# THE COMPLETE BOOK OF TREES OF BRITAIN & EUROPE

THE COMPLETE BOOK OF

# TREES

OF BRITAIN & EUROPE

An authoritative visual reference to more than 550 native and introduced species, with more than 1600 specially commissioned illustrations and photographs

**TONY RUSSELL**

**Photography by Peter Anderson**

HERMES HOUSE

This edition is published by Hermes House

Hermes House is an imprint of Anness Publishing Ltd
Hermes House
88–89 Blackfriars Road
London SE1 8HA
tel. 020 7401 2077; fax 020 7633 9499
info@anness.com



Publisher: Joanna Lorenz
Editorial Director: Helen Sudell
Editor: Simona Hill
Designer: Nigel Partridge
Editorial Reader: Jay Thundercliffe
Production Controller: Pedro Nelson
Photographer: Peter Anderson
Illustrators: Peter Barrett, Penny Brown, Stuart Carter, Anthony Duke, Stuart Lafford, David More, Sebastian Quigley

Parts of this title also appear in *The World Encyclopedia of Trees* and *Trees of the Americas*

10 9 8 7 6 5 4 3 2 1

A CIP catalogue record for this book is available from the British Library.

P1 *Quercus frainetto*
P2 *Salix* x *sepulcralis* 'Chrysocoma'
P3 *Crataegus monogyna*
P4 Left to right: *Crataegus laciniata, Robinia pseudoacacia* 'Frisia', *Cupressus sempervirens, Cercis siliquastrum, Prunus sargentii.*
P5 Left to right: *Abies delavayi, Pseudolarix amabilis, Magnolia* x *loebneri, Juglans ailantifolia, Castanea crenata.*

# CONTENTS

# INTRODUCTION

Trees are the most complex and successful plants on earth. They have been around for 370 million years and quite likely will be around for many millions of years to come. Today, they cover almost a third of the earth's dry land and comprise more than 80,000 different species ranging from small Arctic willows that are just a few centimetres high to the lofty giant redwoods, which stand at an amazing 113m/368ft.

Trees are the oldest living organisms on earth. In the United Kingdom there are yew trees known to be over 4,500 years old. In California, USA, there are Bristlecone pines which are of a similar age. Ever since the first primates appeared in the Palaeocene epoch, 65 million years ago, trees have played an integral part in human development, providing food, shelter, safety, medicines, timber and fuel among other things.

Trees are indeed essential to all life. They reduce pollution by absorbing vast amounts of carbon dioxide from the atmosphere while at the same time replacing it with "clean" oxygen. Each day 0.4 ha/1 acre of trees will produce enough oxygen to keep 18 people alive. Forests of trees help to regulate water flow and can reduce the effects of flooding and soil erosion. They also influence weather patterns by increasing humidity and generating rainfall.

*Above: Palms trees survive in the heat. They usually have large, attractive, compound leaves and a single trunk.*

With their myriad shades of green, trees make our cities and towns more colourful. They increase wildlife diversity and create a more pleasant living and working environment. They provide shade in summer and shelter in winter. It is a fact that post-operative hospital stays are shortened when patients are in rooms with views of trees.

For centuries poets, writers and artists have been inspired by the beauty of trees. Works such as Wordsworth's *Borrowdale Yews* and John

*Left:* Robinia pseudoacacia *trees have been used to create avenues fcr at least 400 years. Pollarding keeps the shape neat and even.*

*Above: Ancient trees are important points of reference in our towns and the countryside, and help to determine the character of an area.*

Constable's majestic elms in *The Hay Wain* will live on long after the original trees depicted have died. Trees help to bring beauty to our gardens and parks. Chosen well, they will provide stunning flowers, foliage, fruit and bark every day of the year. Nothing brings structure and maturity to a garden more successfully than a tree.

With so many obvious values it should be safe to assume that trees are venerated the world over. Unfortunately that is not the case. Over ten per cent of the world's tree species are now endangered. More than 8,750 species are threatened with extinction – some are literally down to their last one or two specimens. Across the world we are losing at least 40 ha/100 acres of trees every minute.

This book is a celebration of trees in all their forms from hardy evergreens and deciduous broadleaves, to sun-worshipping and tender palms. It reveals what incredible organisms trees are and describes the diversity that exists throughout the world and how they each contribute to the planet. The first section describes the origins of trees, how they have evolved, how they live, grow, reproduce and why they die. It looks in detail at their leaves, bark, fruit, flowers, buds, cones and seeds, and details the fascinating role each plays in the life of the tree. Trees inhabit many natural landscapes, from the highest mountain ridges all the way down to sea level, and have adapted to different circumstances. The heat of the tropics, the biting cold of northern lands, the salt and wind of the sea and the pollution of the city have all contributed to the evolution of the tree.

The second section of this book features a comprehensive encyclopedia of the most well-known, unusual, or economically and ecologically important species from around Europe. Each entry provides a detailed description of the tree, its height, habit, colour and leaf shape and whether it produces flowers, fruit or cones. Its habitat and most interesting features are described to aid identification and a map helps to locate wild populations for each entry.

This book aims to bring a greater understanding and appreciation of trees to a wider audience. It should encourage you to look more closely at the diversity of trees in your own locality and if you have the opportunity to visit far-flung countries, to appreciate the diversity that exists on the planet.

*Below: The monkey puzzle tree,* Araucaria araucana, *has a distinctive and instantly recognizable silhouette.*

# HOW TREES LIVE

Trees have three obvious features that together distinguish them from all other living plants. First, they produce a woody stem, roots and branches which do not die back each winter but continue to grow year upon year. This means that from the time a tree begins to germinate until the time it dies it is always visible. Be it the smallest garden apple tree or the largest English oak, this basic principle of growth remains the same.

Second, trees live longer than any other living organism on the planet. It is not exceptional to find living trees that are more than 1,000 years old and many are considerably older. Third, trees are the largest living organisms on the planet. Around the world there are trees in excess of 100m/328ft tall or 1,500 tonnes in weight.

Trees have been growing on earth for 370 million years and today can be found growing almost everywhere from the Arctic Circle to the Sahara Desert. For much of the world, trees are the climax species of all plants – which in simple terms means if land is left untended long enough it will eventually become colonized by trees.

So why are trees so successful? Well, as with all plants, trees need light to survive. Without light, photosynthesis cannot take place and food for growth cannot be made. Trees are superb competitors for light; their woody stem enables them to hold their leaves way above the leaves of any other plant. This means they can absorb vast quantities of light while shading out other plants in the process.

Such is the extensive nature of a tree's root system that it can access moisture from deep in the subsoil – something few other plants can do. As such, trees are well equipped to survive periods of drought, particularly as their structure and size allows them to store food and water for times of deficiency. All in all trees are an incredibly competitive and successful group of plants – which is why they have been around so long. They are also a fascinating group of plants, as the following pages will clearly show.

*Left: Cedars of Lebanon,* Cedrus libani, *in the remnants of a forest in the Bcharre Valley, in Lebanon. This species is known to live for over 2,000 years.*

# THE EVOLUTION OF TREES

*The first trees evolved more than 300 million years ago. By 200 million years ago they were the most successful land plants on earth, growing in all but the most inhospitable places, such as the Polar regions. Their ability to produce vast amounts of oxygen has enabled other life forms, including humans, to evolve.*

The first living organisms appeared on earth 3,800 million years ago. These primitive, single-celled life forms were followed 500 million years later by the earliest cyanobacteria or blue-green algae. Also single-celled, these were the first organisms able to harness the sun's energy to produce food. This process, known as photosynthesis, had an important by-product – oxygen, which gradually began to accumulate in the earth's atmosphere.

### *Archaeopteris*: the first tree

The first known land plant, which was called *Cooksonia*, evolved around 430 million years ago. *Cooksonia* was erect and green-stemmed with a simple underground root system. It was followed about 60 million years later by *Archaeopteris*, the first real tree.

*Below: The timeline below shows the evolution of life forms from the first ammonites of the Devonian period, 417–360 million years ago, through to the development of flowering trees such as magnolias during the Cretaceous period, 144–65 million years ago.*

With a woody trunk up to 40cm/16in across, *Archaeopteris* had branches and a large root system. It also had the ability to produce buds and continue growing year after year. Fossils of *Archaeopteris* found recently suggest that it may have been able to live for as long as 50 years. As forests of *Archaeopteris* spread across the earth, the amount of oxygen in the atmosphere rapidly increased, paving the way for an explosion in the evolution of new land animals.

### The Carboniferous period

During the Carboniferous period, the earth's climate was warm and humid. Great forests and swamps of trees, ferns and mosses covered the land. One of the most common trees was *Lepidodendron*. Known as the scale tree, it reached heights of 30m/98ft and had a trunk more than 3m/10ft across. It looked like a palm tree, but instead of fans of long, thin leaves it had fern-like fronds, each ending with cone-shaped structures containing spores for reproduction.

At the close of this period, the first primitive conifers, or gymnosperms, began to appear. These plants protected their seeds in cones and had a much more efficient reproductive system than their predecessors. None of these early conifers survives today. Their nearest relatives are species of *Araucaria* (monkey puzzle), *Podocarpus* and *Taxus* (yew).

### Pangaea

A vast supercontinent that existed 280–193 million years ago was known as Pangaea. The northern part, called Laurasia, comprised the landmasses of North America, Europe and Asia all joined together. The southern part, Gondwanaland, was made up of South America, Africa, Arabia, India, Australia and Antarctica.

Since they were part of Pangaea the continents have moved. Fossil evidence taken from samples of ice deep in the Antarctic ice cap shows that relatives of *Nothofagus moorei*, the Antarctic beech, grew in that region more than 200 million years ago.

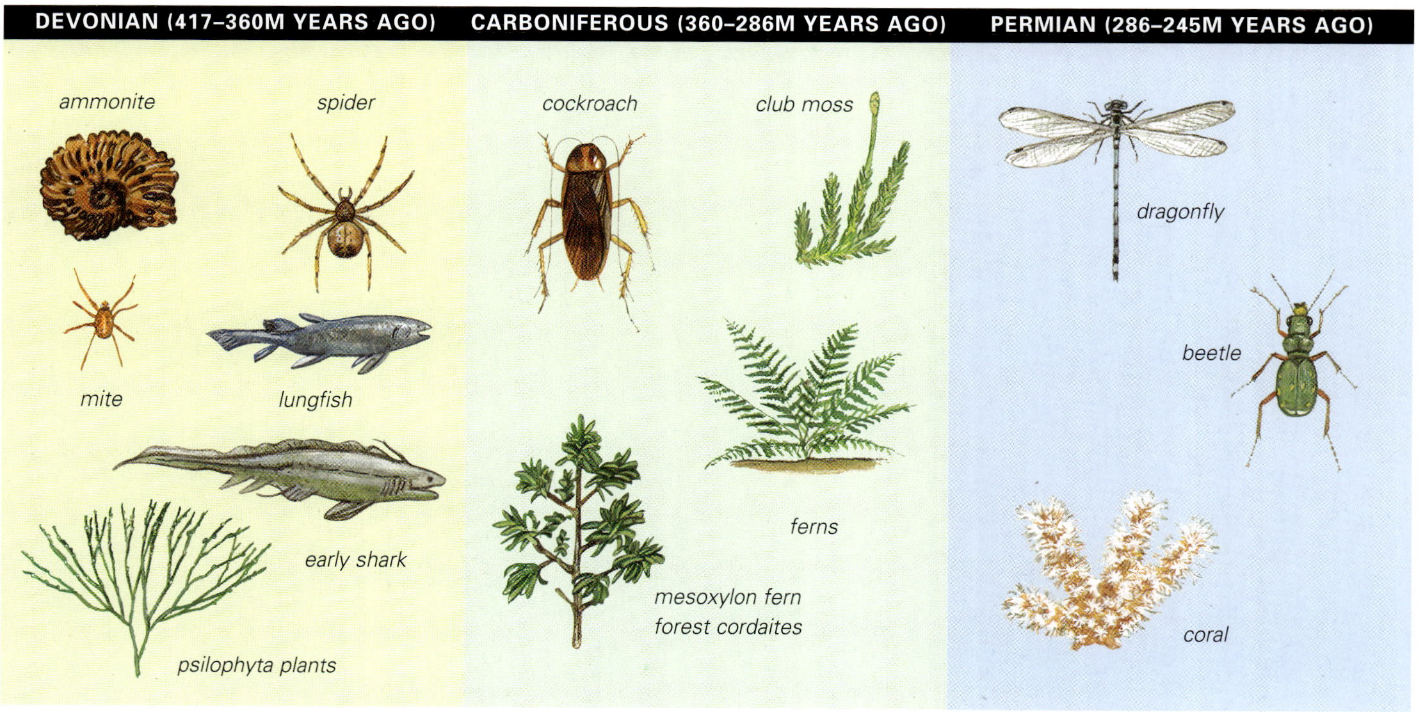

*Above: One very early tree that is still around today is the deciduous* Ginkgo biloba, *or maidenhair tree. It is the last surviving member of a family of trees called the ginkgos; along with conifers, they dominated the land for 250 million years.*

## The Mesozoic era

This era lasted from 245–65 million years ago. It was the age of dinosaurs and saw dramatic fluctuations in world climate. Conifers adapted to these changes so successfully that different species evolved for almost every environment. Today they survive in some of the coldest and hottest parts of the planet.

Ginkgos were also successful: fossils show that they grew throughout the Northern Hemisphere, from the Arctic Circle to the Mediterranean and from North America to China. Fossils of the Jurassic period (208–144 million years ago) also show the dawn redwood, *Metasequoia glyptostroboides*. Previously thought to have been extinct since that time, the dawn redwood was discovered growing in China in 1941. During the Cretaceous period (144–65 million years ago) flowering plants (angiosperms) evolved and began to exert their dominance over conifers. Among the earliest were magnolias, which are common today.

## The Tertiary era

Many of the trees that grew during the Tertiary era (65–2 million years ago) still grow today. The main difference between the Tertiary and the present was the scale of the forests. During the early Tertiary era the planet was warmer than it is today. Europe and North America had a similar climate to that of present-day South-east Asia and vast swathes of forest covered virtually every available piece of land. Oak, beech, magnolia, hemlock, cedar, maple, chestnut, lime and elm occurred alongside tropical trees such as the nypa palm. As the era progressed however, the climate began to cool.

## The ice ages

By 1.5 million years ago the climate had cooled so much that the first of four ice ages began. Trees that we now regard as tropical began to die at the far north and south of their ranges. As the temperature dropped further so more temperate species succumbed. Only those trees close enough to the Equator were able to survive. Each glaciation was interspersed with warmer inter-glacial periods lasting anything up to 60,000 years. During these warmer periods, many trees recolonized their previous ranges. Every continent suffered; however, some fared better than others because of differences in topography. In North America, for example, the mountain ranges all run from north to south. Heat-loving trees were able to spread south as the ice sheets advanced, using the valleys between mountain ranges to reach refuges nearer the Equator. The trees were able to recolonize their old ranges back along these same routes. In Europe, however, recolonization was impossible. The Pyrenees and the Alps, which stretch from east to west, meant that many trees were unable to move south ahead of the ice. Once trapped they perished, leaving Europe with a far less diverse tree flora than that of North America or Asia.

## The modern era

Since the last ice age began to wane 14,000 years ago, the temperature of the earth has gradually increased and trees have begun to recolonize temperate areas of the world. Today there are over 80,000 different species of trees on earth.

# CLASSIFICATION OF TREES

*Classification is the process by which plants or animals are grouped and named according to their specific similarities. The theory and practice of classification is called taxonomy and those that work in this field are known as taxonomists.*

There are over 300,000 different species of flowering plants and gymnosperms or conifers in the world. Botanists have classified them in order to try and make sense of the way that they are related to each other. Rudimentary grouping of trees has occurred for centuries, not always with great accuracy. For example the English oak, *Quercus robur*, and the holm (evergreen) oak, *Q. ilex*, have always been regarded as being closely related because of the fruit they produce. However, the sweet chestnut, *Castanea sativa*, and the horse chestnut, *Aesculus hippocastanum*, which were also once classified on the basis of their fruit, are now thought to belong to two quite different families.

The science of classification starts to become ever more complex as botanists study trees more closely. Where once trees were classified on the basis of just one or perhaps two characteristics, now many more of their features are compared before a degree of relatedness is decided.

*Below: The horse chestnut (left) and sweet chestnut (right) were once thought to be related.*

## Carl Von Linné (1707–78)

Ever since the time of the Greek philosopher Aristotle (384–322BC) it had been recognized that, both in the plant and animal world, there was a natural order where everything had its place and was linked to other species by a common thread. However, it was not until the 18th century that the Swedish botanist Carl Von Linné (also known as Linnaeus – the Latin name that he gave himself) made the first attempt to link all plants by one specific feature. He classified them by the way they reproduced themselves and the make-up of their reproductive systems – in the case of flowering plants, their flowers. As he admitted, his choice of feature for classification was artificial. Linnaeus had not found the common thread, the natural order of all living things. Nevertheless, he did create a system of classification that is still in use today.

Linnaeus invented the principle of using two Latin words to name a species. He chose Latin because it was the language of scholarship, and was understood across the world but no longer used as a spoken language, so the meaning of its words would not change over time. The first of the two words is known as the generic (genus) name and the second the specific (species) name. The generic name gives a clue to the species' relationship with others. Closely related species are given the same generic name but different specific names. For example, the English oak is called *Quercus robur* and the closely related turkey oak is called *Quercus cerris*. All species with the same generic name are said to belong to the same genus.

*Above: Trees often have common names that refer to their place of origin, colouring, or use.*

Similar genera (the plural of genus) are combined into larger groups known as families. For example the oak genus, *Quercus*, belongs to the same family as the beech genus, *Fagus*. This family is called Fagaceae, and is commonly known as the beech family. Similar families are gathered together in turn into larger groups called orders. The beech family, Fagaceae, combines with the birch family, Betulaceae, to make the beech tree

order Fagales. Similar orders are then combined into subclasses. The beech order is part of the hazel subclass, which is called Hamamelidae. In turn, Hamamelidae is combined with all of the other plant subclasses that are characterized by embryos that contain two seed leaves, to form a group that is known as the dicotyledons. This group is then joined together with all plants that have an embryo that contains only one seed leaf (monocotyledons) into one group that contains all flowering plants – the Magnoliophytina. Finally, this is gathered together with all of the other groups of seed-producing plants and then combined with the non seed-producing plants, such as ferns, into the Plant Kingdom.

## Charles Darwin (1809–82)

The "common thread" or natural order of all living things was left for Charles Darwin to discover. Darwin recognized that plants, or animals for that matter, were usually alike because of their common ancestry.

Trees alive today can be classified in terms of their relatedness because they have all evolved over time from a single common ancestor that existed millions of years ago. The science of the ancestry of all living things is called phylogeny and it goes hand in glove with taxonomy.

Once the interrelatedness of all plants was understood, scientists began to trace back the evolution of trees. In many ways this process is similar to tracing back one's own family tree. The major difference is that fossil records are used. The different characteristics of trees living today compared to fossils of those from the past reflect the evolutionary changes that have occurred to the common ancestral line over millions of years. Each evolutionary change has been in response to a different environmental condition and has resulted in a different tree.

For most of us, classification only becomes pertinent when we are trying to identify a species.

*Below: The cork oak (left) and common beech (right) look different, but in fact they are both members of the beech family, Fagaceae.*

*Above: It is possible to recognize trees that belong to the same family by certain obvious characteristics. For example hazel (above left), alder (above centre), birch (above right), and hornbeam (not shown) all belong to the birch family, Betulaceae, and all produce catkins.*

For botanists and taxonomists however, classification is an everyday procedure and a frequent cause of disagreement. It is now more than 200 years since Linnaeus developed his system of classification and 150 years since Darwin announced his theory of evolution. Nevertheless taxonomists still move species from one genus to another and some botanists cast doubt on whether plant classification should be based upon the evolutionary process at all.

# ROOTS

*Tree roots provide anchorage, ensuring that the tree does not fall over. They obtain water, the lifeblood of any tree, by sucking it from the soil. Roots provide the tree with minerals, which are essential for growth. They also store food, such as starch produced by the leaves, for later use.*

Roots have the ability to influence the size of a tree. Around 60 per cent of the total mass of any tree is made up by its trunk. The remaining 40 per cent is split evenly between the branches and the root system, each having a direct relationship with the other. If there are not enough roots, the canopy and leaves will not be able to obtain enough water and branches will start to die back. In turn, if branches are damaged or removed and there are fewer leaves to produce food, a tree's roots will begin to die back.

## A shallow existence

Contrary to popular belief, tree roots do not penetrate deep into the soil. In most cases the roots of even the tallest tree seldom reach down more than 3m/10ft. The overall shape of a tree is like a wine glass, with the roots forming a shallow but spreading base.

More than three-quarters of most trees' roots can be found within 60cm/24in of the surface. They seldom need to go deeper: the top layers of the soil are normally rich in organic material, minerals and moisture, which are just the ingredients that roots require.

However, roots do spread outwards considerably within the upper layers of the soil. The bulk of a root system will be found within 3–4m/10–13ft of a tree's trunk. However, very fine roots may spread anything up to twice the radius of the canopy, which in a large tree can mean anything up to 30m/98ft away from the trunk.

*Tap root*

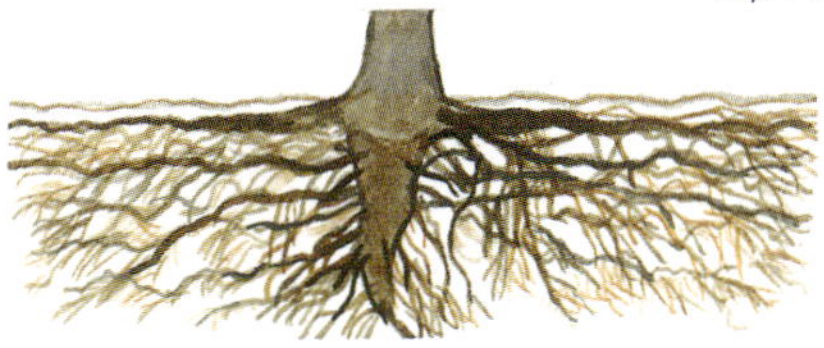

## Tap roots

The first root that every tree grows from its seed is called a tap root. Tap roots grow straight down and from day one have the ability to extract moisture and minerals from the soil. Within days of the tap root emerging from a seed, side roots (known as laterals) grow off the tap root and begin to move horizontally through the top layers of the soil. On some trees, such as oak, the tap root persists for several years. In most species, it withers and the lateral roots take over.

*Lateral roots*

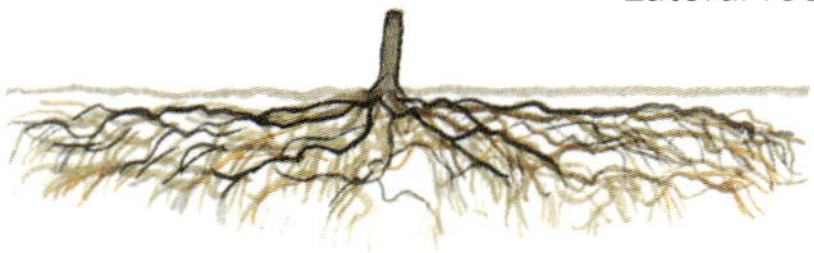

## Lateral roots

Most lateral roots stay close to the surface for the whole of a tree's life. Sometimes they may develop from the tap root or grow directly from the base of the trunk; in the latter case they can be more than 30cm/12in wide. Within 1m/3ft of the trunk they taper to around 10cm/4in across, and at 4m/13ft away they are usually under 5cm/2in in diameter and far more soft and pliable.

*Stilt roots*

## Stilt roots

Mangroves grow throughout the tropics on coastal mudflats. Many species have stilt-like roots that arch from the main stem down into the mud. Once these have taken root, they help to anchor the tree so that it remains stable in the constantly moving mudflat silt. The roots graft together, creating a three-dimensional framework that holds and supports the mangrove tree clear of the mud.

## Pillar roots

The weeping fig, *Ficus benjamina*, and the banyan, *F. benghalensis*, have roots that grow and hang down from the

*Left: Few trees can survive indefinitely in waterlogged conditions such as these. In swampy or boggy ground most trees will survive for a time until their roots become waterlogged. Many trees will happily grow alongside watercourses though. Willow, ash and poplar all thrive near water.*

*Left: With some trees, such as this rainforest giant near Belem in Brazil, it is difficult to know where the root stops and the trunk begins. These wedge-like buttress roots prevent the tree from toppling over.*

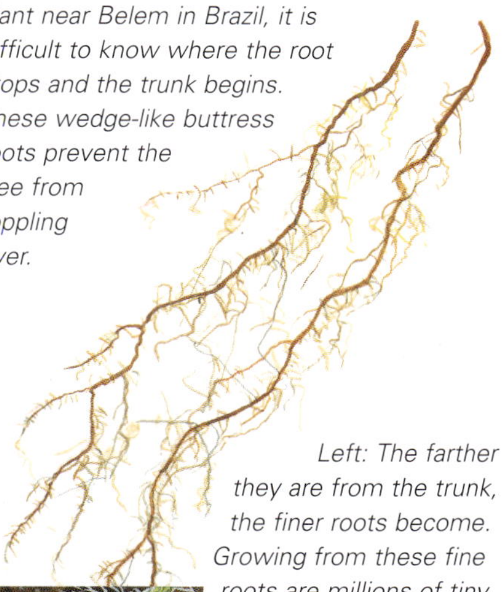

*Left: The farther they are from the trunk, the finer roots become. Growing from these fine roots are millions of tiny hairs, each one made up of a single cell. It is these root hairs that collect the necessary ingredients, such as nitrogen and potassium, for a tree to grow. Each is in contact with the soil particles around it and is able to absorb both the moisture and the diluted minerals that surround each particle. Root hairs have a lifespan of no more than a few weeks, but as they die, new ones are formed.*

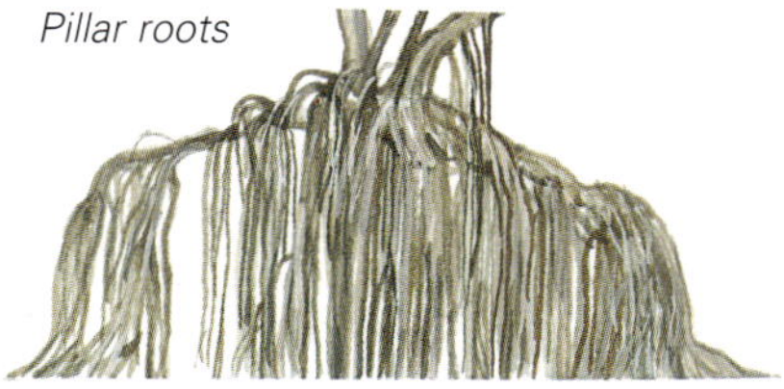

*Pillar roots*

branches. These roots grow quickly – up to 1cm/½in a day – and once anchored in the soil they form prop-like pillars, capable of bearing the weight of the spreading branches they grew from originally. This system enables the tree to continue to grow outwards almost indefinitely. A banyan tree planted in the Royal Botanic Garden of Calcutta in 1782, for example, now covers an area of 1.2ha/3 acres and has 1,775 pillar roots.

## Symbiotic associations

Within the soil, tree roots come into contact with the living threads, or *hyphae*, of numerous fungi. Quite often this association is beneficial to both the tree and the fungus. Usually the tree acquires hard-to-obtain nutrients such as phosphorus from the fungus and the fungus gets carbohydrates from the tree. The structures formed between tree roots and fungi in these mutually beneficial associations are known as mycorrhiza. Sometimes, however, contact with fungus can be damaging for a tree.

## Roots and water

Tree roots require water to survive but they also need to obtain a supply of oxygen. It is important that they have water readily available, but roots will not do well if they are continually submerged. In constantly waterlogged conditions roots will not be able to obtain enough oxygen and a tree will effectively drown.

## Roots and oxygen

The swamp cypress thrives in wet conditions. To counter the lack of soil oxygen, its roots have developed strange knobbly growths, called knees. These grow out of the water or wet ground to gain access to the air, and therefore to a supply of oxygen. Swamp cypress knees can reach a height of 4m/13ft. They not only absorb oxygen, but also provide support to the tree, making it less likely to blow over in strong wind.

*Below: The breathing roots of this mangrove protrude through the sand on Mafia Island, near Tanzania.*

*Below: Honey fungus,* Armillariella mellea, *is one of the biggest killers of trees in the temperate world. Once it has made contact with a tree's roots, it rapidly spreads through the entire vascular system of the tree, killing tissue as it goes.*

# TRUNK AND BARK

*What makes a tree different from all other plants is the tough, woody framework it raises above the ground: a framework, made up of a trunk and branches, that lasts for the entire life of the tree. As each year passes, this framework gets bigger as the trunk and branches expand upwards and outwards.*

The main purpose of the trunk is to position the leaves as far as possible from the ground. The higher they are, the less competition there is from other plants for light. Without light trees die. The trunk supports the branches and the branches support the leaves.

The trunk and branches have two other functions. They transport water, which has been collected by the roots, up through the tree to the leaves. Second, they move food, which is produced in the leaves, to every other part of the tree, including the roots.

Considering the importance of the functions that the trunk and branches perform, it is extraordinary that more than 80 per cent of their mass are made up of dead cells. The only living cells in a tree's trunk and branches are those in the area immediately beneath the bark. It is here that all of the activity takes place.

## The inner tree

A tree's bark is like a skin. It is a corky waterproof layer that protects the all-important inner cells from disease, animal attack and, in the case of redwoods and eucalyptus, forest fires.

Some barks, such as that of the rubber tree, exude latex to 'gum up' the mouths of feeding predators. Pine trees have a similar defence mechanism, exuding a sticky resin which can literally engulf a whole insect. Some trees, such as the South American quinine tree, *Cinchona corymbosa*, produce chemicals in their bark which are poisonous to attackers.

Bark is perforated with millions of tiny breathing pores called lenticels, which pass oxygen from the outside atmosphere through to the living cells beneath. In cities and along busy roads these lenticels get clogged up with dirt and carbon. Some trees, such as the London plane, *Platanus* x *hispanica*, have adapted by regularly shedding their old bark. All trees are constantly growing and their girth expanding. This is reflected in the cracks and crevices that appear in the bark of many trees. As bark splits, new corky cells are produced to plug the gap.

Beneath the outer bark is the inner bark, or phloem. This is a soft spongy layer of living tissue that transports sap – sugary liquid food – from the leaves to the rest of the tree.

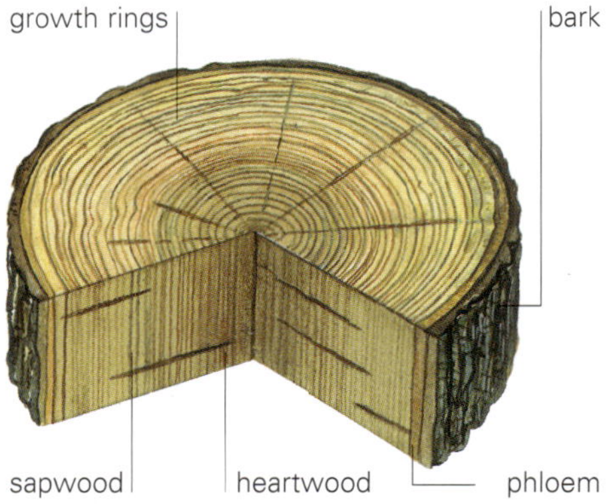

*Above: A section through the trunk of a larch tree showing the darker heartwood and the lighter sapwood.*

Beneath the phloem is a thin tissue known as the cambium. Although it is only one cell thick, the cambium is extremely important. It is here that all tree growth takes place. Cambium cells are constantly dividing, producing phloem cells on the outside and on the inside wood cells, or xylem.

Xylem has two parts: the sapwood, made up of living cells, and the heartwood, composed of dead cells. The sapwood transports water and minerals from the roots to the leaves. Most of these are carried in sapwood made by the cambium during that year. The heartwood forms the dense central core of the trunk, supporting the tree and giving the trunk rigidity. The two main constituents of xylem are cellulose and lignin. Cellulose, a glucose-based carbohydrate, makes up three-quarters of the xylem and is used in the construction of cell walls. Lignin comprises most of the remaining quarter and is a complex organic polymer. It is lignin that gives wood its structural strength. If water and air reach the heartwood as a result of damage to the outer layers of the trunk, decay will occur and in time the tree may become hollow.

*Left: Most trees over 500 years old are hollow. Eight people sitting around a table can fit inside the trunk of this tree.*

*Banyan tree*

*Cola nut*

*Kapok*

*Papaya*

*Flame of the forest*

*Tembusu*

## Bark invaders

While bark exists to provide a protective barrier over the living tissue of a tree's trunk and branches, there are plenty of creatures capable of penetrating that barrier. Bark beetles and wood-boring insects eat cellulose and excavate breeding chambers and galleries for egg-laying purposes. Often the damage inflicted by insects is much greater than just the physical effects of their mining. Insects may carry fungal diseases. Beetles that bore into infected trees become coated with fungal spores, which they carry to other, healthy trees. Once underneath the bark, the fungus quickly blocks the cells transporting food and water, leading to the tree's demise.

*Below: Trees grow from terminal and lateral buds positioned towards the tip of the branches.*

## How we use bark

Bark not only forms protection for trees, it can also be incredibly useful to us. Much of the wine that we drink is sealed in bottles with bark from the cork oak tree, *Quercus suber*. In Mediterranean regions, cork oaks are grown in orchards. Every ten years or so, the outer corky bark is carefully removed, leaving the cambium layer intact. The cambium then produces more cork cells to replace the bark that has been harvested. Bark also provides us with food and medicine. The spice cinnamon is made from the dried and ground bark of the Sri Lankan cinnamon tree, *Cinnamomum ceylanicum*, while the bark of the Pacific Yew, *Taxus brevifolia*, contains a substance called taxol, which has been highly effective in the treatment of some forms of cancer. Some trees have very attractive bark, making them ideal ornamental plants for parks and gardens. The Tibetan cherry, *Prunus tibetica*, has highly polished mahogany-red bark, for example, and the Himalayan birch, *Betula utilis*, has bark the colour of freshly fallen snow.

*Himalayan cherry*

*Indian horse chestnut*

*Floss silk tree*

*Eucalyptus*

*Paperbark maple*

*Birch 'Snow Queen'*

# BUDS

*Buds act as protective sheaths for the growing tips of trees during the coldest months of the year. In winter, even though deciduous trees will have shed their leaves, they can still be readily identified by their buds.*

For trees to grow they need water, minerals, nutrients and the right growing conditions, namely sunlight and warmth. In parts of the world where there is little seasonal variation, such as the tropics, favourable climatic conditions may allow growth to continue all year round. However, even in tropical rainforests very few trees grow non-stop. The normal pattern for most trees, particularly those in temperate regions, is for a period of growth followed by a period of rest. The period of rest coincides with the time of year when the climate is least favourable to growth. Across Britain, Europe and North America this is during the cold and dark of winter.

Throughout the winter resting period, the growing tips of a tree, known as the meristem, are vulnerable to cold winds and frost. Prolonged low temperatures can very easily damage or even kill the meristem. Trees have therefore evolved ways to protect this all-important tissue.

### Protective sheath

During early autumn, as the growing season approaches its end, the last few leaves to be produced by the tree are turned into much thicker but smaller bud leaves, known as scales. These toughened leaves stay on the tree after all of the other leaves have fallen off and form a protective sheath around the meristem. This sheath is known as a leaf-bud. Its thick scales are waterproof and overlap each other, creating a defence system able to withstand the onslaught of winter. Often a coating of wax, resin or gum is used to strengthen these defences.

*Below: Some buds contain all of the cells needed for the whole of the following year's growth. Others contain just enough to start growing in spring and then produce more growth cells once the leaves have emerged from the bud.*

*Above: A lime tree breaking bud.*

*Above: A sycamore breaking bud.*

### Inside the bud

Winter buds contain all that the tree will need to resume growing once the days lengthen and the temperature increases in spring. Inside is a miniature shoot, miniature leaves all carefully folded over one another and, in some species, such as the horse chestnut, *Aesculus hippocastanum*, miniature flowers.

### Trees without buds

Not all trees produce buds, even in temperate regions. Some, such as the wayfaring tree, *Viburnum lantana*, have "naked buds" with no bud scales. At the end of the growing season in this species, the last leaves to be formed stop growing before they are fully developed. A dense layer of hair then forms on them to protect them from the cold and they proceed to wrap themselves around the meristem. When spring arrives the protective leaves simply start growing again from where they left off.

Eucalyptus trees also have "naked buds" but as back-up they produce tiny concealed buds beneath the leaf base. These are only activated if the growing tip gets damaged.

*Below: An Indian horse chestnut bud opening to reveal long, thin, down-covered leaves.*

Some conifers, such as western red cedar, *Thuja plicata*, and lawson cypress, *Chamaecyparis lawsoniana*, have no distinct buds at all; instead they produce little packets of meristematic cells, which are hidden beneath the surface of each frond of needles.

## The growing season

As spring arrives, buds open and the leaves begin to emerge. For all trees the trigger for this to happen is increasing warmth and light. Individual species each have their own trigger point, which is determined by their geographical origins. Species that originated in colder regions, such as birch or willow, burst bud earlier than those such as horse chestnut or sweet chestnut, which evolved in warmer parts of the world. Birch instinctively knows that northern European summers are relatively short affairs, and that it needs to get going as quickly as possible to make the most of the growing season. Sweet chestnut, on the other hand, instinctively expects a long, Mediterranean summer, so is in less of a rush to get started.

**Horse chestnut bud**

*Left: The sticky buds of horse chestnut will open over a period of three days in springtime.*

## BUD ARRANGEMENTS

Even in winter, when deciduous trees display bare branches, trees can still be identified by the shape, size, colour and arrangement of the buds on the twigs.

### Opposite buds

The buds of trees such as sycamore and ash are said to be opposite – that is, in pairs on each side of the twig, exactly opposite each other. Ash buds are easily recognizable by their distinctive black colouring.

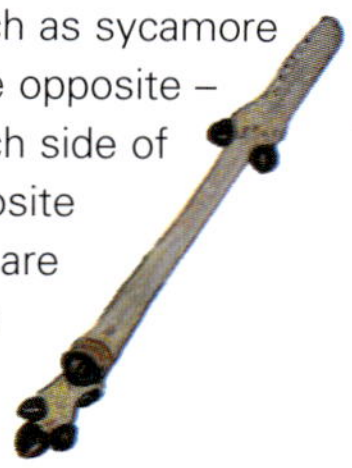

### Alternate buds

The buds of trees such as beech and willow are arranged alternately on different sides of the twig. Willow buds are generally longer and more slender than those of beech.

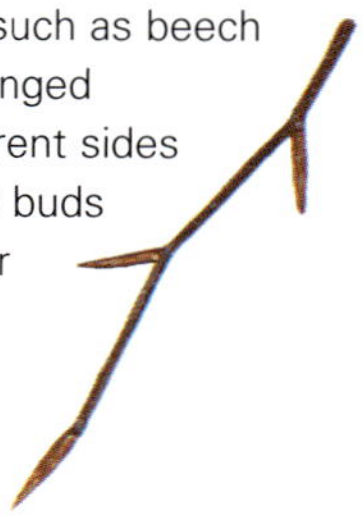

### Hairy buds

Magnolia buds are very distinctive and easily recognized by their covering of thick grey fur. Magnolia buds are some of the largest found on any tree.

### Clustered buds

Oak buds appear almost randomly on the twig, but always with a cluster of buds at the tip. Cherries also adopt this clustered approach.

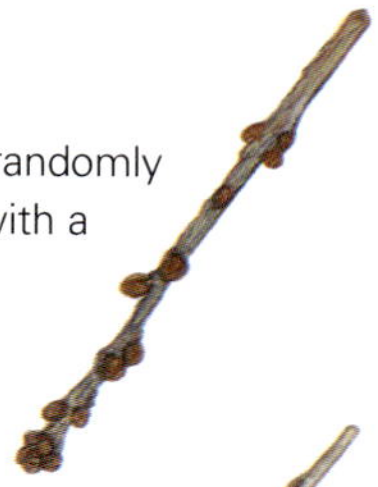

### Whiskered buds

As well as being clustered, some oaks, such as the turkey oak, also have thin whiskers surrounding the buds.

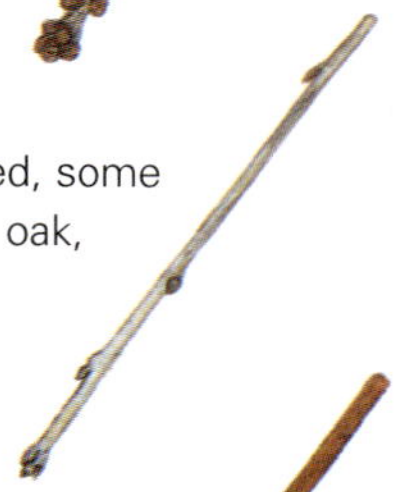

### Naked buds

The wayfaring tree does not have a true bud. Instead it has immature hairy leaves which surround the growing tip to protect it from the cold.

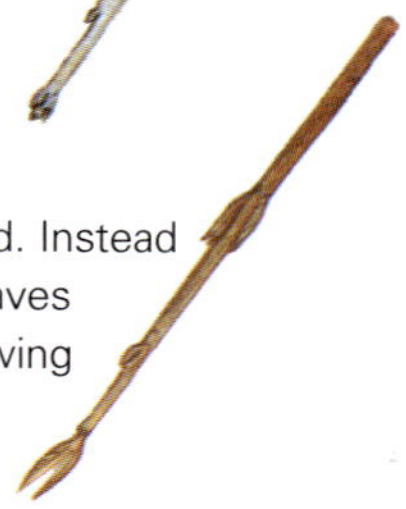

Trees that have everything for the coming year's growth pre-packaged inside the bud tend to have a single growth spurt immediately after their leaves emerge. This can mean that they achieve virtually all their growth for the whole year within the first four weeks of spring. Those trees that over-winter with just enough growth cells in the bud to aid emergence in spring grow more slowly but grow for a longer period of time. In some instances these species may continue growing for more than 100 days. However, by the end of the season the overall growth of each will be similar.

*Below: Sweet cherry buds.*

*Below: Wingnut bud.*

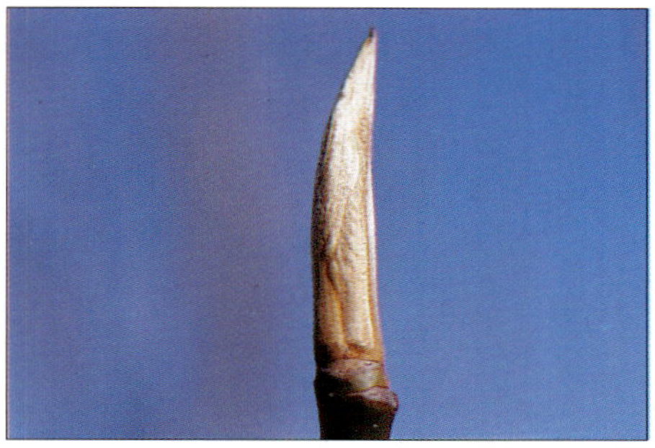

*Below: Magnolia bud.*

*Below: Horse chestnut bud.*

# LEAVES

*Each leaf on a tree is a mini power station, generating food, which the tree uses to provide the necessary energy for living and growing. The process by which leaves produce food is called photosynthesis. During this process the leaves absorb carbon dioxide and emit oxygen.*

Leaves contain a green pigment called chlorophyll, which absorbs light energy from the sun. This energy is used to combine carbon dioxide, which the leaf absorbs from the atmosphere, with water taken from the soil. The resulting products are glucose and oxygen. Glucose provides the energy to run the tree and can be turned into starch for storage or cellulose, which forms the tree's cell walls. The oxygen is released by the leaf back into the atmosphere. A mature tree can produce the same amount of oxygen every year as that used by ten people.

Above: When chlorophyll production stops and any residue decays, other pigments are seen.

## Leaf structure

Each leaf is covered by a skin of tightly packed cells known as the epidermis. This skin is coated by a waxy covering called the cuticle. The cuticle acts as waterproofing, preventing the leaf from losing any more water than is necessary. The transfer of oxygen and carbon dioxide to and from the atmosphere takes place through tiny holes in the cuticle known as stomata. Stomata are concentrated on the underside of the leaf away from the direct heat of the sun to minimize water loss. The cells around the stomata have the ability to enlarge and decrease the size of the hole. Despite this, water is lost from the leaf through the stomata. This loss of water is known as transpiration. Even though stomata normally cover less than one per cent of a leaf's total area, the amount of water lost in this way can be astonishing. A large deciduous tree can lose up to 300 litres per day in summer. The lost water is usually replaced by water drawn from its roots. In times of drought however, the amount of water lost may exceed that available to the roots. When this happens the leaves wilt and die, stopping the tree from producing food.

Inside the leaf cells, the chlorophyll is contained in millions of tiny cell-like vessels called chloroplasts. Most of these are found in the upper part of the leaf, which receives the most light. Beneath the chloroplasts are the vascular tissues that make up the xylem, and which transport the raw ingredients for photosynthesis, such as water and minerals, all the way from

Left: Tiny breathing holes in the leaf are known as stomata.

Below. Cross section of a leaf.

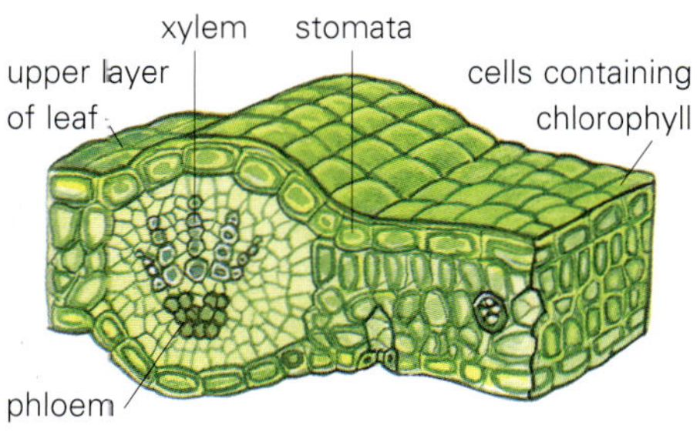


**Life cycle of leaves**

Above: In spring new leaves form.

Above: Summer.

Below: The changing tones of autumn.

Above: Winter profile.

the roots to the leaf. Alongside the xylem is the phloem, which transports the sugary products of photosynthesis from the leaf to all other parts of the tree. Both vascular systems rely on a process called osmosis to move liquid. Osmosis is a process whereby liquid moves from one cell to another. The catalyst for this to happen is the fullness, or turgidness, of each cell. As one cell becomes full, so the liquid within it permeates through the cell wall into a neighbouring cell that is less turgid. Once this cell is full, liquid starts to permeate from it into the next empty cell and so on.

## Leaf size and shape

One of the most interesting things about leaves is the incredible range of shapes and sizes. The smallest so-called broad leaf is produced by the Arctic willow, *Salix nivalis*. This tundra species has leaves less than 5mm/¼in long. Some conifers have needles that are even smaller.

All broad-leaved tree leaves have one thing in common: a network of visible veins, which spread out across the leaf from its base. It is within these veins that the xylem and phloem are found. The veins join together at the leaf base to form the stalk, or petiole.

## Simple leaves

These come in a wide variety of shapes. At their most basic they may be entirely round or heart-shaped, or have no indentation around the leaf edge. Many leaves, such as those of cherry trees, are oval in shape and have small serrations around the edge. On others the serrations may be more pronounced, as with the sweet chestnut, *Castanea sativa*. Some trees, such as the oak, produce leaves with distinctive lobing. These lobes may be rounded, or more angular.

## Compound leaves

At first glance the leaflets of compound leaves look like separate leaves growing off the same stalk. However, closer inspection of a new compound leaf reveals that the whole stalk and its leaflets all emerge from the same leaf bud. It is in essence all one leaf. Many of the trees in our cities have compound leaves. One of the most easily recognized is the horse chestnut, which has seven or nine large leaflets all attached to the same point of the main leaf stalk. The golden-leaved robinia, *Robinia pseudoacacia* 'Frisia', has paired leaflets that come off the leaf stalk opposite each other (pinnate leaflets), as does the European ash, *Fraxinus excelsior*. Occasionally the leaf stalk to which the leaflets are attached may sub-divide, producing side stalks and a bipinnate leaf. One of the best examples of a tree with bipinnate leaves is the Japanese angelica tree, *Aralia elata*, which has leaves in excess of 50cm/20in long.

## Evergreen leaves

A deciduous tree keeps its leaves for only part of the year; they grow in the spring and fall off in the autumn. By contrast, evergreen trees, which include most conifers and trees such as holly, box and laurel, have leaves all year round. This does not mean the same leaves stay on the tree for the whole of its life. Evergreen leaves fall from trees and are replaced throughout the year. The real difference between evergreen and deciduous trees is that the leaves of deciduous trees all fall at around the same time, while those of evergreens do not. On average, evergreens keep their leaves for between three and five years, although on some firs and spruces the needles may be retained for up to ten years.

## Needles

Pines, firs, larches, spruces and cedars all have needles, as do yews and redwoods. Although visually quite unlike other leaves, needles are in fact just compact versions of simple leaves, and do the same job of producing food for the tree. Needles lose far less water than the leaves of broadleaf trees. They are therefore better equipped to survive in areas where water is in short supply, such as northern temperate regions where the ground is frozen for months at a time.

*Below: The soft, feathery needles of the western red cedar.*

*Below: The 1m/3ft-long leaves of the tropical breadfruit tree.*

*Below: The fine pencil-like leaves of* Eucalyptus champmaniana.

# FLOWERS

*Flowers contain the tree's reproductive organs. Some trees, such as cherry, have both male and female reproductive organs within the same flower. Others, such as hazel, have separate male and female flowers on the same tree. Some trees only produce flowers of one sex.*

Flowers are the sex organs of a tree. What happens in them determines the ability of the tree to reproduce itself. Trees are passive organisms; they cannot actively go out and search for a mate, so they have to engage in sex by proxy. Each tree needs a go-between to get its pollen either to another tree or from the male to the female part of its own flowers. Depending on the species of tree, this go-between may be wind, water or an animal, such as a bird or insect. Over countless generations each species has developed its own flower to suit a specific go-between. The African baobab tree, *Adansonia digitata*, for example, has developed large flowers that produce vast quantities of nectar at night. These flowers attract bats, which feed on the nectar and in the process get covered in pollen. The bats transfer that pollen from flower to flower and tree to tree.

## Inside the flower

There are almost as many different forms of tree flower as there are trees. Indeed the whole classification system for trees (and other flowering plants) is built around the design of the flowers. Although outwardly tree flowers may look very different, their basic components are all the same. Most flowers have four main parts: the stamen, which is the male reproductive organ and produces the pollen; the stigma, which receives the pollen; the style, which links the stigma to the ovary; and the ovary, which contains ovules that, after fertilization, develop into seeds. A few tree flowers have only male or female parts.

If both male and female components are present in the same flower, then the flower is said to be "perfect". The tree is then capable of self-pollination and it is known as an hermaphrodite. Self-pollination is far from ideal, and can lead to genetic weaknesses in the same way as inbreeding does in animals. Cross-pollination with another tree is better because it enables different genes to mix. Trees that are hermaphrodites include cherry, laburnum and lime.

To avoid self-pollination, some trees have developed separate male and female flowers. Such trees are known as monoecious and are particularly common where the main vector for pollination is the wind. Monoecious trees include beech, birch and hazel.

Some species only produce male or female flowers on any one tree. These species are "dioecious". This division of trees into sexes overcomes the problem of self-pollination but raises a new problem. Trees of opposite sexes must be relatively close together to have any chance of breeding at all. Trees that are dioecious include yew, holly and the New Zealand kauri pine. Holly berries are found only on female trees and then only when there is a male tree not too far away.

Below: Magnolias are insect-pollinated.

Below: Italian alders are wind-pollinated.

**Life cycle of a flower from bud**

*Left: In winter the flowers are protected within buds.*

*Left: The flowers emerge as the temperature rises in spring.*

*Right: Once fully open the flowers are pollinated by insects.*

*Left: Fertilized flowers produce berries in summer.*

*Right: Birds eat the berries and the seeds they contain are dispersed within the birds' droppings.*

## Welcoming guests

Tree flowers come in all manner of sizes, shapes and colours. Much of this diversity is linked to the pollinator. In general, flowers that are pollinated by animals tend to be larger and showier then those that are pollinated by wind. The wind is indiscriminate but animals need to be attracted. Some animal pollinators are attracted to flowers of certain colours and a few trees actually alter the colour of flowers once they have been pollinated to discourage further visitors. For example the colour of the markings inside the flowers of the horse chestnut, *Aesculus hippocastanum*, changes from yellow to red after pollination. To a bee, red looks black and very unattractive, so it visits a flower that has yet to be pollinated instead.

Sometimes tree flowers themselves may be quite inconspicuous but are surrounded by showy sterile flowers or leaf bracts to draw pollinators to them. The pocket handkerchief tree, *Davidia involucrata*, from China, for example, has large white bracts that guide pollinating moths to its flowers.

*Above:* Magnolia grandiflora *has some of the largest flowers borne by any tree.*

## Gone with the wind

Most wind-pollinated trees evolved in places where there was a shortage of insects. Wind pollination is common in the colder northern temperate regions of the world. All conifers are wind pollinated and most produce such large amounts of tiny-grained pollen that on breezy days, clouds of the stuff may fill the air around them. Conifer stamens are positioned at the tips of the branches to aid dispersal. Those of pine trees are bright yellow and stand upright from the needles like candles.

Alder, birch and hazel are also wind pollinated. Rather than having erect stamens like those of pines they have drooping catkins, each containing millions of pollen grains. In many places, these catkins are one of the first signs of spring. They are made all the more conspicuous by the fact that they appear before the tree comes into leaf. Oak also has pollen-bearing catkins but these are hardly ever seen because they open in late spring after the tree's leaves have emerged.

*Below: Burmese fish tail palm is pollinated by insect and by wind.*

## Insect pollinators

Pollination by insects is by far the most common method of reproducing among trees. More than 60 per cent of tree species in equatorial regions are pollinated by some kind of insect. Trees that use insects as pollinators tend to produce flowers with copious amounts of sugar-rich nectar. Their pollen grains are larger than those of wind-pollinated species and also quite sticky so that they adhere to insects' bodies.

*Below: The flowers of the persimmon tree.*

## Birds and mammals

In the tropics, birds are important pollinators of tree flowers. Flowers that are pollinated by birds tend to be tubular in shape (to keep the nectar out of reach of other animals), brightly coloured and unscented, since most birds have a poor sense of smell. Hummingbirds use their long beaks to reach inside the flowers of trees such as the angel's trumpet, *Brugmansia*, from Brazil. In Australia and South Africa, bottle-brush trees, *Banksia*, have masses of protruding pollen-covered stamens, which brush against birds' feathers as they collect nectar. Few temperate trees are pollinated by birds. The giraffe is the most unique pollinator of all. It transfers pollen between the flowers of the knobthorn acacia, *Acacia nigrescens*, which grow high up in the tree's branches.

*Below: The flowers of* Prunus *x* yedoensis.

# SEEDS

*Seeds are the next generation of tree. They contain all that is necessary for the creation of a mature tree virtually identical to its parents. Seeds come in a variety of forms; they may be contained within nuts, fruit or berries, or have "wings" to aid dispersal by the wind.*

Every tree seed has the potential to develop to become part of the next generation. "From little acorns mighty oak trees grow" is a well-known and accurate saying, although it has to be said that a tiny proportion of all acorns will ever have the opportunity to become mighty oaks. A mature oak tree can produce up to 90,000 acorns in a good year, but fewer than 0.01 per cent will grow to become anything like as mighty as their parent. Most acorns will be eaten by mammals or birds (a wood pigeon can eat up to 120 acorns a day), or simply land in a spot where germination and growth are impossible. It is because of this low success rate that the oak produces so many acorns. In terms of seed production 90,000 is actually quite modest; alder trees will produce around 250,000 seeds a year.

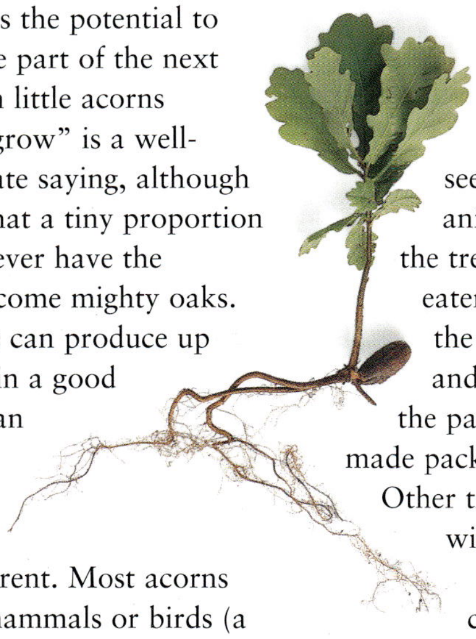

*Left: The first year's growth from an acorn.*

## Different seed types

Seeds are produced from the female part of the flower once it has been pollinated and one or more of its ovules fertilized. Just as tree flowers have evolved over millions of years in their quest to find the most effective method of pollen dispersal, so tree seeds also take on many different forms. The fundamental problem facing pollen is exactly the same as that for seeds; trees cannot move, so they have to find other ways of distributing what they produce.

Some trees wrap their seeds inside brightly coloured, sweet tasting fruits or berries. The fruit or berry has two roles; firstly to protect the seed and secondly to tempt animals to take it away from the tree. After a fruit or berry is eaten, the seed passes through the animal's digestive system and is excreted, often far from the parent tree, in its own ready-made package of fertilizer.

*Left: Apple seeds are contained within an edible, fleshy, protective fruit.*

Other trees enclose their seeds within tough outer casings as nuts. Once again, these casings help to protect the seed, but in this case it is the actual seed inside that is the attraction. Squirrels will collect and hoard the nuts, eating some in the process, but many of the nuts are never eaten and wherever the squirrel has stored them they will proceed to germinate and grow.

Conifer seed is known as "naked seed" because each individual seed is produced without a protective coat or cover. Conifer seeds are encased together in a cone but the scales of each cone can be bent back to reveal the unprotected seed inside. Each seed is often equipped with light, papery wings, which enable it to "fly" away from the parent tree on the wind.

*Below: Crab apples contain seeds.*

Other seeds, such as those of alder, are contained in cones, but rely on water for their distribution.

Whatever the method of dispersal there is always one aim: to get the seed as far away from the parent plant as possible. There is no point competing for living space with one's own progeny. Dispersal also reduces the risk of cross-pollination between parent and offspring in years to come.

## Matters of time

Most ovules are fertilized within days of pollen landing on the stigma. How long seed takes to ripen varies from tree to tree. Elm seed can be ready for dispersal less than ten weeks after fertilization.

Most temperate broad-leaved trees disperse their seed in the autumn of the year in which their flowers were fertilized. In many conifers, on the other hand, seed takes two years to develop. This is because fertilization is delayed for a year after pollination. Some conifers will hold seed in a sealed cone for many years after it has ripened, waiting for a special event to trigger its release. For the giant redwood this trigger is forest fire, which kills off all competing vegetation and provides a thick bed of nutrient-rich ash for its seeds.

*Below: Douglas fir seeds are paper-thin.*

*Above: Tamarind pods grow to 18cm/7in long and contain a soft pulp.*

*Right: The seeds of sweet chestnut are contained within a spiny casing to protect them from predation.*

## Berries and fruit

Most fruit and berries are brightly coloured to attract birds. Bright red rowan berries are loved by starlings, while red holly berries attract waxwings and fieldfares. The berries of hawthorn provide a vital source of food for many different birds in winter.

Normally, the flesh of berries is digested but the seed is not, and it gets passed out in the bird's droppings.

Fruits range in size from large tropical varieties, such as mango and papaya, to the small, glossy, black berry of the European elder tree. Many fruits are eaten by humans, and some trees, such as apple and olive, are farmed specifically for their fruit. Some fruits only become good for human consumption as they begin to rot, such as the fruit of the medlar tree, *Mespilus germanica*.

## Nuts and other seeds

Essentially, nuts are edible seeds. Some, such as hazelnuts, are encased in a woody shell. Others, like chestnuts, are surrounded by an inedible but more fleshy outer coating. They are distributed by birds and mammals. Squirrels and jays bury those nuts that they are unable to eat straight away. Some of the store is never returned to and these may germinate. Some seed casings are impenetrable to all but the most determined of foragers. The Brazil nut has one of the toughest cases of all but it is staple food for the agouti, a cat-sized rodent. Agoutis collect Brazil nuts and bury them, just as squirrels do in temperate forests.

Many dry seeds rely on wind for their dispersal. Eucalyptus seed is like fine dust and can be borne considerable distances on the wind. Some heavier seeds also ride on the wind. Those of maple and ash trees have extended wings known as keys, which help to keep the seed airborne. Sycamore seeds have paired keys.

Alder trees grow alongside rivers and watercourses. Each of their seeds is attached to a droplet of oil, which acts like a tiny buoyancy aid. After falling from the tree into the water, the seed floats downstream until it is washed ashore. Wherever it lands it will attempt to grow.

*Above: Seeds such as this of the sycamore are attached to wings to aid dispersal.*

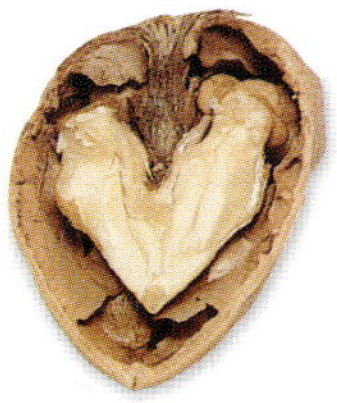

*Right: Walnut seeds are protected within a hard wooden casing.*

*Above: The nuts or seeds of the cream nut tree are edible, but difficult to find because they are also irresistible to monkeys.*

*Above: Some seed heads are incredibly attractive, such as these remarkable magnolia seed capsules.*

The world's largest seed comes from the coco de mer palm, *Lodoicea maldivica*, which is found in the Seychelles. It looks like an enormous double coconut and takes ten years to ripen. The heaviest of these seeds can weigh up to 20kg/45lb.

## Germination

Inside every ripe tree seed are the beginnings of a root, a shoot and two specialized leaves, which are known as cotyledons. If a seed arrives in a suitable location it will germinate straight away or wait until conditions become right for it to do so. In temperate areas this is in spring, when air and soil temperatures begin to rise.

### Germination to seedling

*Below: The first thing to emerge from the seed is the root. No matter which way the seed is lying, the root will instinctively grow downwards into the soil. Once the root has become established and is providing additional food and moisture, the two cotyledons emerge and begin the process of photosynthesis. Shortly afterwards, true leaves appear from a bud between the cotyledons and the tree begins to grow.*

# LIFE CYCLE OF TREES

*The life cycle of a tree is a fascinating, and in many cases very long, process of change and development. The initial struggle is coupled with rapid growth, while a sapling establishes itself. It then goes through a middle period of relative inactivity, to an eventual slow decline into old age and death.*

There is a saying that "an oak tree spends 300 years growing, 300 years resting and 300 years dying". Although these time spans may be optimistic for some oaks and very optimistic for most other tree species, there are, in fact, several important truths within this statement.

There is no doubt that trees have the potential to live for a very long time. They include by far the oldest living organisms on earth. The oldest tree in the world is a bristlecone pine, *Pinus longaeva*, which is growing 3,050m/10,000ft up in the White Mountains of California and has been verified as 4,700 years old. Close on its heels is Scotland's Fortingall yew, which is estimated to be somewhere between 3,000 and 5,000 years old.

Trees go through various stages of growth in much the same way as humans. In our early years we develop and grow at a relatively fast rate. By the end of our teens, growth slows down and stops and our bodies stay pretty much the same for the next 40 years or so. Then, as our three score years and ten approaches, we begin our decline into old age and eventual demise. This is similar to a tree's life cycle, the only real difference being the amount of time that it takes.

*Below: This sweet chestnut is in the final stages of its life cycle. It is still alive even though its trunk is hollow.*

## So how do trees grow?

As with any living organism, it all begins with a birth. In the case of trees it is the germination of a seed. However, it can also occur naturally when a piece of an older tree breaks away, develops its own root system and grows as a completely new tree. This frequently happens with willows growing along riverbanks. When the river floods, a lower branch may be broken off by the force of the water and swept downstream. Eventually this branch will come to rest and from it roots will develop, grow down into the mud and a new willow tree will grow. A tree grown in this way is known as a cutting. Cuttings have the same DNA as the tree they were once part of.

*Above: In old age some trees need help to retain their branches.*

Seeds do not have the same DNA as the tree that produces them. A seedling tree will develop its own genetic identity, taking on characteristics from both its male and female parent or, in the case of a self-fertilized tree, the characteristics contained in the genes of the male and female sex cell that initially produced it.

Once a seed or cutting has put down roots and sprouted its first leaves, the process of growth begins. The first few years, known as the establishment years, are critical in the life of any tree and the odds are stacked against survival. A young tree is vulnerable to being eaten or trampled by animals, its root system may not be able to withstand drought and it is far more vulnerable to forest fire than a larger tree. Other major threats include long periods of frost or waterlogging, which a fully grown tree would survive more easily.

## The growing years

Once a tree is established, it can get down to some serious growing. Trees grow upwards, downwards and outwards. The rate of growth will be determined by many factors, including the availability of water, light levels and climatic conditions.

## Upward growth

There is a popular misconception that trees grow from the bottom up and are continually moving skywards. In other words, if you were to go to any tree and paint a ring around it 2m/6½ft above the ground and then return to it five years later when the tree was 2m/6½ft taller then the ring would be 4m/13ft above the ground. Well, this is not the case; the tree may well be 2m/6½ft taller but the painted ring will still be 2m/6½ft above the ground. Growth occurs year on year only from the tips of the previous year's growth.

At the tip of each branch are growing cells. As these divide, they make the branch grow longer, so the tree becomes taller and wider. How fast these cells divide will depend on the species and many other external factors, such as the availability of water and light. Some plants, such as bamboo, can grow more than 50cm/20in a day, but there are no trees that grow at anywhere near this rate. The fastest-growing trees come from tropical parts of the world, simply because there are no seasonal changes and so growing conditions remain good throughout the year. One species of tropical eucalyptus from New Guinea, *Eucalyptus deglupta*, can grow 10m/33ft in just over a year, as can *Albizia falcata*, another tropical tree, from Malaysia. Willow is one of the fastest growing temperate trees. When it is coppiced (the stem is cut back down to the stump) it can grow more than 3m/10ft in a year.

Growth in any tree is affected by age; as trees get older their growth rate decreases until they eventually stop growing altogether.

*Below: Tree growth is determined by the amount of sunlight the leaves can absorb and the uptake of water through the roots.*

## Downward growth

There is a direct relationship between growth put on above ground by the branches and that achieved below ground by the roots. This relationship is known as the root:shoot ratio. The leaves on the branches provide food for the roots and in turn the roots provide water and minerals for the leaves. As a tree grows, it produces more leaves. These require more water and minerals, so the root system needs to grow in order to provide these minerals. To do that it needs more food from the leaves. All parts of the tree must work in harmony in order to continue the growth of the tree. The tree roots must develop to provide sufficient anchorage for the tree. The balance is a fine one; if leaves or roots fail, the tree will suffer and may die.

## Outward growth

As the branches grow longer, so the trunk, branches and roots become thicker. In temperate regions a mature tree trunk increases in diameter by about 2.5cm/1in every year. This growth is a result of the need for the tree to be able to transport increasing amounts of water and food to and from its branches. This process occurs immediately below the bark surface in the vascular system, which contains the phloem and xylem. Throughout a tree's life, the cambium constantly produces new phloem and xylem cells, which cover the inner wood. As these cells are added, so the tree's girth expands. In tropical regions this growth continues throughout the year. In temperate areas, growth only occurs in the spring and summer.

*Above: Without competition for light, saplings will establish much more quickly than those trying to grow in another tree's shade.*

## Growth rings

The cycle of growth in a temperate tree can be clearly seen when the tree is cut down. Each year the new cells that are produced under the bark create a new ring of tissue, visible in a cross section of the trunk. Each ring has light and dark sections. The light tissue is less dense and is made up of cells produced in the spring when the tree is growing fastest. The dark part of the ring is composed of cells laid down in the summer when the rate of growth has slowed. These rings are known as growth rings. By counting them it is possible to work out the age of a tree.

## Old age

As a tree gets older, so its rate of growth slows down and eventually it stops. In theory, provided that the root:shoot ratio remains stable the tree should live for many years. However, as a tree ages and its growth slows, so it also loses the ability to defend itself from attack. Opportunistic fungi will exploit this and eventually disease and decay upset the root:shoot ratio and the tree starts to decline.

## Rejuvenation

Some trees respond to hard pruning or coppicing. The re-growth is effectively young wood and displays all the characteristics of a young tree. Coppicing carried out on a regular basis can extend a tree's life almost indefinitely. In England, there is a coppiced small-leafed lime, *Tilia cordata*, at least 2,000 years old, which is still growing as a juvenile.

# TREES AND WEATHER

*Climate is the main controlling influence over where and how trees grow. Throughout time, climate changes have dictated the pattern of tree distribution and evolution across the world. In times of intense cold, such as the ice ages, billions of trees perished.*

The relatively settled climate of the last 12,000 years has resulted in fairly static patterns of tree distribution over that time. However, even minor changes in the earth's climate now, perhaps due to the greenhouse effect, could have a dramatic effect on future patterns of tree distribution and growth. An increase in the mean temperatures of just 2°C/35°F would result in a significant northward migration of temperate trees in the Northern Hemisphere. Thousands of acres of sugar maple plantations in New England would disappear as the climate became too warm for them. Spruce would have difficulty surviving in the United States, Great Britain and central Europe for the same reason. Deserts would expand into the Mediterranean regions of the world, threatening the natural diversity of trees in California, Spain and France. In the Southern Hemisphere more than half of the rainforest of northern Australia would disappear, along with vast areas of rainforest in central Africa and South America.

*Below: In areas of severe exposure trees will grow away from the direction of the prevailing wind.*

## Influencing weather patterns

Whereas the climate controls tree distribution on a global scale, trees actually influence weather patterns on a regional or local level. The process of photosynthesis raises humidity in the air. Where trees are found in large numbers, such as in equatorial rainforests, this humidity has an effect on daily rainfall. In the morning the sun warms up the forest and warm, moist air rises from the trees. As the air rises, it cools and condenses into water droplets, causing clouds to form, and it begins to rain. This process is repeated daily throughout the year all around the world's equatorial regions.

## Reducing the effects of weather

Ever since man evolved, trees have been used to reduce the effects of cold and wind exposure. Forests and woodlands provided natural shelter, and many original human settlements were created in clearings cut from the forest. Trees were also used to shelter stock. The practice of "wood pasture", grazing cattle or sheep within a forest, has been going on in Britain since the 2nd century. Timber from trees has been used to build shelters and dwellings for thousands of years and early man discovered that wood from trees could be burnt to provide heat.

*Above: Palm trees are well equipped to cope with the heat and drought of the tropics.*

Today, our use of trees to control the extreme effects of the weather has become far more sophisticated. We now know which are the best species to include within wind shelter belts, for example. We know how tall, how wide and how dense the belt should be. We also know how far away it should be from the area we wish to shelter. On average, a shelter belt 20m/66ft wide and 20m/66ft tall will provide wind protection on the leeward side for a distance of 400m/1,312ft. Such protection can increase cereal crop production by as much as 20 per cent.

Trees also help reduce the effects of frost. If tree shelter belts are planted across a hillside, cool air descending the slope will become "trapped" by the trees. Frost "ponds up" above the trees rather than travelling farther down the hillside or into the valley bottom. The same principle applies to snow. Strategic planting of trees on lower mountain slopes dramatically reduces the chance of avalanches occurring and their effects if and when they do occur.

Another important function trees have is the stabilization of soil and prevention of erosion. Tree roots help bind soil to the ground and soak up rainfall, while leaves and branches reduce the effects of wind on the ground. The latter is particularly important in areas of low rainfall where soil is often dry and loose. One of the biggest causes of soil erosion is deforestation. Once trees have been felled, fragile topsoil becomes exposed to both wind and rain, and is soon washed or blown away. Once the soil has gone so has the opportunity to grow food crops. Trees are now being re-planted in the Sahel region of Africa to try to reduce the effects of soil erosion and also the expansion of the Sahara Desert.

Trees also help protect against the effects of heavy rainfall and flooding. Those planted in water catchment areas soak up excessive amounts of rain, enabling the soil to release smaller volumes of water into the watercourses gradually, thus reducing the possibility of flash-flooding farther downstream. Trees such as willow, planted alongside riverbanks, reduce the effects of riverbank erosion when water levels are high.

*Above: Reducing temperatures in autumn will trigger an explosion of colour as the leaves begin to die.*

*Below: Many conifers have adapted to regular heavy snowfall by developing weeping branches, which are able to shed snow.*

*Above:* Welwitschia mirabilis *has adapted successfully to the Namibia Desert.*

### Indicators of climate change

In many parts of the world where there are seasonal differences in rainfall and temperature, trees form clear annual growth rings in their trunks. The width of these rings varies depending on the growing conditions in any one year. In cold, dry years, tree growth is slow, producing a narrow ring. In warm, wet years, tree growth is faster and the ring produced wider. As tree rings build up, they provide a year-by-year record of changes in climate. Because trees live for such a long time, these records may cover hundreds or thousands of years.

# TREES AND POLLUTION

*Trees are the air filters of the world. They absorb carbon dioxide from the air and replace it with oxygen. They also trap airborne particle pollutants, which are one of the main causes of asthma and other respiratory problems in humans.*

The process by which trees produce food and thereby harness energy for growth is called photosynthesis. As part of the process, trees absorb vast amounts of carbon dioxide from the atmosphere and break it down. The carbon is effectively locked up within the trees' woody structures of roots, trunk and branches. A healthy tree can store about 6kg/13lb of carbon a year. On average, 0.4ha/1 acre of trees will store 2.5 tonnes of carbon per year. Trees are the most effective way of removing carbon dioxide from the atmosphere and thereby reducing the effects of global warming.

When trees die naturally, the carbon they contain is gradually released back into the atmosphere as carbon dioxide. This happens so slowly that the gas can be reabsorbed by the next generation of trees growing alongside. However, when trees, or coal (fossilized wood), are burnt, the carbon they contain is released much more quickly. Living trees and other plants are only able to reabsorb some of it – the remainder stays in the atmosphere. Continual burning means a continual build-up of carbon dioxide.

The practice of "slash and burn" agriculture, carried out in tropical rainforests to create agricultural land, releases hundreds of thousands of tonnes of carbon dioxide back into the atmosphere. Even more serious is the large-scale burning of fossil fuels, such as coal and oil, in the West. The carbon dioxide produced traps more of the sun's energy than normal inside the atmosphere and so contributes to global warming.

*Below: Across the world 40ha/100 acres of forest are felled every minute.*

*Above: An ongoing threat to the tropical rainforests is the expansion of agriculture.*

During the photosynthesis process trees not only remove carbon dioxide from the atmosphere, they also replace it with oxygen, effectively producing clean air. Every day 0.4ha/1 acre of trees produces enough oxygen to keep 18 people alive.

## Biological filters

As well as removing carbon dioxide from the atmosphere, trees absorb sulphur dioxide produced by the burning of coal; hydrogen fluoride and tetrafluoride released in steel and phosphate fertilizer production; and chlorofluorocarbons, which are produced by air-conditioning units and refrigerators. Trees also trap other particle pollutants, many of which are by-products of the internal combustion engines in cars. These particles are one of the main reasons for the increasing incidence of asthma and other respiratory illness in people across the world. Research has shown that trees act as excellent biological filters, removing up to 234 tonnes of particle pollutants every year in cities the size of Chicago.

## Trees cause pollution

Some trees emit large amounts of certain volatile organic compounds (VOC), which react with nitrogen oxides and sunlight to form ozone – a significant ground-level air pollutant. Volatile organic compounds exist in fossil fuels, such as petrol. Most petrol nozzles are fitted with filters to stop the VOC from escaping into the atmosphere. It is of course impossible to stop trees from emitting high rates of VOC, but some tree species produce more than others. Scientists suggest that these trees should not be grown in large quantities where high levels of nitrogen oxides already exist, such as in and around towns and cities.

Trees that produce high levels of VOC include eucalyptus, oaks and poplars. The blue haze often seen over the Blue Mountains near Sydney, Australia, is in part caused by the release of VOC by eucalyptus trees. Ten thousand eucalyptus trees will emit about 10kg/22lb of VOC an hour, which is equivalent to that released by the spilling of 54 litres/12 gallons of petrol an hour.

There is evidence to suggest that certain tree species, particularly conifers such as spruce and fir, increase acidification of streams, rivers and lakes. This increased acidification can cause the decline of freshwater flora and deplete stocks of freshwater fish. The evidence for this effect is not clear-cut however. Acid deposition from the atmosphere (acid rain) can increase acidification in freshwater, and therefore the decline of freshwater flora and fauna may be attributable to that. There is much debate over whether conifer plantations in water-catchment areas actually do increase the level of acidification. Long-term studies are currently underway in North America, Great Britain and Scandinavia to establish the truth. In the meantime, forest policy in Great Britain at least has been revised so that coniferous tree species are no longer being planted directly adjacent to lakes and rivers.

## Trees are subject to pollution

Although trees can act as natural "air filters", ideally they need clean air to live and grow. Photosynthesis becomes more difficult for trees in areas of high air pollution. In highly polluted cities such as Mexico City, it is estimated that less than ten per cent of the tree population is healthy. Some trees, such as the London plane, *Platanus* x *hispanica*, are able to cope with relatively high levels of air pollution, but it is estimated that more than half of the trees in large cities are in decline due to air pollution. In New York City the average lifespan of trees is less than 40 years.

In many parts of the world the air is now highly polluted. Pollutants such as sulphur dioxide reach high into the atmosphere where they vaporize and mix with other chemicals and moisture to form acid rain. The damage caused by acid rain affects both coniferous and broad-leaved trees. The effects are more obvious on evergreen trees than deciduous ones because their needles or leaves are replaced less often. Discoloration of foliage is the first sign of acid rain damage. This is followed in extreme cases by defoliation and death. Nutrients are stripped from the leaves as acid rain falls through the canopy and the roots are slowly killed as the acid soaks into the soil.

There are a range of other sources of pollution that affect trees. Too much ozone disrupts the process of photosynthesis and can sterilize pollen, so reducing seed production. Particles of soot are also harmful because they coat leaves and thus prevent vital sunlight getting through. Even salt that is spread to de-ice roads can affect the chemistry of the soil around the roots of roadside trees.

*Below: Views of greenery make travelling by road less stressful for motorists, but the pollution emitted by vehicles is ultimately damaging to the trees.*

# GIANT TREES

*Trees are by far the largest living organisms on earth. Some of the tallest specimens would dwarf the Leaning Tower of Pisa in Italy, or Big Ben in London. A single banyan tree in India covers an area that is larger than a football pitch.*

Not only are trees the oldest living things on earth, they are also the largest. The world's biggest trees include the most famous individual trees of all. Some of these arboreal giants are local celebrities, others nationally famous and a few known about around the world.

Almost every country has its dendrologists (tree experts) and tree measurers, who can always be readily identified by their measuring tapes and skyward gaze. Countries such as Britain and the United States even have their own tree registers, which detail the largest specimen of just about every tree species that grows in that country. Books are written about the biggest trees and photographs taken. Champion trees are big news and interest in them is growing.

## What is a champion tree?

A champion tree is the tallest or fattest living example of a species. In order to be proclaimed champion it must have been accurately measured and those measurements recorded in an agreed way. The height is taken to be the distance from the ground to the top of the tallest living part of the tree. Girth is considered to be the distance around the trunk, and is read at 1.3m/4ft 3in from the ground.

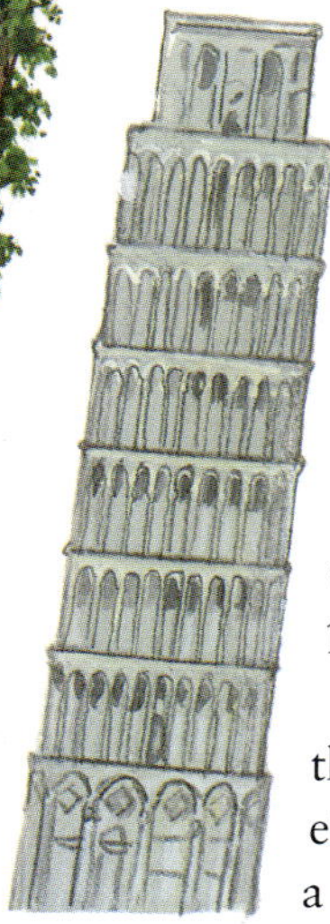

*Right: Compare the height of trees to the Leaning Tower of Pisa, which stands 58m/190ft tall. From the left: Montezuma cypress, 35m/115ft; New Zealand kauri, 51m/167ft; giant redwood, 83m/272ft.*

## Tropical giants

The tallest tropical tree, which is called *Araucaria hunsteinii*, is a relative of the monkey puzzle and grows in New Guinea. When last measured the largest specimen was 89m/293ft tall. In Africa, Dr David Livingstone (1813–73), camped under a baobab tree, *Adansonia digitata*, which had a girth of 26m/85ft. This tree appears not to exist now and the largest baobab alive today is 13.7m/45ft in circumference.

One of the largest trees in the world is found in the Calcutta Botanic Garden in India. It is a banyan tree, *Ficus benghalensis*, that was planted in 1782. In not much over 200 years it has grown into an arboreal titan with vital statistics that are simply astounding. The tree covers an area of about 1.2ha/3 acres and can provide shade for more than 20,000 people. It has 1,775 "trunks" (pillar roots) and an average diameter of more than 131m/430ft.

## Temperate giants

For sheer volume, the largest single living thing on earth is a giant redwood, *Sequoiadendron giganteum*, called General Sherman. The tree, which stands in the Sequoia National Park, California, has a diameter of 17.6m/58ft, is 95m/311ft tall and weighs 1,200 tonnes.

*Above: One of the largest trees in South America is this* Fitzroya cupressoides.

General Sherman is not the tallest tree in the world, however. This accolade belongs to a specimen of its cousin, a coastal redwood, *Sequoia sempervirens*, which goes by the simple but appropriate name of "Tall Tree". It grows on the Californian coast and when last measured, in October 1996, was 112.2m/368ft tall. If transported to London and placed next to the Houses of Parliament, this tree would be more than 14m/46ft taller than Big Ben.

General Sherman is not the fattest tree in the world either. That title is held by a Montezuma cypress,

*Taxodium mucronatum*, growing in the grounds of a church at Santa Maria del Tule, near Oaxaca in southern Mexico. This enormous tree is made even more impressive by its very close proximity to the church and other buildings, which take on toy-town proportions in its shade. The Santa Maria del Tule Montezuma cypress has a girth of 36.3m/119ft, outstripping even the mighty African baobabs.

Two trees from New Zealand also deserve a mention. They are the kauri, *Agathis australis*, which grows in the north of the North Island, and the totara, *Podocarpus totara*, which grows on both the North and South Islands. Both trees are antipodean giants, reaching ages approaching 2,000 years, girths of 13m/43ft and heights approaching 60m/197ft. They hold great religious significance for the Maori people, who believe that important spirits live within the trees. Both species have suffered at the hands of the loggers over the last 200 years and many of the biggest specimens have gone. Those that remain are protected within special sanctuaries, such as Waipoua State Forest, north of Auckland.

The tallest tree ever recorded was an Australian eucalyptus called the mountain ash, *Eucalyptus regnans*, measured in 1872 in Victoria. Unfortunately it never qualified as a champion tree because when it was measured it was already on the ground. It was 132.6m/435ft tall at the time, and thought to have been over 150m/500ft tall when it was at its peak. At one time giant mountain ash clothed the valleys that run from Melbourne to Tasmania. Today, only remnants of this mighty forest remain. There are still some big eucalyptuses in Australia, but nothing approaching these dimensions. There are now no trees over 100m/328ft tall.

There are no world-record-breaking trees in Britain but there is plenty of time for that situation to change. Britain has a good climate for tree growth. It is moist with few extremes of temperature as a result of its proximity to the Gulf Stream. A large number of exotic trees have been introduced into Britain in the last 200 years and many of them are world-beaters in their native habitats. The British examples are still babies but their growth rates so far suggest that some have the potential to develop into record-breaking giants. The title of tallest tree in Britain is currently shared between two Douglas firs, *Pseudotsuga menziesii*, both growing in Scotland. Each measures 62m/203ft tall – taller than the Leaning Tower of Pisa, which stands at 58m/190ft.

*Below: Giant redwoods can attain heights in excess of 100m/328ft.*

# THE FOLKLORE OF TREES

*Trees have been worshipped by people since at least the beginning of recorded time and most likely long before that. Like all ancient cultures in different parts of the world, the first inhabitants of Britain and Europe depended on trees for shelter, fuel, food and to cure their ills.*

Trees were by far the largest living things ancient people encountered and, as such, inspired awe and admiration and acquired great spiritual importance. They were believed to hold mystical powers, to be able to foretell the future and even to bestow everlasting life. We may dismiss such extravagant ancient beliefs as superstitious nonsense, but much tree folklore is based upon a deep factual understanding of how trees grow and respond to the changing seasons.

## The mystical yew

Out of all the native trees of Europe it is the yew, *Taxus baccata*, that attracts the most folklore. It has long been associated with darkness and death. In 1664 Robert Turner wrote: "*If the yew be set in a place subject to poisonous vapours, the very branches will draw and imbibe them, hence... the judicious in former times planted it in churchyards on the west side, because those places were... fuller of putrefaction and gross oleaginous vapours exhaled out of the graves by the setting sun.*" Such ideas seem extreme, but there are actually some strands of fact here. Yew trees are commonly found in churchyards and of those that are newly planted many are on the west side of the church.

So why are there so many yew trees in churchyards, particularly in Britain? Several explanations (including that of yew trees collecting bad air) have been given. A central ready supply of yew wood in the village could be quickly turned into long bows in time of strife by the parishioners, and that the walls around most churchyards would stop roaming stock from eating the yew's poisonous foliage, are two of the more plausible ones. However, the most likely reason and the one most tree experts agree on is that the trees predated the churches. The yew was not always associated with death. The Druids believed that its evergreen foliage represented eternal life, a sacred gift that would pass to anyone who slept beneath its boughs. So the locations where yew trees grew became sites of worship, places where people congregated. Centuries later, when Christianity arrived, Christians chose to build their churches on these sacred sites, symbolically linking the tree to Christ's resurrection. Slowly but surely the original reason for the yew's presence in early churchyards faded and it was simply accepted that yew trees would be planted alongside new churches as they were built.

Below: Druids believed that the evergreen foliage of the yew represented eternal life.

Above: The rowan's red berries symbolized Christ's blood.

## A protective rowan

The rowan or mountain ash, *Sorbus aucuparia*, is also found near churches and quite often around dwellings, especially in regions of Europe inhabited by the Celts. It is said that the wood was used to make the cross upon which Christ was crucified – its bright red berries symbolizing Christ's blood. It was also believed that the presence of a rowan in a churchyard would prevent the "slumbers of the dead from being disturbed". Rowans were widely believed to have protective powers, particularly against witchcraft. A rowan planted by a house door would prevent evil spirits entering.

Other trees were considered to have the opposite effect. At one time many people believed that if the white flowers of hawthorn, *Crataegus*

Above: The Druids followed their rites in groves of oak.

*monogyna*, were taken into the house, a death would surely follow. The same was believed of the elder, *Sambucus nigra*. If a sleeper inhaled the scent of elderflowers, the unlucky person would drift into a deep coma and death would eventually follow.

### A mighty oak

The oak, *Quercus robur* or *Q. petraea*, has always been considered the king of the forest, the greatest and noblest of trees. It measures its lifespan in centuries and produces hard, strong timber, which has been used to build everything from churches and houses to ships. In Roman Europe citizens who had achieved some great deed were crowned with oak wreaths, and even today, oak leaves are still included in the design of military decorations.

Oak leaves always feature in depictions of the mythical Green Man, who is in essence the spirit of all trees, but in particular of oak, and an ancient fertility symbol. In ancient British belief, people were the children of the oak, and so the Green Man was seen as the source of humankind. Carvings of the Green Man can be found in churches across Europe, often placed high in the roof, so as not to be in conflict with Biblical references.

It is widely documented that King Charles II of England hid in an oak tree in an attempt to avoid capture after his defeat at the Battle of Worcester in September 1651. His father, Charles I, had a similar adventure in an oak tree in 1646. Henry VI, head of the House of Lancastrians, sought shelter under an oak tree before fleeing to Muncaster in northern England in 1464 during the course of the War of the Roses. In Britain, 29 May was once widely celebrated as Oak Apple Day, marking the anniversary of Charles II's triumphal return to London on his restoration in 1660. Anyone not wearing a sprig of oak leaves on that day traditionally risked having their legs whipped with stinging nettles.

### Prediction and country lore

Many trees were believed to predict the future. Ash, *Fraxinus excelsior*, was thought to be helpful in romantic predictions. Ash leaves placed beneath the pillow of sleeping unmarried girls would offer them a glimpse in a dream of their future husbands. In parts of Europe, girls gathered crab apples on Michaelmas Day (29 September) and arranged them on the ground to form the initials of the one they wished to marry. On 10 October the apples were checked and the girl whose initials were best preserved was deemed most likely to achieve her aim.

Some old sayings reflect the deep understanding that farmers and others who worked the land had of their environment. "When the elmen (elm) leaves are as large as a farden (farthing, an old English coin), It's time to plant kidney beans in the garden," is an example of how trees were used as a natural calendar, indicating the gradual warming of the soil in spring by their response to the changing seasons and weather conditions.

"When the oak is out before the ash...we are in for a splash; when the ash is out before the oak we are in for a soak" is an old country way of forecasting the British summer weather. There is some confusion over just what is "out". Some consider it to be the flowers, and yet oak flowers are almost invariably produced up to two months after ash flowers. Others consider it to be the leaves, yet oak leaves are almost always well formed long before the bare branches of ash even begin to green up.

Some folklore shows what now seems an astute understanding of the healing properties of trees. Those suffering from aches and pains – or hangovers – were often advised to eat willow bark. This is not as strange as it may seem, for the bark of white willow, *Salix alba*, contains salicylic acid, which is a natural pain-reliever and a component of the drug aspirin.

Below: Historically ash has been associated with healing and protective properties.

# ANCIENT TREES

*The oldest living things in the world are trees. The life span of most is measured in centuries rather than years and there are some that have existed for millennia. Temperate trees generally live longer than tropical trees; although there are baobabs in South Africa said to be more than 3,000 years old.*

Until recently we knew more about the ages of trees in temperate than tropical regions (because there are no annual growth rings to count in tropical trees), but now evidence suggests that tropical trees can live just as long as their temperate counterparts. For many years it was thought that the rapid growth and decay that occurs in tropical rainforests meant that tropical trees rarely lived for more than 200–300 years. However, recent advances in carbon-dating have clearly shown that many tropical trees are capable of living for more than 1,000 years.

## Ancient tropical trees

There is speculation that some tropical trees may be more than 1,500 years old. The oldest tropical tree recorded with any certainty is a *castanha de macaco* (monkey nut), *Cariniana micrantha*, which is related to the Brazil nut. One specimen of this Amazonian rainforest tree is known to be 1,400 years old. The cumaru tree, *Dipteryx odorata*, from Brazil, is also known to live for more than 1,000 years. One of the best known and largest of all tropical rainforest trees, the Brazil nut, *Bertholletia excelsa*, regularly attains heights in excess of 50m/164ft, but none of those carbon-dated so far has been found to be more than 500 years old.

Africa's oldest known tree is a baobab, *Adansonia digitata*, growing in Sagole, in South Africa's Northern Province. Near its base it is 13.7m/45ft in diameter, and it is thought to be more than 5,000 years old.

The oldest tree in the world with a known and authenticated planting date is a fig tree, *Ficus religiosa*. It grows in the temple gardens in Anuradhapura, Sri Lanka, and was planted as a cutting taken from another fig tree given to King Tissa in 288BC. King Tissa planted it and prophesied that it would live forever: over 2,000 years later it is still going strong.

*Below: Africa's oldest known tree is a baobab growing in Sagole, South Africa. It could be over 5,000 years old.*

*Above: English oaks,* Quercus robur, *have been known to live for more than 1,000 years.*

## The world's oldest?

People have a fascination for the oldest and biggest. Over the years ancient trees have grabbed their fair share of the headlines. An 11,700-year-old creosote bush, *Larrea tridentata*, was said to grow in California's Mojave Desert. A 10,000-year-old huon pine, *Dacrydium franklinii*, was "discovered" in Tasmania. The most incredible is a king's holly, *Lomatia tasmania*, also from Tasmania, which was reported to be up to 40,000 years old. These are all great stories, but are they true?

Close scrutiny reveals that these are all clones that have grown from plants that were on the same site before. In terms of the age of the growth that can be seen above ground today, none is any older than 2,000 years. Whether or not they qualify for the title of oldest living trees is debatable. They have the same genetic material as the seedlings that first grew on the same spot all those millennia ago, but then all living things that reproduce asexually have the same genetic make-

up as their ancestors. We would not consider a female aphid produced asexually to be the same animal as its mother. No doubt the veracity of these claims will continue to be debated for some time to come.

## Ancient temperate trees

The oldest living tree in the temperate world is the bristlecone pine, *Pinus longaeva*. Bristlecone pines originated in the White Mountains of eastern California. The oldest are found in an area known as Schulman Grove, named after Dr Edward Schulman, who spent more than 20 years studying the trees there. In 1957 he discovered that many of them were over 4,000 years old and one of them, which he christened Methuselah, was considerably older. These trees all have solid centres, so Schulman was able to bore right into the centre of each tree and take out pencil-thick radial cores, from which he was able to count the growth rings. Carbon-dating since then has confirmed Dr Schulman's original age for these trees and Methuselah is verified as being 4,700 years old.

*Below: Olive trees will live for centuries; the oldest is believed to be 2,000 years old.*

For many years the giant redwoods, *Sequoiadendron giganteum*, were assumed to be the oldest trees because they were the biggest. We now know that isn't the case. The oldest redwood is a giant known as General Sherman, which stands in the Sequoia National Park, California, and is approximately 2,700 years old.

The oldest tree in Europe is believed to be the Fortingall Yew, which stands in a churchyard in Perthshire, Scotland. Although much of its trunk has rotted away, its girth suggests that it is at least 4,000 years old. There are many contenders for the oldest oak tree, *Quercus robur*, and in truth we will probably never know for certain which is the oldest. Oaks have a habit of looking more ancient than they actually are. England and Wales have possibly the best collection of ancient oaks in western Europe. There are several oaks in Britain and across Europe that are believed to be up to 1,000 years old, but their exact age is anyone's guess. Three oaks are locally proclaimed as being 1,500 years old: one in Brittany, France, another one in Raesfeld, Germany, and a third in Nordskoven, Denmark. The oldest olive tree, *Olea europaea*, grows in the Garden of Gethsemane, at the foot of the Mount of Olives in Jerusalem. It is said to have been planted at the time of Christ.

*Above: Methuselah, the world's oldest bristlecone pine.*

# UNDERSTANDING BOTANICAL NAMES

*Botanical names offer an international language, understood by gardeners, botanists, horticulturalists and foresters all over the world. Although they may appear confusing and difficult to remember, botanical names can tell us much about a plant once we understand the basics of plant nomenclature.*

Botanical names are sometimes referred to as "scientific names" or "Latin names", even though some are derived from Greek, Arabic and other languages. It was the Swedish botanist Carl von Linné (1707–78), who published his work under the Latinized form of his name, Carolus Linnaeus, who created a referable system of nomenclature that became the foundation of the system used today. He classified organisms by the way they reproduced themselves and the make-up of their reproductive systems.

## The binomial system of nomenclature

Linnaeus employed what is known as the "binomial" system of nomenclature, which gives each organism a two-word name. The first word is the generic (genus) name, which is always a noun. Most generic names are derived from older Latin, Greek or Arabic names. Some are based on characters in Greek mythology, such as Daphne, the nymph who was turned into a laurel tree by Apollo, while others commemorate people such as botanists; *Buddleia*, for example, is named after the Reverend Adam Buddle, who described it in the early 18th century. The generic name gives a clue to the relationships between organisms, because closely related species are given the same generic name but different specific names. For example, the English oak is called *Quercus robur* and the closely related Turkey oak is called *Quercus cerris*. All the species with the same generic name are said to belong to the same genus.

The second part of the name is the specific (species) epithet and is normally a qualifying adjective. The same descriptive word may be applied to quite different plants in many different genera: for example in the names *Magnolia sinensis*, *Camellia sinensis* and *Nyssa sinensis*, *sinensis* means "from China", and quite often specific names indicate the geographical origin of a plant in this way. Others describe a particular feature, such as leaf colour. *Quercus alnifolia*, for example, means "golden-leaved oak". Specific epithets also commemorate people, and many species, such as *Magnolia wilsonii* or *Pterocarya forrestii*, are named in honour of the prolific plant hunters (in these cases the Englishman Ernest Wilson and the Scotsman George Forrest) who brought the first propagating material to Europe.

Each two-word name, known as the binomial, can only ever represent one plant. The advantage of using old languages such as Latin or ancient Greek is that they are recognized universally but are no longer used as a spoken language, so that the meanings of the words should not change over time. If a continually evolving language, such as English, were used, there is every chance that names would be corrupted or changed over time and cause confusion. We all think we know what we mean by "dandelion", for instance, but across Europe this name represents over 1,000 different plants.

*Above: The English oak,* Quercus robur, *is known all over Europe, but is one of many species in the* Quercus *genus. All species of oak have common features.*

Quercus dentata

Quercus petraea

Quercus suber

Quercus frainetto

## Subspecies, varieties and cultivars

Occasionally, other words and names follow the binomial name. These reflect the fact that the plant is not exactly the same as the named species. A subspecies, such as *Magnolia campbellii* subsp. *mollicomata*, occurs when a regionally isolated population of a species shows distinct and regular characteristic differences to the species as a whole. This may be because it is growing on an island, on the other side of a mountain range, or in another part of the world, and the difference may result from climatic or physical changes in the environment. In evolutionary terms, a subspecies may be on its way to becoming a different species, but the differences are such that reclassification is not justified.

A variety is a plant that occurs naturally in the wild and regularly displays different characteristics to those that are normal for the species,

*Above:* Sorbus americana.

*Above:* Tilia americana.

*Above:* Fraxinus americana.

but not across whole populations, or not great enough to warrant it being named as a different species. For example the Christmas box, *Sarcococca hookeriana* var. *digyna*, is the same as the species, *S. hookeriana*, except that it regularly produces narrower leaves and two stigmas in each flower, whereas the species has three stigmas.

A cultivar exhibits a difference that does not occur regularly in the wild but is maintained only through horticultural techniques such as continual selection or grafting. An example is *Quercus rubra* 'Aurea': the species, *Q. rubra*, has deep green leaves, whereas the cultivar 'Aurea' has bright yellow leaves in spring, which gradually turn green. It is the result of a chance mutation, which was noted and propagated; once its reproduction was perfected it was made commercially available.

*Above:* Taxus baccata, the original species.

*Below:* Taxus baccata *'Semperaurea', a variety of the original species.*

## Name changes

The natural conclusion from all of this is that once a plant is named under the binomial system that name will remain the same for ever, and everyone will have a clear understanding of what plant that name represents. Most of the time this is true, but for botanists and taxonomists, some botanical names are a subject of constant disagreement. Taxonomists are still regularly moving species from one genus to another, and when this happens the consequence is that the generic name changes, and quite possibly the specific name does too.

### What do they mean?

It can be useful to learn the meanings of some of the most commonly used specific names. Many provide information about a tree's origins, habit or appearance, and can be helpful clues to the conditions they will need in order to grow well.

**Geographical**

*atlantica* = of the Atlas mountains of North Africa (*Cedrus atlantica*)
*europaeus* = of Europe (*Euonymus europaeus*)
*lusitanica* = of Portugal (*Prunus lusitanica*)
*orientalis* = eastern (*Thuja orientalis*)
*sinensis* = of China (*Nyssa sinensis*)

**Habitat**

*aquatica* = growing by water (*Nyssa aquatica*)
*campestre* = of the fields (*Acer campestre*)
*maritima* = by the sea (*Prunus maritima*)
*montana* = of the mountains (*Clematis montana*)
*sylvatica* = of the woods (*Fagus sylvatica*)

**Habit**

*arboreum* = tree-like (*Rhododendron arboreum*)
*fastigiata* = erect, upright (*Taxus baccata* 'Fastigiata')
*nana* = dwarf (*Betula nana*)
*pendula* = weeping, pendulous (*Betula pendula*)
*procera* = very tall, high (*Abies procera*)
*repens* = creeping (*Salix repens*)

**Leaves**

*cordata* = heart-shaped (*Tilia cordata*)
*decidua* = dropping its leaves, deciduous (*Larix decidua*)
*incana* = grey and/or downy (*Alnus incana*)
*latifolia* = broad-leaved (*Ilex latifolia*)
*macrophyllum* = large-leaved (*Acer macrophyllum*)

**Flowers**

*floribunda* = free-flowering (*Dipelta floribunda*)
*paniculata* = flowers in panicles (*Koelreuteria paniculata*)
*parviflora* = small-flowered (*Aesculus parviflora*)
*racemosa* = flowers in racemes (*Sambucus racemosa*)
*stellata* = star-like flowers (*Magnolia stellata*)

**Colours**

*alba* = white (*Populus alba*)
*aurea* = golden (*Pinus sylvestris* 'Aurea')
*bicolor* = two-coloured (*Picea bicolor*)
*carnea* = flesh pink (*Aesculus* x *carnea*)
*nigra* = black (*Pinus nigra*)
*rubra* = red (*Quercus rubra*)

# THE PLANT HUNTERS

*For centuries man has sought out new tree species; firstly for food, medicine and timber and latterly for ornamental purposes. Over the last 200 years plant hunters have scoured the world in search of previously unknown species to introduce to gardens and arboreta.*

Ever since they first appeared on earth, trees have spread to new places by natural means. Sometimes their movements have been the result of large-scale geological events, such as the break up of the supercontinent Pangaea, which began to occur around 193 million years ago. Sometimes they have been in response to changes in global climate. For example successive ice ages saw a migration of plants away from the poles and towards the Equator. Interglacial warm periods saw a movement back towards the poles but also away from low-lying land, which became flooded as the polar caps melted and sea levels rose. These movements occurred over many plant generations and sometimes took tens of thousands of years.

Later, the development of human civilization brought an accompanying quickening in the rate at which trees migrated and species settled in entirely new areas. Humans quickly realized that trees were useful. They could be used to make basic tools, produce food and provide shelter and fuel. As humans moved around the earth, they started using the trees that surrounded them and took parts of other trees with them, sometimes in the form of their fruits and berries. Inadvertently to begin with, but then consciously, humans became the vehicle for seed distribution and then, ultimately, moving trees. Trees took on new importance, becoming symbols for pagan worship. As time passed, their medicinal properties became better understood too.

Above: The Oriental plane was introduced to much of Europe by the Romans.

## The early plant hunters

The first record of plant hunting and tree collecting dates from 1495BC. Queen Hatshepsut of ancient Egypt sent out expeditions to Somalia to collect the incense tree, *Commiphora myrrha*, which produced a resin that was burned in Egyptian temples.

The Romans sped up the process of distribution, taking many trees with them as their empire expanded. Later, during medieval times, monks were also responsible for moving trees right across Europe, as they developed a network of monasteries from Russia to Portugal.

As civilization developed, so did the aesthetic appreciation of trees. Trees were considered an integral part of garden creation and from the 16th century onwards, European plant hunters started to look outside their own continent for new introductions. One of the first trees to be brought in was the horse chestnut, *Aesculus hippocastanum*, which was introduced into Vienna and then into France and England from Constantinople by the Austrian botanist Clusius in 1576. It was followed shortly afterwards by the Oriental plane, *Platanus orientalis*, introduced into Britain from Greece in the late 1590s.

It was in the early 1600s, as the exploration of North America began, that plant hunting really started to take off. Stories of amazing new trees swept through Europe and everyone with influence and money wanted their own collection. John Tradescant, (1570–1638) and his son, also called John (1608–62), were the first organized plant hunters. After starting out as gardeners for the rich and famous (including King Charles I), they introduced a phenomenal range of plants including dogwoods, *Cornus* species; lilac, *Syringa* species; red maple, *Acer rubrum*; the tulip tree,

Below: The Royal Botanic Gardens at Kew, England, has a fine ornamental tree collection.

*Liriodendron tulipifera*; and the false acacia, *Robinia pseudoacacia*, from North America into Britain.

## The past 200 years

As travel became easier, European plant hunters ventured farther and farther afield in search of ever more exotic and wonderful trees. North America, South America, the Himalayas, China, Japan, Australia and New Zealand all contained huge, largely unexplored tracts of land, ripe for discovery. Expeditions were funded by wealthy landowners or botanical institutions such as London's Kew Gardens, keen to build up their collections of botanical rarities. Arboreta sprang up all over Europe. Among the first was the magnificent Westonbirt Arboretum in England, created by Robert Holford in 1829.

## David Douglas (1799–1834)

Born in Perthshire, Scotland, David Douglas was probably one of the greatest tree collectors of all time. From an early age he displayed a great interest in all things horticultural. By the time he was ten he was apprentice gardener to the Earl of Mansfield at Scone Palace. In his early twenties he was commissioned by the Horticultural Society in London (later to become the Royal Horticultural Society) to collect for them in North America. Over the next ten years Douglas walked almost 10,000 miles, exploring the Pacific coast of North America. Along the way he collected over 200 species never seen in Europe before, which included the Monterey pine, *Pinus radiata*, from Southern California; the noble fir, *Abies procera*; the grand fir, *Abies grandis*; and perhaps the finest tree of them all, the Douglas fir, *Pseudotsuga menziesii*.

*Below:* Liriodendron tulipifera, *was introduced into Europe by John Tradescant (junior) in 1650.*

## William Lobb (1809–64)

A Cornishman, William Lobb was the first plant hunter employed by Veitch and Sons, nurserymen of London and Exeter. His first journey for the company in 1840 took him to the South American Andes, where, among other things, he collected more than 3,000 seeds from the monkey puzzle tree, *Araucaria araucana*. The seed was dispatched to Veitch and Sons, and by 1843 the first seedlings were on sale. In 1849 Lobb was sent on his second trip, this time to North America, with the aim of picking up from where Douglas had left off 20 years before. It was on this trip that he discovered the western red cedar, *Thuja plicata*, and collected seed from the coastal redwood, *Sequoia sempervirens*. Lobb's third trip in 1852, again to North America, was the one for which he is best remembered. It was on this trip that he discovered the largest tree in the world, the giant redwood, *Sequoiadendron giganteum*. Lobb arrived back at Veitch and Sons with seed from this remarkable tree just before Christmas in 1853. The tree was immediately named 'Wellingtonia' in honour of the Duke of Wellington, who had recently died. The Victorians fell in love with Wellingtonia and virtually overnight it became the most sought-after tree for estates across the British Isles.

## Ernest Wilson (1876–1930)

Another employee of Veitch and Sons, Ernest Wilson was sent to China in 1899 to find what had been described as "the most beautiful tree in the world" – the pocket handkerchief tree, *Davidia involucrata*. When he arrived in China, he was presented with a scruffy piece of paper with a map on it. The map covered an area of roughly 51,200km²/20,000 miles². On it was marked the rough position of a single pocket handkerchief tree. Amazingly Wilson found the tree's location, but all that was left was a stump. The tree had been cut down and its timber used to build a house. Undaunted, Wilson continued to search the area and eventually found another pocket handkerchief tree, from which he collected seed to send back to England. In 1906 Wilson left the employment of Veitch and Sons to became a plant hunter for the Arnold Arboretum in Boston, USA. From there he carried out further trips to China and to Japan. During his plant-hunting career Wilson introduced more than 1,000 new species to the western world.

*Above: The giant redwood species was introduced to Europe by William Lobb.*

## Charles Sargent (1841–1927)

Born in Boston, USA, Sargent created the Arnold Arboretum. A botanist and plant hunter, he collected mainly in North America and Japan. Sargent had several plants named after him. Perhaps the best-known is the Chinese rowan, *Sorbus sargentiana*.

# TREE CULTIVATION

*People grow trees for several reasons. Foresters plant on a large scale to produce trees for timber, while farmers and landowners cultivate them for their fruit. Many trees are grown for pure ornament, to brighten up parks or add structure to gardens.*

Ever since people first appeared on earth they have lived in a world dominated by trees. When modern humans finally arrived 35,000 years ago, trees covered more than two-thirds of all dry land on the planet. The other third was covered by ice or occupied by grassland or desert.

Modern humans have grown up alongside trees and forests. People have used them for shelter, as a source of food, fuel and all the necessary implements for life. In the beginning, they would have used whatever tree happened to be near to them. Gradually, however, awareness grew that certain tree species were better used for specific purposes. Harvesting trees from the wild had one serious disadvantage however – the more that were taken, the farther people had to travel from home to find the right tree for the job.

Above: Fruit trees have been cultivated in orchards for more than 2,000 years.

Below: The first roadside tree planting was carried out by the Romans, to provide shade for their marching legions.

## Semi-natural forests

Eventually people learnt that trees could be "managed". Seed could be collected, seedlings grown and trees planted in more convenient locations closer to human settlement. These were the first artificial plantations. At the same time, people realized that some trees did not die once they had been cut down, but re-grew from the stump. They learned that regular cutting (coppicing) provided a ready supply of thin, straight sticks, which could be used for a variety of purposes. The trees were all native to the area, and had been part of the original natural forest, but now that people were managing them in these ways, the forest had become semi-natural.

Above: The para rubber tree is grown in plantations throughout South-east Asia, where it is widely cultivated for its harvest.

## Early plantations

The idea of growing specific trees together in one place for a clearly defined purpose dates back almost to prehistory. Ancient Egyptians grew plantations of sandalwood so that they could have a ready supply of incense, for example. Both the ancient Greeks and Romans planted groves of olives, as well as orchards of cork oak. Fruit, such as apples and pears, has been grown in orchards in much of western Europe since early medieval times. In Britain from the 16th century, forests of oak were planted to supply timber to build wooden ships for the navy. One large warship could consume as many as 3,000 trees.

With the advent of the Industrial Revolution came the need for vast quantities of raw materials, many of which came from trees. One tree that typifies this change is the para rubber tree, *Hevea brasiliensis*, which originated in the Amazon rainforest. As early as the 15th century, rubber was being extracted from this tree to make shoes, clothes and balls. In 1823, Charles Macintosh, a Scottish inventor, coated cloth with rubber and invented the raincoat we now know as the mackintosh. Just 16 years later, an American, Charles Goodyear, discovered that heating rubber with sulphur caused the rubber to stabilize. This led to many new uses for rubber, including the manufacture of tyres. In 1872, Joseph Hooker, director of the Royal Botanic Gardens at Kew in England, sent plant hunter James Wickham to Brazil to collect seed from the para rubber tree. There, Wickham collected more than 70,000 seeds. Out of these seeds, 9,000 grew into young saplings, which were then shipped to Sri Lanka and Singapore. These saplings began rubber growing on an entirely new continent and formed the basis for plantations that now cover more than a million ha/2,500,000 acres across South-east Asia.

## Plantation forestry

The form of tree cultivation that we are most familiar with is the growing of trees in man-made forests for timber. Around 1.2 billion ha/3 billion acres of the temperate world are now covered with commercial timber plantations. More than 80 per cent of these are plantations of softwood trees – fast-growing conifers, such as the Monterey pine, *Pinus radiata*, which is widely planted in New Zealand, Australia and South Africa, and sitka spruce, *Picea sitchensis*, which is grown in the Northern Hemisphere. Both of these species have the potential to grow more than 1m/3ft in height per year. They are both harvested at 20–60 years of age, depending on growth rates, for use in construction or to create pulp for papermaking. Once harvested the forest is re-planted and the whole process begins again.

This "sustainable" forestry has far less impact on the environment than the wholesale destruction of natural forests, such as that which occurs in many parts of the tropics. Even so, it does have its disadvantages. Conifer plantations are normally made up of just one species, often not native to the country it is being grown in. The trees are planted close together in rows, thus creating poor habitats for wildlife. Some organizations have begun to acknowledge that this is unacceptable, and have taken steps to remedy the problem. In Britain, the Forestry Commission, for example, now ensures that at least 20 per cent of the ground is left unplanted and that other, native tree species are also grown within the conifer plantation.

Plantation forestry is not restricted to temperate countries, though. Large areas of Java are now planted with teak, for instance, to provide wood for furniture-making. Like the softwoods from conifer plantations, this timber is sustainably produced – the trees are replaced with new saplings immediately after they are cut down. Sustainable forestry is now being developed all over the world and is actively encouraged by organizations such as the Forestry Stewardship Council (FSC), based in Oaxaca, Mexico. Consumers are also better informed about where wood comes from – most products made from timber grown in certified sustainable forests now carry an FSC label.

*Below: In many areas, timber-producing, fast-growing conifers have replaced traditional broad-leaved woodlands.*

*Above: There are more than 20 different varieties of fir which are now cultivated for Christmas trees.*

## Christmas trees

Growing Christmas trees is now big business, with thousands of acres being devoted to their cultivation. These are "short rotation crops" being harvested normally in less than ten years from planting. Traditional Christmas trees, such as the Norway spruce, *Picea abies*, now have competition from many other species, such as the Nordman fir, *Abies nordmanniana*, which has citrus-scented needles that remain on the tree for longer after it has been felled.

# TREES FOR TIMBER

*Wood is humanity's oldest natural resource. It has provided us with food, fuel, weapons, shelter and tools for thousands of years. Wood can be easily shaped, it has great strength and is durable, hard-wearing and naturally beautiful.*

If you look around any room, you will see several things that are made of wood. Furniture, panelling, doors and window frames are the most obvious, but even the paint on the doors, the paper this book is printed on and the photographic film the photographs were taken with, all have a proportion of wood in them.

Wood is unique because it is the one basic natural resource that mankind can renew. When all of the world's oil, gas and coal has been exhausted, there will still be trees and we will still have wood – that is, as long as we manage our forests and woodlands properly.

## What is wood?

Wood is a type of tissue produced within trees by a specialized cell layer known as the cambium. Cambium encircles a tree, producing on its outside phloem cells, which transport food manufactured in the leaves to other parts of the tree, and on its inside xylem cells, or living sapwood, which transport water and minerals from the roots to the leaves. This sapwood is constantly being renewed, overlaying the existing sapwood and so enlarging the core of the tree. As each growing season passes, so the core of the tree gets larger. Only the xylem cells in the current year's sapwood are able to transport water and minerals; the cells beneath gradually die. As they die the old cells undergo a chemical change, turning drier, harder and normally darker. It is this change that creates the visually distinctive banding in a sawn log, demarcating the boundary between the young, soft sapwood and the older, harder inner wood, or heartwood.

*Below: Teak produces a strong, durable timber.*

## Is there a difference between wood and timber?

No, both refer to the woody cells that make up the structure of a tree. The difference between the two words is a matter of timing. When a tree is standing and growing, its bulk is referred to as wood. Once the tree has been cut down and sawn up, that bulk becomes timber. We buy planks from a timber merchant, but once the planks have been turned into something, the object that they have been turned into is generally referred to as being made of wood, rather than timber.

## A global business

Timber is produced by almost every country in the world. In general, softwood timber, such as spruce, larch and pine, is more likely to have been grown in temperate regions, whereas hardwood timber, such as teak, mahogany and ebony, is more likely to have come from the tropics. Some countries, such as Canada and Brazil, are virtually self-sufficient in timber supplies. Others, such as Great Britain and Japan, must import up to 90 per cent of their timber requirements. Over the last 100 years or so, the global trade in timber has increased dramatically. For many developing countries, timber is financially by far their single most important export. Unfortunately, this reliance on timber has resulted in the destruction of millions of hectares of natural forest. It is estimated that 80–200 million hectares/200–500 million acres of natural forest were destroyed in the last ten years of the 20th century.

## Timbers of the world

The world's most famous types of timber are household names. But, perhaps surprisingly, not every type is sourced from trees of just one species. Ebony, for instance, may come from any one of five different trees.

### Mahogany

Ever since the 16th century, when it was first brought to Europe by the Spanish, mahogany has been the most prized wood for cabinet- and furniture-making in the world. Mahogany is the collective name for the timber of several species of tree in the genus *Swietenia*, which originate from Central and South America. The most

*Below: The red-brown colouring of mahogany timber is valued for furniture production.*

*Above: Pine is used for construction work and was traditionally used for economy furniture.*

favoured species is *S. mahoganii,* but because this tree is now very rare most commercial supplies of mahogany now come from *S. macrophylla.* Mahogany has distinctive, rich red-brown colouring, complemented by dark figuring. As well as being beautiful, it is also very durable and is quite impervious to rot and woodworm.

### Teak

Indigenous to India, Burma and Indonesia, teak, *Tectona grandis,* has been introduced to Central America, where it is widely planted. Its timber has beautiful golden brown heartwood and is extremely strong and durable. Teak timber is used to make all manner of things, including furniture, boats, staircases and sea defences.

### Ebony

Certain species of *Diospyros* provide the timber known as ebony. There are two main types: African ebony, produced by trees that originated in West Africa and Madagascar, and East Indian ebony, produced by trees from Sri Lanka and southern India. Both types have a distinctive almost jet-black colouring. Ebony has always been used for furniture and sculpture, but it is best known as the timber used to make the black keys of pianos.

### Oak

There are more than 450 species of oak, most of which occur in temperate regions. The most important group for timber production is known as the "white oaks" and includes the English oaks, *Quercus robur* and *Q. petraea*; the American oak, *Q. alba;* and the Japanese oak, *Q. mongolica.* Timber from white oak has a creamy fawn sapwood and yellow-brown heartwood with silver-grey veining. It is one of the world's most popular timbers. Oak beams were used in the construction of many of the most important old buildings in western Europe, including the majority of tithe barns, churches and cathedrals.

### Spruce

This is a group of 20 evergreen conifers found growing naturally in most of the cool temperate regions of the Northern Hemisphere. Of those, only two are commercially important: Norway spruce or 'whitewood', *Picea abies*; and sitka spruce, *Picea sitchensis.* Norway spruce occurs in the wild throughout much of northern Europe, while sitka spruce originates from the Pacific coast of North America. Both timbers are widely used for interior building work, general joinery and the manufacture of pallets. Sitka spruce produces a significant amount of the world's virgin pulp supply for newspapers.

### Pine

European redwood, red deal and Scots pine are just three of the names given to the timber of *Pinus sylvestris,* a tree that occurs right across Eurasia from Spain to Siberia. Pine is one of the heaviest softwoods and has attractive pale red-brown heartwood. It is often used in the manufacture of economy furniture, as well as for general building work. In Britain, pine has been used for many years for making railway sleepers (railroad ties).

### Elm

The elm occurs naturally throughout northern temperate regions of North America, Europe and Asia. Although different species grow in different regions, the characteristics of the timber are broadly similar. The heartwood is dull brown with a reddish tinge and has prominent, irregular growth rings, which give an attractive figuring. Elm is very water-resistant – in Roman times it was used as a conduit for water, the heartwood being bored out to create a basic drainpipe. The Rialto bridge in Venice stands on elm piles. Sadly, because of Dutch elm disease, elm timber is in short supply across much of Europe and elm trees are much rarer than they once were.

*Below: Oak was tonce used for building ships. It has a straight grain and is very durable.*

# GENERAL USES OF TREES

*It is easy to take wood for granted because it features in almost every area of life. Humans have had a long association with trees and consequently there is an impressively wide range of useful products that can be obtained from trees.*

Over the centuries and to the present day, tree products have found their way into the larder, medicine cupboard, wine cellar, paint store, garage, garden shed, wardrobe, bathroom, library and jewellery box.

## Medicinal uses

Although the bark and wood of trees is seldom edible, extracts from them have given rise to some of the world's most important medicines. Malaria is said to have killed more people than all of the wars and plagues in history combined. Oliver Cromwell and Alexander the Great are two of the better-known people to have died at its hands. For centuries the only known treatment was quinine, an alkaloid found in the bark of the evergreen cinchona tree, which grows in the tropical forests of Peru and Bolivia. Quinine was first used to treat malaria by the Quechua Indians and in the 16th century the Spanish Conquistadors realized its potential. Called the "miracle cure" when it finally arrived in Europe, it was used to cure King Charles II, King Louis XIV and the Queen of Spain, among countless others. Quinine has been chemically reproduced since the 1940s; however, in recent years some forms of malaria have developed resistance to synthetic quinine and the cinchona tree has once again become the centre of attention.

*Below: Aspirin was originally derived from the bark of the white willow,* Salix alba.

*Above: Extracts of the leaves of* Ginkgo biloba *have been used to improve memory loss.*

If you have ever had a headache then the chances are that you will have reached for a bottle of aspirin, the world's most widely used drug. Before aspirin came in bottles, aches and pains could be cured by walking to the nearest river and finding a piece of willow bark to chew on. Aspirin is a derivative of salicylic acid, which comes from the bark of the white willow, *Salix alba*. Nowadays aspirin is produced synthetically.

The last remaining member of a family that existed when dinosaurs roamed the earth, the maidenhair tree, *Ginkgo biloba*, has long been used for medicinal purposes. The leaves have traditionally been a staple of Chinese herbal medicine and used to treat everything from asthma to haemorrhoids. Now maidenhair tree leaves have found their way into western medicine and are used to treat memory loss and coronary conditions. Fluid extracted from the leaves helps to improve blood circulation. It relaxes blood vessels, enhancing blood flow throughout the body but in particular that going to the brain.

More than 2,000 different trees are currently used for medicinal purposes. Many, such as the Pacific yew, *Taxus brevifolia*, are helping in the fight against cancer. *Castanospermum australe,* the Australian Moreton Bay chestnut, contains an unusual alkaloid called castanospermine, which is able to help neutralize the Aids virus HIV. Witch hazel, *Hamamelis virginiana*, is a tree with strong antiseptic qualities. Native American tribes such as the Cherokee made a "tea" of the leaves, which they used to wash sores and wounds. Another important medicinal tree species is *Eucalyptus globulus*. Its leaves contain the oil cineol, which is very effective in the treatment of coughs, sore throats, bronchitis and asthma.

## Trees in the home

One of the world's favourite drinks – coffee – is made from the seeds (beans) of three small evergreen trees, *Coffea arabica*, *C. canephora* and *C. liberica*. Now cultivated extensively throughout the tropical world, they originate from the montane forests of Ethiopia, where they grow to approximately 6m/20ft tall.

Products made from the Amazonian tree *Hevea brasiliensis* have found their way into just about every home in the world. Better known as the para rubber tree, its cultivation accounts for about 90 per cent of the world's raw rubber supply. *Hevea brasiliensis* produces a gummy, milky white sap beneath its bark as a natural defence against attack from wood-boring insects. This sap, known as latex, is tapped and collected once the tree reaches seven years old. An experienced tapper can harvest about

Above: A mature cork oak may produce up to 4,000 bottle stoppers per harvest.

Above: Olive trees provide an important crop of fruit for many Mediterranean regions.

450 trees a day. *Hevea brasiliensis* is cultivated on more than 7 million ha/17 million acres of land across the tropics. These plantations yield about 6.5 million tonnes of natural rubber every year.

Cork comes from the outer bark of the cork oak tree, *Quercus suber*. An evergreen tree, the cork oak is grown in Mediterranean countries, such as Portugal, Spain and Italy. Cork is a great insulator and it protects the tree's inner bark from forest fires and hot dry summer winds. It is also resistant to moisture and liquid penetration. The Romans used cork to insulate their houses and beehives, as soles for their shoes, stoppers for bottles, jugs and vases, floats for fishing nets and buoys for navigation purposes. Today its main use is in the wine industry. The cork oak is not stripped of its bark until it reaches 25 years old. After that, the cork is harvested every nine to twelve years, giving the tree time to grow a new "skin". Cork oaks are long-lived trees, regularly exceeding 200 years old. A mature tree provides enough cork to make 4,000 bottle stoppers per harvest.

Much of the food that stocks our supermarket shelves comes from trees. Citrus fruits, such as oranges and lemons, are produced by evergreen trees of the *Citrus* genus, originally from South-east Asia. The species that yields Seville oranges, *Citrus aurantium*, was introduced to Spain in the 12th century and its fruit became a valuable provision on long sea voyages, helping to prevent scurvy among the sailors. Today the orange is the most widely grown fruit in the world – every year more than 70 million tonnes are harvested.

Olive trees, *Olea* species, have been grown for their fruit for more than 5,000 years. Originally from Europe's Mediterranean region, they are now cultivated across the world, from Australia to California. The fruit is either eaten whole or pressed for its oil, which has significant health benefits. A ripe olive is about 20 per cent oil.

Even when we brush our teeth we are using products from trees. Toothpaste contains carboxymethal cellulose, which is basically pulped up wood. In Africa, small sticks made from the wood of a tropical tree called *Diospyros usambarensis* are chewed to clean teeth. The wood contains anti-fungal bacteria, which help to combat gum disease and tooth decay.

Below: Timber has been used for boat building.

# CHOOSING AND PLANTING TREES

*"The right tree for the right spot" may sound rather simplistic, but time spent researching just what is the right tree for a location is likely to yield dividends both in terms of maintenance and the tree's survival. Simple actions can help give a tree a good start in life and maintain its health.*

*Above: Bare-rooted plants need planting immediately so that they do not dry out or become damaged. This is the most economical method of buying hedging.*

*Above: Buying small and established trees that have a protected root ball is a good idea, as long as the soil around the root doesn't become deficient in nutrients.*

Although trees are generally more resilient than other plant forms, most still require specific conditions of soil, shelter and light to thrive. Climatic changes may also affect a tree's ability to establish and grow satisfactorily. It is therefore important to ascertain the key facts about a location before selecting, obtaining and planting a tree. For example, trees such as the Japanese maples, *Acer* spp, prefer to be grown in dappled shade, while others such as the Judas tree, *Cercis siliquastrum*, thrive only in full sun. Most willows, *Salix* spp, prefer moist soil, and some tender trees such as the Loquat, *Eriobotrya japonica*, require a sheltered warm location to survive.

The golden rule is to identify the conditions of your location first and then find a tree to suit it. Acquiring the tree first and then trying to find a suitable spot may lead to compromise and ultimately to growing difficulties. An important point on species choice: find out the maximum height of any tree you are considering and check that you have room to plant at least that distance away from any buildings.

## Soil type

It is important to discover the alkalinity or acidity of the soil – measured on the "pH scale", which runs from 1–14. Soil-testing kits are available from garden centres and nurseries. Neutral soil gives a reading of around 7, below 7 indicates that the soil is on the acid side, and from 7–14 indicates an alkaline soil. Once you have this information you can begin to identify suitable trees for the site. There are some general rules that apply to choosing trees: for example, cherry trees and other fruit trees, such as plums, thrive in alkaline soil, while most conifers and tree rhododendrons require acid soil to grow well. Observation and identification of the trees that are already growing well in the locality is also helpful. Changing the soil to suit the tree is not really an option for most gardens and it is better to work with the soil that you have.

*Right: Nurseries and garden centres stock a whole array of trees to choose from, and should offer advice about the suitability of the purchase for its intended site, as well as aftercare instructions.*

*Below: Electronic meters are one way of indicating the acidity and alkalinity of soil. Garden centres also sell inexpensive soil-testing kits.*

## Purchasing trees

Trees can usually be obtained from nurseries, garden centres, do-it-yourself stores and even supermarkets. They can also be purchased by mail order and over the internet. Your objective should be to obtain the best quality tree, wherever it comes from. Even when you cannot see the actual tree before purchasing there are certain questions worth asking: Has the tree been container-grown or is it bare-rooted? Container-grown trees can be planted at almost any time of year, except in high summer or when the ground is frozen. Bare-rooted trees should ideally be planted in the period from mid-autumn to early winter or, failing this, from late winter to mid-spring. Bare-rooted trees are likely to be cheaper.

Avoid "pot-bound" trees. These have been grown for too long in a pot that is too small for the root system. A telltale sign is roots growing out of the drainage holes in the bottom of the pot. Moss or other plant material growing on the soil surface is also an indication that the tree has been in the pot for a long period. It is perfectly in order for you to lift the tree from the pot to examine the root system. As well as identifying pot-bound trees, this can also be helpful in establishing if the tree has been containerized – that is dug up from the soil and simply placed in a pot for sale. If this is the case the soil may simply fall from the root system as you lift it from the pot.

It is recognized that smaller trees, below 1m/3ft in height, generally survive and establish more easily than larger, more developed, trees. They are also less likely to require staking and are less costly to purchase.

## Tree planting

**1** Choose your site away from buildings and other trees. Strip off the surface vegetation or turf over an area of about 1m/3ft diameter.

**2** Dig a hole in the centre of the stripped area at least 15cm/6in wider and deeper than the root system of the tree or the container holding it. Heap the soil you dig out on a plastic sheet. Break up the walls and bottom of the planting hole to ensure that the roots can penetrate the surrounding soil.

**3** Apply water to the hole, the soil heap and the roots of the tree. In soils that have either a high clay content or are extremely free-draining, mix some well-rotted organic material, such as leaf mould or garden compost, into the soil to improve its quality.

**4** If you have chosen a container-grown tree, remove it from the pot and gently tease the roots on the outside and bottom of the root-ball away from the compost. Place the tree in the centre of the hole, ensuring that the top of the root-ball is level with the surrounding ground level.

**5** Replace the heaped soil around the roots in the pit, firming gently with your heel as you proceed. Fork over the soil lightly and water well.

**AFTERCARE**

It is worth considering introducing a short length of plastic pipe, such as kitchen waste pipe, into the hole before planting. This should be at a 45 degree angle, with the lower end positioned directly beneath where the root system will be, and the upper end just above the ground surface, approximately 60–90cm/24–36in from where the stem of the tree will be. Once planted, the tree can be watered through this pipe, ensuring that the water gets to where it is most easily absorbed by the roots with minimal waste or evaporation.

Place a 1m/3ft diameter mulch mat (a piece of carpet underlay or thick blanket will do) around the tree. Finally, apply a mulch of treated pulverized or chipped bark over the matting to a depth of 5cm/2in. This will suppress weed growth and help to retain moisture in the soil.

# BEST TREES FOR ALKALINE SOIL

*Alkaline soils normally overlie limestone or chalk. They are quite often shallow, stony soils but have the advantage of being free-draining and warm quickly in the spring. Several trees from the Mediterranean region thrive in these conditions including the Judas tree,* Cercis siliquastrum.

The acidity or alkalinity of soil is reflected by its pH reading, which basically measures the level of calcium in the soil. Calcium is an alkaline mineral that can be leached from soil through rainwater percolation, but where the soil overlies calcium-rich chalk or limestone the levels of calcium will remain high. Alkaline soil is often shallow and stony, but has the advantages that it is free-draining and warms up quickly. Alkaline soils will normally dry out more quickly in times of drought. Many of the trees that grow well in these conditions originate from the Mediterranean region and other warm countries with similar climates and free-draining soils. They therefore thrive in sunny positions, where the warmth helps to ripen the wood, producing more flowers.

*Below: Olive trees are found on shallow, stony, alkaline soils. Trees that are hundreds of years old still produce good crops of olives.*

*Above: In warm regions of Europe* Albizia julibrissin *grows into a beautiful 'mimosa-like' tree with clusters of salmon-pink flowers.*

**Silk tree** *Albizia julibrissin*
Also known as pink siris, this broadly spreading Asian tree is sometimes confused with mimosa, *Acacia dealbata*, because of its finely divided, delicate, fern-like leaves. However, when in flower it is easily distinguished from mimosa by its clusters of beautiful, fluffy flowers with salmon-pink stamens, which are borne on the tree from late summer to early autumn. The flowers are slightly fragrant with a scent of freshly cut hay. The tree enjoys long warm summers and is best grown in a sheltered sunny position. In autumn the leaves turn yellow and orange before falling.

**Japanese crab apple** *Malus floribunda*
This is one of the most popular of all the flowering crab apples and one of the most floriferous, producing masses of flowers over its small, rounded crown in mid- to late spring – usually a full two weeks before most other flowering crabs. When in bud, each flower is a deep rich pink, opening to pale pink and then white flushed pink as it matures. Japanese crab apple was introduced into Europe from Japan in 1862, but intriguingly there is no known wild population in

*Above: Japanese crab apple produces masses of pale pink flowers in mid- to late spring.*

Japan. After flowering, small pea-sized, rounded fruits, yellow flecked with red, are borne from late summer to autumn.

### Variegated box elder *Acer negundo* 'Flamingo'

This sparsely branched, fast-growing maple, also known as ash-leaf maple because of its pinnate, ash-like foliage, is one of relatively few maples that actively thrive in alkaline soil. 'Flamingo' is a Dutch clone developed in the 1970s. It has soft leaves that are pink in bud and open pale green, with creamy-white to pale pink variegation around the margin, which may extend in places to the midrib of the leaf. The shoots have a blue-green bloom.

*Right: The tall candle-like spikes of horse chestnut flowers are a familiar sight in parks and gardens right across Europe in spring.*

*Below: The variegated box elder* Acer negundo *'Flamingo' is one of very few maples that thrives in alkaline soil.*

### Common horse chestnut *Aesculus hippocastanum*

Horse chestnuts thrive in any soil, including alkaline, and are common in cultivation in parks and gardens across Europe. They are among the most ornamental of all spring-flowering trees. A large specimen grown in an open position is a spectacular sight, especially when covered with a profusion of erect, candle-like spikes of flowers, white with blotches of red and yellow, in late spring.

Horse chestnut fruits contain an oil which was once used to make horses hooves and fetlocks shine: this is one explanation for the common name.

# BEST TREES FOR ACID SOIL

*Calcium is an alkaline mineral which is found in all soils. It can be lost from soil through rainwater percolation. Where the soil overlies sandstone or granite, the level of calcium falls and the soil becomes acidic. The trees detailed below cannot grow in calcium-rich soil because they become chlorotic.*

The acidity or alkalinity of soil is reflected by its pH reading, which measures the level of calcium in the soil. Neutral soil (a balanced mix of calcium and other minerals) has a pH reading of seven. A pH level below seven indicates an acid soil with less calcium. The trees described here will thrive in soil with pH less than seven.

## Red oak *Quercus rubra*

This North American oak has been widely grown in Britain and Europe since its introduction in 1724 and will grow well (and quickly) on both acid and alkaline soils. It is a tree for parkland, arboreta, and open spaces. However, it will produce vibrant red autumn leaf colour only on acid soils – elsewhere the colour tends to be a dull yellow or brown. Even on acid soil red oak rarely colours well in milder regions of Europe. This is because the tree needs the dramatic night-time temperature reductions, which occur as autumn approaches in northern Europe, to stimulate the leaf colour change.

*Above: One of the most recognized "acid loving" genera is* Rhododendron. *There are several species that grow to tree-like proportions in Europe, including* R. arboreum.

*Below: In the right conditions the red oak will produce stunning autumn leaf colour displays.*

## Sweet gum *Liquidambar styraciflua*

In Britain and Europe this large deciduous tree has yet to exceed 30m/100ft, even through it has been grown there since the 17th century. It thrives on moist, acid soils, where it produces autumn leaves of every hue

*Below: The leaves of the sweet gum produce a vibrant autumn colour on acid soils.*

from orange through red to deep plum purple, whereas on shallow alkaline soils it seldom colours well. Two of the best cultivars for autumn colour are 'Lane Roberts' and 'Worplesdon'.

### *Sassafras albidum*

This is an attractive tree with distinctive variable-shaped leaves that may be heavily lobed and similar in outline to leaves of the common fig. It belongs to the same family as cinnamon, *Cinnamomum* spp., and bay, *Laurus nobilis*, and has aromatic bark. It was long used to flavour non-alcoholic root beer and in Creole cooking in its native south-eastern United States, until sassafras oil was found to be carcinogenic in the 1960s (most sassafras flavouring is now produced synthetically). The tree grows well only on acid soils and has leaves that turn yellow, orange or purple in autumn.

### *Eucryphia glutinosa*

In cultivation in Europe this beautiful small tree, with pinnate leaves, is deciduous to semi-evergreen, but in its native Chile it is evergreen. It grows best in a sheltered position with dappled shade and in moist acid loam.

*Below:* Rhododendron arboreum *is one of the largest of all the tree rhododendrons.*

Given these conditions it will produce masses of fragrant white flowers with rose-pink stamens throughout the late summer months, followed by beautiful autumn leaf colours.

### Tree rhododendron
*Rhododendron arboreum*

Sometimes known as Cornish red, this magnificent rhododendron is known to reach heights in excess of 15m/50ft in its native habitat in the Himalayas. It is widely grown in temperate northern and western regions of Europe, especially Britain and Ireland. Like most rhododendrons it will not tolerate lime in the soil and grows best in moist but well-drained acid loam.

### Japanese stewartia *Stewartia pseudocamellia*

This is one of the most beautiful trees for growing on acid soil. It has red-brown flaking bark and a graceful spreading habit. Its white camellia-like flowers have bright orange stamens and appear in late summer, when few other trees are in flower. It also has excellent autumn leaf colour, which varies from bright orange to deep wine-purple. Originally from Japan, it has been cultivated in Europe since the 1870s but despite its beauty it is not widely grown possibly because it doesn't tolerate drought and wind.

*Below: The beautiful camellia-like flowers of the Japanese stewartia are borne in summer.*

*Above: The evergreen Chilean fire bush is one of the most beautiful small trees.*

### Rauli *Nothofagus nervosa*

This fast-growing attractive tree from Chile and Argentina is known as southern beech because its leaves are reminiscent of European beech. In fact they show more resemblance to common hornbeam because of their prominent leaf veins. On moist acid soils the new leaves emerge in spring a warm bronze and in autumn turn a rich orange-marmalade colour.

### Chilean firebush *Embothrium coccineum*

The Chilean fire bush grows wild from the Pacific Coast to high in the Andes Mountains in Chile and southern Argentina. It is a beautiful small evergreen tree that will thrive only in moist acid soil and prefers to be grown in dappled shade. In spring it produces clusters of large brilliant orange-red flowers that look like glowing red embers along the branches.

### Red maple *Acer rubrum*

This is one of the most striking of all American maples. It produces bright red showy flowers before the leaves emerge in early spring and, on acid soil, scarlet-red autumn leaf colours. It is a fast-growing, handsome large tree, ultimately reaching 30m/100ft in its native eastern North America. It will grow on neutral to slightly alkaline soils but seldom thrives and rarely produces vibrant autumn colour.

# BEST GARDEN TREES

*Such is the diversity of trees that grow in Europe that it doesn't matter how big or small the plot, how acid or alkaline the soil, or how sheltered or exposed the location, there is bound to be at least one tree that is exactly right for a site.*

In addition to their beauty, trees bring structure, height and permanency to a garden. They provide welcome shade on hot summer days. They screen us from views we would rather not look at and shelter other garden plants from cold wind and torrential rain. Trees produce oxygen, help to clean the air and lock up carbon, and can even reduce the effects of flooding and soil erosion.

Trees are a prominent part of the permanent structure of a garden, and isolated specimens are often planted as focal points of the design. The most popular choices put on a show-stopping performance at some time in the year, such as spectacular blossom or vivid autumn colour. They also need to be able to look good in all seasons to earn their place in the garden.

For garden planting close to buildings there are certain trees that are best avoided because their invasive roots can cause structural damage. These include some species of poplar, willow and oak. It is difficult to give a rule-of-thumb, "no go" distance when planting trees near to buildings because there are so many factors to be considered, such as the species, soil type and fluctuations in the water table. The answer is to research each location and potential species carefully.

There are trees for every season, such as cherries for spring, dogwoods for summer, maples for autumn and golden pines for winter. Some trees, such as the Asian birches, have it all: great bark, stunning autumn colour, graceful weeping form and delightful catkin flowers.

### Mount Fuji cherry *Prunus* 'Shirotae'

The Japanese name 'Shirotae' roughly translates as "snow white" and is said to be a reference to the way that this tree's pure white pendulous flowers hang – looking like drifts of snow in the deep gullies that surround Mount Fuji. It is a beautiful tree and one of the finest cherries for spring flowering in any garden. The Mount Fuji cherry has a fresh look to it: the flowers are well set off against bright apple green leaves, which emerge at the same time as the flowers. It has been cultivated in Europe since the early 20th century and is sometimes sold as *Prunus* 'Mount Fuji'.

*Below: As well as adding height, colour, fruit, scent and definition to the garden, trees help reduce sound and pollution.*

*Above: The Mount Fuji cherry,* Prunus 'Shirotae', *has long been considered one of the best flowering cherries for cultivation.*

**Willow-leaved pear** *Pyrus salicifolia*

This small tree can withstand long periods of ice and snow, prolonged drought and temperatures in excess of 32°C/90°F. It is also extremely attractive. The tree is fairly dense and slightly mounded, with the main branches more or less horizontal and younger branchlets drooping from them. In ideal growing conditions it may reach a height of 9m/30ft; more commonly 6–7m/20–23ft is a good size. The narrow silver-grey leaves are willow-like (hence the name) and emerge in mid-spring, covered in a beautiful soft white down that extends to the young branches and buds. These are followed in mid-summer by small brown, pear-shaped fruits.

*Below: In late spring the leaves of the willow-leaved pear tree are joined by pure white, closely packed lightly scented flowerheads.*

**Wedding-cake tree** *Cornus controversa* 'Variegata'

For unadulterated flamboyancy, one small tree is outstanding: the wedding-cake tree brightens even the dullest spring day, bringing to a garden a sense of light and sustained ornamentation that is rarely achieved by other trees. Each branch grows away from the main stem in an ordered, horizontal plane, reducing in length towards the top of the tree. The result is perfectly symmetrical and reminiscent of the tiers of a wedding cake, hence the name. The icing on this particular cake is the leaves, which are bright fresh green bordered by a rich cream margin. In spring the leaves are joined by clusters of cream flowers, and in autumn small purple-black berries shine out among purple-red leaves. This variegated form was cultivated in Japan in the mid-1800s. It was introduced into Europe in 1889.

**Himalayan birch** *Betula utilis* var. *jacquemontii*

The bark colour of this beautiful Asian birch is very variable and, depending on the location from which seed is collected, may vary from copper-brown through pink to pure white. The variety *jacquemontii* has stunning white bark flecked with orange-brown lenticels. Although attractive all year round, it is in winter that this tree really stands out: when planted against an evergreen backdrop the effect is sensational. It is a graceful medium-sized tree, eventually reaching 15–20m/50–65ft.

*Below: This beautiful wedding-cake tree is a variegated cultivar of the dogwood,* C. controversa, *which originated in Japan, China and the Himalayas.*

# TREES FOR SPRING

*More trees flower in spring than at any other time of year. Magnolias and cherries predominate, but unusual and dramatic trees such as the handkerchief tree are well worth seeking out. Many public gardens are renowned for their springtime flowering displays of trees and shrubs.*

As lengthening days and rising temperatures encourage their plump buds to open, deciduous trees usher in the spring with an explosion of fresh colour that makes them the centre of attention, both in the wild and in parks and gardens. Before they have come fully into leaf the trees allow sunshine to filter through their young foliage, creating dappled, luminous effects. The absence of a heavy leafy canopy also means that early-flowering bulbs and herbaceous plants can flourish on the ground beneath, completing the seasonal picture.

Some of the world's finest ornamental trees flower in springtime. Catkins are an endearing feature of trees such as willow, hazel and alder early in the year, and the flowers of many species open before the leaves appear, or while they are still very small. From the large flamboyant flowers of the Asian magnolias to more delicate blossoms such as those of the North American snowy mespilus, *Amelanchier lamarckii*, trees herald the spring with a remarkable variety of floral colour and form.

Flowers are not the only spring attraction trees have to offer. Some species have richly coloured new leaves, creating a display that can be just as striking as a covering of blossom, and many cultivars have been bred to accentuate such features. One of the loveliest springtime displays is that of *Sorbus aria* 'Lutescens', whose new leaves open in globular buds of shining silvery white. Unusual leaf colour may be spectacular yet fleeting, as in the sunrise horse chestnut, *Aesculus* x *neglecta* 'Erythroblastos', with bright salmon-pink leaves, or a more lasting feature, such as the soft yellow foliage of the golden false acacia, *Robinia pseudoacacia* 'Frisia'.

*Above: Some of the earliest magnolias flower in mid-winter and continue until late spring.*

*Below: Along with cherries, one of the first small trees to flower in spring is the snowy mespilus,* Amelanchier lamarckii.

### Snowy mespilus

*Amelanchier lamarckii*

The snowy mespilus has a rather spreading habit, quite often with several stems growing from the base, which adds to its attractiveness. The overall form is light and airy, with plenty of space between the slender branches. In spring it produces masses of small, white star-shaped flowers in open, spreading racemes, which contrast superbly with the warm copper-bronze of the newly emerging leaves. First described in 1783 from a plant cultivated in France, the snowy mespilus is believed to be native to Canada, from where it was introduced into France in the 17th century. It has been widely cultivated in Europe since the early 19th century and has naturalized in parts of southern England, as well as in Belgium, Holland and north-west Germany.

**Judas tree** *Cercis siliquastrum*
The Judas tree is one of the loveliest of all spring-flowering trees. It is also one of the most curious, because it produces lilac-pink, pea-shaped flowers, as if by magic, from every twig and branch, and even growing straight out of the main stem. In a good spring the whole tree is covered in a lilac-pink cloud of blossom. Attractive, delicate, blue-green heart-shaped leaves emerge just as the flowers are fading in late spring. It has been cultivated in western Europe for more than 350 years, but originates from the eastern Mediterranean and western Asia. It flowers at its very best in hot and sunny regions.

**Voss's laburnum** *Laburnum x watereri* 'Vossii'
This beautiful small tree makes a spectacular garden feature in late spring. It is sometimes known as the golden chain tree, and when fully in flower there is really nothing quite like it. It has weeping chains, up to 30cm/12in long, of large, fragrant, bright yellow pea-flowers, which drip from every spreading branch. Raised in Holland late in the 19th century, it has superseded all other forms of cultivated laburnum in Europe.

*Below: Voss's laburnum is one of the most popular garden trees in Europe. It produces beautiful long chain-like racemes of fragrant, golden-yellow pea-flowers in late spring.*

*Above:* Magnolia campbellii *is one of the earliest magnolias to come into flower.*

**Great white cherry** *Prunus* 'Tai Haku'
The great white cherry is one of the most beautiful flowering cherries in cultivation. It also has one of the most curious stories. 'Tai Haku' was an old Japanese cultivar, grown in the Kyoto region until the 1700s. It was thought to be extinct until, in 1923, the English plantsman Collingwood Ingram found a moribund specimen in a garden in England, and progeny from the original tree was taken back to Japan and planted in the Kyoto Botanical Gardens. Today it is widely cultivated in Japan and throughout Europe and North America. In mid-spring it produces a profusion of delightful, pure white single flowers, with dusky pink central stamens. These flowers, which drip from every branch and twig, are accompanied by bronze young leaves newly emerged from their winter buds.

**Handkerchief tree** *Davidia involucrata*
This large, stately tree is one of the highlights of spring. Its flower clusters are small, but each one is guarded by a pair of long white, or creamy-white, bracts, which hang in profusion from every bough, like miniature sails. The whole effect is dramatic, particularly when a slight breeze sets the bracts fluttering. The tree is native to Sichuan and Hubei provinces in western China, where the French Jesuit Missionary Abbé Armand David discovered it in 1869. When Ernest Wilson, the English plant collector, went to find it in 1903, he described finding a grove of around 20 mature trees "growing on a precipitous slope…their crowns one mass of white…and most conspicuous as the shades of night close in".

**Campbell's magnolia** *Magnolia campbellii*
One of the most remarkable of all flowering trees, this magnificent large magnolia is covered in pale pink to deep rose-pink flowers, which may be up to 30cm/12in across, in early spring. The flowers resemble large pink water lilies: the inner tepals are closed in bud-like formation while the outer tepals become wide-spreading as they mature, creating a cup-and-saucer shape. After its introduction into Europe around 1865 from its native Himalayas, the tree is believed to have flowered first in County Cork, Ireland, in 1885. Specimens 20m/65ft tall are now common in western Europe.

# TREES FOR SUMMER

*Trees appear to dominate the landscape in summer more than at any other time of the year, with their broad canopies adding a rich tapestry of colour and texture to town and countryside. When foliage abounds, trees are at their most easy to identify.*

As the leaves expand to create a canopy, deciduous trees take on their characteristic silhouettes and create strong blocks of colour and texture. Trees are a valuable addition to the summer garden, forming excellent backdrops for other, smaller plants as well as providing much-needed shade and shelter.

Although not so numerous as spring-flowering trees, those species that flower in summer can be just as attractive. The early summer flowers of the horse chestnuts, *Aesculus* spp, are held above the leaves in spectacular white, pink or red spikes, or "candles", effectively set off by the dark green foliage behind them. While the spring-flowering magnolias bear their flowers before the leaves, their summer-flowering cousins, such as *Magnolia grandiflora,* benefit from the beautiful contrast between their huge waxy white blooms and glossy dark foliage. The cup-shaped flowers of *M. wilsonii* hang face-down, so that they can be enjoyed from under the tree. The starry flowers of the Japanese snowbell tree, *Styrax japonica*, also hang under the branches so that they are not lost among the neat, bright green leaves.

The fragrance of summer-flowering trees travels far on a breeze or in the still air of warm evenings. Limes, *Tilia cordata* and *T. tomentosa*, have abundant, richly scented flowers, which are much visited by honeybees. Aromatic leaves such as those of the myrtle, *Myrtus luma*, are also at their most powerful in hot weather.

*Above:* Robinia pseudoacacia 'Frisia' *holds its yellow colour into summer.*

### **Golden-leaved robinia** *Robinia pseudoacacia* 'Frisia'

The pinnate leaves of this graceful, wide-spreading tree are a bright, intense, almost glowing, golden yellow. The tree looks its very best when the early morning or late evening summer sun shines through its translucent foliage. The vibrancy of the leaves is at its most intense in early summer, when the tree's beauty is heightened by clusters of white pea-like flowers interspersed with the foliage. 'Frisia' is a cultivated form of the American black locust tree, *R. pseudoacacia.* It was raised in 1935 and has become one of the most popular trees for ornamental planting in western Europe.

### **Golden rain tree** *Koelreuteria paniculata*

This is one of the most stunning of all summer-flowering trees. Each flower is made up of four small bright yellow petals, around a central cluster of orange-red stamens. Individually they are not particularly striking, but as there are about a hundred on each long panicle, with several hundred panicles distributed throughout the tree's canopy, the effect is stunning. The abundant flowers, coupled with the attractive bright green pinnate leaves, make it easy to see why this tree has become a firm favourite for planting in the summer garden. Native to northern China and southern

*Above:* Koelreuteria paniculata *produces a spectacular display of flowers in summer.*

Mongolia, it was introduced to Europe as long ago as 1763 and is believed to have been cultivated at Croome Park, Worcestershire, England, one of the first landscapes to be designed by Lancelot "Capability" Brown.

**Snowdrop tree** *Halesia carolina*

This is a beautiful, early summer-flowering tree. In a good year every branch is laden with pendulous, creamy-white, snowdrop-shaped flowers, the centres of which are filled with a mass of bright orange stamens. It has oval pointed leaves which, when they first emerge in early summer, are covered in soft white down. They make the perfect landing strips for bees, which love to forage for the sweet nectar within the flowers. This spreading, medium-sized tree originates from south-eastern United States and was introduced into Europe in 1756, having been identified by Stephen Hales, an English clergyman and amateur botanist for whom the genus was named, but it is still surprisingly uncommon in cultivation.

*Below: The small bell-shaped, white flowers of* Halesia carolina *are resistant to pests and diseases making it a good garden tree.*

**Honey locust** *Gleditsia triacanthos* 'Sunburst'

As its name suggests, the frond-like, almost feathery, new leaves of this attractive small tree are bright golden-yellow, and contrast superbly with the older dark green foliage. The effect is carried right through summer as fresh leaves are produced. This is the ideal tree to brighten up a dull corner in the garden. The honey locust is native to central and south-eastern United States and had arrived in Europe by 1700, when Bishop Henry Compton planted a specimen from Virginia in his garden at Fulham Palace in London. The cultivar 'Sunburst' originated in 1953 as a sport of the American form, *G. triacanthos inermis.*

**Tulip tree** *Liriodendron tulipifera*

This is one of the finest and largest of all summer-flowering trees. A mature specimen is truly majestic and never more so than in summer, when its broad shapely crown is covered with tulip-like flowers. Each cup-shaped flower is bright yellow-green with orange markings on the inside. *Liriodendron tulipifera* is native to eastern North America and is believed to have been introduced into Europe by John Tradescant, gardener to King Charles I of England, in the 1640s. By 1688 specimens were being grown in the grounds of Fulham Palace, London.

**Manna ash** *Fraxinus ornus*

Ash trees are not renowned for their flowers, as most are small, inconspicuous and borne in early spring, but the manna ash produces very attractive, large pendulous panicles of creamy-white fragrant flowers, which hang in numerous fluffy clusters from each branch in late spring and early summer. This beautiful flowering ash grows wild in south-western Asia and southern Europe and has been widely cultivated throughout central and western Europe since around 1700.

**Golden-leaved Indian bean tree** *Catalpa bignonioides* 'Aurea'

There can be no better tree to have in the garden in summer than the golden-leaved Indian bean tree, sometimes called the golden catalpa. It has just about the most radiant golden leaves of any tree. The leaves are slightly translucent and appear to intensify the light, so that even on cloudy days walking beneath the canopy makes it feel as if the sun has come out. *Catalpa bignonioides* 'Aurea' is a distinct form of the Indian bean tree, native to the eastern United States from Mississippi to Florida. It was cultivated in Europe in 1877 and has been maintained in cultivation by propagation ever since.

*Below:* Fraxinus ornus *is a small compact tree with a showy display of stunning white flowers in summer.*

# TREES FOR AUTUMN

*There are few trees that flower in autumn. Instead an incredible kaleidoscope of colour is produced by the leaves of many tree species commonly grown in Europe. The myriad shades of green foliage fade in many species to leave pigments in a startling array of colours such as red, scarlet, orange and yellow.*

*Above: Acer species are renowned for their spectacular autumn foliage.*

Autumn is the time of year when trees really do take centre stage, bringing a hundred different shades of fiery colour to the countryside and the garden. During spring and summer, leaves contain a green pigment known as chlorophyll, which absorbs light energy from the sun. The energy is used to carry out the process called photosynthesis, which produces the food trees require to live and grow, and this is stored in leaves in the form of starches and sugars. In autumn, as day length reduces and night-time temperatures begin to fall, most trees begin a period of dormancy when they do not need food. Any remaining chlorophyll gradually breaks down and the green pigment begins to disappear, so that other pigments, which have been present in the leaves all the time, are revealed. These range from purple through red and orange to gold and yellow, representing the varying levels of starches and sugars that have been stored in the leaves.

Many trees also produce fruits that ripen to beautiful, glowing colours in autumn. Some, such as apples, pears and walnuts, are edible straight from the trees; others can be turned into jams or wine, while many provide food for birds and other creatures. The red and orange berries of many species harmonize with the autumn tints of the surrounding foliage, but some autumn fruits are truly spectacular. The berries of the Kashmir rowan, *Sorbus cashmiriana*, are pure white, in startling contrast to the orange leaves. The fruits of the spindle trees, *Euonymus* spp, are even more dramatic, as their lipstick pink lobes split open to reveal vivid orange seeds inside.

### Japanese maple *Acer palmatum*

One of the finest small trees for autumn leaf colour is the Japanese maple, *Acer palmatum*, which, despite its name, is also native to China and Korea. It has been a favourite for cultivation in European parks and gardens since its introduction in the

1820s from Japan, where it was grown in temple gardens. Today, literally hundreds of cultivars of *A. palmatum* are available. They vary according to the colour and vibrancy of their autumn foliage, as well as the shape and size of their palmate leaves. The cultivar 'Osakazuki' is one of the most reliable for autumn colour, turning brilliant red every year. 'Burgundy Lace' has fine, dissected purple palmate leaves.

**Katsura tree** *Cercidiphyllum japonicum*
The Katsura tree is one of the most beautiful of all autumn trees. It has been described as the "queen of the forest" and the title is quite apt, for there are few trees that can match its regal splendour in autumn. It has delicate heart-shaped leaves that turn a clear butter yellow in mid-autumn. When seen against a blue autumnal sky the effect is stunning. As a bonus, the leaves also emit a sweet caramel-like fragrance as they turn colour. The tree is native to China and Japan and was introduced into Europe in 1881. Today, it is frequently found planted in parks, gardens and arboreta right across western Europe.

**Persian ironwood** *Parrotia persica*
This medium-sized tree is remarkable in autumn for the way its leaves have the ability to turn almost any colour, from deep purple to bright orange, sometimes within the same leaf. In a good year the whole effect is reminiscent of a giant, sprawling bonfire. Persian ironwood is one of the few autumn-colouring trees that perform well in soil with high lime content. In its native habitat, from northern Iran to the Caucasus, it has a relatively upright habit, whereas in cultivation in Europe it tends to be broad and spreading.

**Chinese mountain ash** *Sorbus vilmorinii*
The fruits are unequalled for their beauty and make this graceful small tree worthy of growing in any garden. The fruits are pea-sized, at first deep pink, then gradually fading to blush-white. They are borne in open pendulous clusters from mid-autumn, while the pinnate leaves become plum-coloured. The fruits persist on the tree well into winter.

**Rowan** *Sorbus aucuparia*
Otherwise known as mountain ash, this small tree, which is native to most of northern and western Europe, is widely grown for its fruits, which are prominent from early autumn onwards. They are glossy bright orange-red, about the size and shape of a pea, and are borne in large pendulous clusters. The sight of a rowan covered in berries is unforgettable. However, these fruits are short-lived once ripe because they are loved by birds, who will strip a tree of its fruit overnight in their quest for the sweet fleshy pulp inside. The mountain ash is a common tree of the uplands and, as its alternative name suggests, can often be found growing high on mountain slopes where few other trees would survive. It is, however, a popular tree for planting in urban areas and brings colour to the streets in autumn.

*Below: A typical autumnal parkland scene includes leaf fall and different fading foliage colours.*

# TREES FOR WINTER

*It is in winter that conifers and other evergreen trees come to the fore, retaining their solid presence and form in the landscape when deciduous species are bare and skeletal. It is also the best time of year to enjoy some of the beautiful bark that many trees, such as the Tibetan cherry, produce.*

The elegant form of many large conifers makes them ideal for planting as isolated specimens in key locations, where they become even more prominent in winter. Their foliage contributes texture and colour to the winter scene, and as hedges and screens they provide formal structure and shelter from wind. Their sculptural forms look particularly effective in frosty or snowy weather.

Some deciduous species also have much to offer in winter. Those that retain their fruits, such as the crab apples, *Malus* spp, and the strawberry tree, *Arbutus unedo*, look ornamental and are of value to visiting birds. Some, such as witch hazel, *Hamamelis* spp, and Persian ironwood, *Parrotia persica*, produce flowers in winter, though these are few and far between. Once stripped of their leaves it is the trunks and branches of deciduous trees that are visually important, and there are a number that have interesting contorted or weeping branches, such as the corkscrew hazel, *Corylus avellana* 'Contorta'. These are at their best in winter, when their unusual forms are not obscured by foliage. Among the cherries, maples and birches, there are several species with interesting and unusual bark, polished and glowing in the winter sun, or peeling to reveal fresh new colours beneath. Of course the bark is there all year round, but it is in winter that it really comes into its own.

### Tibetan cherry *Prunus serrula*

This popular Asian cherry, which has been in cultivation in Europe for almost 100 years, seldom grows more than 10m/33ft tall. It is grown mainly for its beautiful bark, which resembles highly polished mahogany, particularly when seen as the waning late afternoon sunshine burns deep into its gleaming surface, lending a warmth and depth to its red-brown colouring that is unique in nature. As the tree matures, horizontal fawn banding develops on the trunk, separating and adding contrast to the bark in a way that seems to intensify its rich colour.

*Below: With their evergreen foliage conifers add so much to the winter landscape.*

*Below: Even in late winter, when deciduous trees are stripped of their foliage, their trunks and branches add height and structure to the surrounding landscape.*

*Above: The smooth bark of* Prunus serrula *is this tree species' most identifiable feature.*

**Corkscrew hazel** *Corylus avellana* 'Contorta'

This small European tree has a twirling, twisted habit, which extends from the main stem to the tips of the uppermost branches. It is what might be called a "structural tree", as it provides interesting form and outline. It is at its most visually striking in winter when the foliage has fallen to reveal its remarkable corkscrew-like skeleton. It is best seen in sunshine on a cold frosty winter morning, against an azure sky. In late winter the tree drips with butter-cream catkins. These are the male flowers, known as "heralds of spring", which will soon release copious amounts of pollen into the air. A closer look reveals the female flowers – tiny ruby-red stars waiting for the males to perform.

**Kilmarnock willow** *Salix caprea* 'Pendula' (*S. caprea* 'Kilmarnock')

The Kilmarnock willow is a weeping form of the European native goat willow, which grows alongside rivers and in wet soils throughout western Europe. It is a neat, umbrella- or mushroom-shaped small tree with attractive silver-grey weeping branches, which may reach to the ground on mature trees. The effect is best seen in winter, when there are no leaves to obscure its sculptural shape. Towards the end of winter each bare weeping branch becomes covered with beautiful silver-grey "pussy-willow" catkins that gradually turn golden yellow.

*Right: The corkscrew hazel has incredible twisted branches, which are at their most visible in winter.*

**Paperbark maple** *Acer griseum*

This popular Chinese tree, cultivated in Europe since its introduction in 1901, has the finest bark of all the maples. It is a striking, vibrant cinnamon colour, very smooth to the touch and wafer thin; it flakes away in papery strips to reveal fresh bright orange bark beneath. It is seen to best advantage when low winter sunlight backlights each translucent flake. The paperbark maple is occasionally planted alongside some of the white-barked birches, and the striking contrast between them can be the highlight of a winter garden.

**Koster's blue spruce** *Picea pungens* 'Koster'

No winter garden is complete without conifers, and one of the finest for winter colour is Koster's blue spruce. It is a cultivar of the North American Colorado blue spruce, and was raised in Holland in 1901. It is now a popular tree in the parks of western Europe, where it is widely planted for its striking silvery-blue needles, which look as if a layer of frosted icing has been carefully laid upon them when viewed from a distance.

*Above: The bark of* Acer griseum *invites touch. It appears rough, though its texture is thin and papery, shedding easily.*

**Golden Scots pine** *Pinus sylvestris* 'Aurea'

Scots pine is native to all of Europe, from Scotland to eastern Russia, and is widely cultivated for its straight-grained timber from its long trunk, as well as for its ornamental warm red bark. In 1876 a yellow-needled variety, *P. sylvestris* 'Aurea', was raised, and is now one of the best conifers for the larger winter garden. For nine months of the year the foliage is very similar to that of ordinary Scots pine, *P. sylvestris*, but in response to reducing temperatures and light levels, the previous year's growth turns a beautiful bright golden-yellow almost overnight. The effect is remarkable, as the bright young foliage is offset by the darker, older needles.

# GLOBAL DISTRIBUTION OF TREES

*The natural distribution of trees around the world is influenced by the weather. Over millions of years each tree species has adapted to a particular set of climatic conditions and so their distribution is limited to where those conditions exist.*

Trees in different parts of the world all function in much the same way. They all require the same things to survive, namely water, minerals, air and light. They all have leaves, roots and a persistent woody stem containing a vascular transport system, which takes water and minerals from the roots to the leaves and food from the leaves to the rest of the tree. That, however, is where the similarity ends.

Throughout the world, trees have adapted to the climate that surrounds them. The amount of rainfall, the temperatures they have to endure, the number of daylight hours and the angle of the sun all influence both the behavioural patterns of trees and their natural distribution across the planet.

Trees growing in the tropics look very different to those found in temperate parts of the world. In a large number of cases they represent very different groups of plants. In general, conifers dominate the colder and drier areas of the world, and broad-leaved trees are more common in warmer and wetter regions.

*Below: The world is broken up into zones that experience different climatic conditions. Individual tree species seldom occur within more than one zone.*

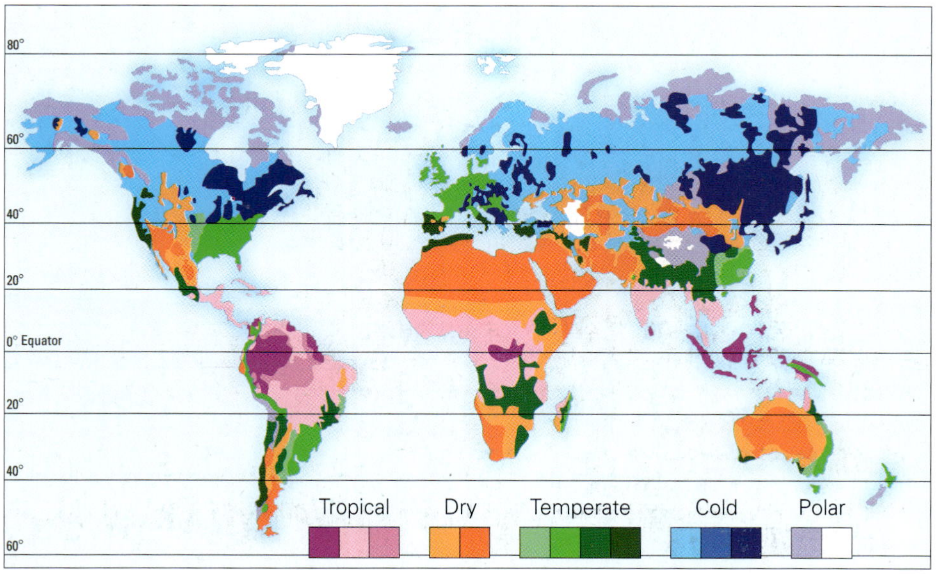


## Equatorial rainforest

Five degrees latitude north and south of the Equator is the area where Equatorial rainforest exists. The conditions in these rainforest areas are perfect for tree growth: the morning sun heats up the vegetation, causing water to evaporate from the leaves. Warm, wet air rises from the trees, forms clouds and produces rain in the afternoon. This happens on every day of the year and there are no major seasonal changes. Numerous trees thrive here, among them rosewood, *Dalbergia nigra*, and the gaboon, *Aucoumea klaineana*.

## Monsoon forest

Moving away from the Equator, the climate becomes drier. Within 5 and 25 degrees north and south of the equator there is a marked dry season during the winter months when the air is colder and clouds do not form. Trees can only grow during the summer months when warm air allows clouds to form and causes rain to fall. This seasonal change is known as monsoon and the forest that grows in these regions is monsoon forest. Monsoon forest covers a vast proportion of the Indian subcontinent, parts of Central America, East Africa, Madagascar and south-eastern China. Trees of the monsoon forest include Indian rosewood, *Dalbergia latifolia*, and East Indian ebonys, among them *Diospyrus melanoxylon*.

## Savannah and desert

Between 25 and 35 degrees of latitude, clouds seldom form, rain rarely falls and the climate becomes progressively drier. Savannah grassland, which borders the monsoon areas, eventually gives way to desert. Few trees can survive in this harsh environment. Those that do include the giant saguaro cactus, *Carnegiea gigantea*, from North America, and the dragon's blood tree, *Dracaena cinnabari*, from Yemen.

## Mediterranean forest

Beyond latitudes of 35 degrees, the conditions for tree growth gradually improve. At 40 degrees from the Equator, the Mediterranean forest region begins. This region contains most European Mediterranean countries, California, Chile and parts of Australia. Typically, the climate is characterized by hot, dry summers, and winters with moderate rainfall. Mediterranean trees include the holm oak, *Quercus ilex*, and the olive tree, *Olea europaea*.

## Temperate forest

Between 40 and 50 degrees of latitude the climate becomes damp and windy, with cold temperatures in winter months restricting tree growth. This temperate region covers central and western Europe (including the British Isles), central North America, New Zealand, Japan and parts of China. The natural tree cover of this area is primarily broad-leaved. Trees that thrive here include oak, beech, ash, birch and maple.

*Above: Conifers are particularly well adapted to cold conditions.*

## Boreal forest

From 50 to 70 degrees of latitude, the length of the tree growing season diminishes and winter lengthens. Known as the boreal region, this area covers by far the greatest landmass of all the forest regions. It includes most of central Canada, northern Europe and Russia, right across to the Pacific coast. The natural tree cover of this region is primarily conifer and includes Scots pine, *Pinus sylvestris*, and sitka spruce, *Picea sitchensis*. The tree density of this region is greater than in temperate or Mediterranean regions, but less than both Monsoon and the Equatorial regions.

## Tundra

Above 70 degrees, winter lasts almost all year and very few trees are able to survive. Known as tundra, this area includes northern Canada, Iceland, Greenland and the far north of Europe and Russia. One tree that does survive is the dwarf willow, *Salix reticulata*.

## Micro climates

There is always some blurring at the edges of every climatic forest region. Land that is close to the sea will generally be warmer than that which is landlocked. Consequently a greater diversity of tree species will grow here than for the same latitude inland. The west coasts of Britain and Ireland benefit from the Gulf Stream, which brings warm, moist air from the Caribbean. This allows trees that grow naturally in the Mediterranean forest region to survive and sometimes to flourish. One Mediterranean forest tree that tends to grow well in gardens in Cornwall, the Isles of Scilly and the west coast of Scotland is the Chilean azara, *Azara lanceolata*.

## Montane forest

In mountainous areas, trees typical of regions farther from the Equator thrive. Because of their latitudes, both the Alps of central Europe and the Rocky Mountains of North America are technically within the temperate region. But because of their high altitude, which decreases average temperatures and effectively shortens the summers, the tree cover is more typical of boreal forest. In the Alps, Norway spruce, *Picea abies*, is the dominant species.

## Tree zoning

Whether or not a tree will survive in any region, given its basic requirements of water, minerals, air and light, depends on the lowest temperatures it will have to endure. Over the years, through trial and error, botanists and horticulturalists have identified the average annual minimum temperatures that individual tree species can withstand. Maps of the world have been produced that put countries or regions into zones, according to the average annual minimum temperatures that occur in them. Most of Britain is suitable for tree species rated at zone eight – trees that are capable of surviving average annual minimum temperatures of around 5°C/41°F. Tree species rated at zone nine would find this average fairly chilly. They prefer the average temperatures not to fall below 10°C/50°F.

*Below: Savannah grassland is the harsh intermediate zone between Mediterranean forest and desert. The closer conditions are to a desert, the fewer trees exist.*

# TEMPERATE TREES

*Temperate trees are found in the bands 40 to 50 degrees north and south of the Equator. These areas include most of North America, Britain and Europe, southern Russia, northern China, Japan, New Zealand, Tasmania, southern Argentina and Chile.*

In temperate regions the climate is suitable for tree growth for six months of the year, when temperatures average more than 10°C/50°F. There are well-defined seasons but few extremes in either temperature or rainfall.

Although the temperate regions are suitable for both deciduous and evergreen trees, it is deciduous broad-leaved trees, such as oak, which predominate. Many trees that live in the windy conditions of the world's temperate regions are wind pollinated.

## Temperate diversity

There is far less tree diversity within temperate regions than in the tropics. This is partly because the climate is less favourable and partly due to historical climatic changes.

Temperate trees have been forced to migrate towards the Equator and back again several times during the last two million years because of successive ice ages. Inevitably these mass movements had casualties. Some tree species perished as they were unable to successfully disperse their seeds with enough speed to escape the freezing conditions expanding outwards from the polar regions. Other species became extinct because their escape routes were blocked by high mountain ranges, such as the Alps and Pyrenees.

## Temperate pioneers

The density of temperate woodlands is such that light is rarely in short supply and there are few other plants, such as climbers, that have the ability to stifle tree growth. Temperate pioneer trees have large canopies and their branches and leaves are free to grow right down the trunk. Their wood is light in colour. Birch is one of the most successful temperate pioneer tree species. It will colonize land far more readily than any other species and is quite often found growing on disused industrial sites, spoil heaps, landfills and railway embankments. Willow, poplar and pine are also early colonizers of inhospitable land.

## Other temperate species

The temperate tree species that has been around longer than any other is the maidenhair tree, *Ginkgo biloba*. Today it grows wild in a small area of Chekiang province, China, although it has been widely planted elsewhere.

There are more than 450 species of oak tree across the temperate world. In Europe the two main species are the English oak, *Quercus robur*, and the

1 Oak
2 Beech

*Above: Oak is the predominant tree species in temperate regions of the world.*

sessile oak, *Q. petraea*. The holm or evergreen oak, *Q. ilex*, originates from the Mediterranean but also grows well in southern temperate regions of Europe and America.

Close to 80 species of oak are native to North America, including the red oak, *Q. rubra*, which has large, sharply pointed leaves that turn red in autumn.

One of the most recognizable temperate trees is the monkey puzzle or Chile pine, *Araucaria araucana*. This hardy evergreen grows up to the snow line in its native Andes Mountains. It has rigid, spiny and prickly leaves.

One temperate tree that looks like it belongs in the tropics is the tree fern, *Dicksonia antarctica*. Native to Tasmania, it grows well in warm, moist temperate regions, such as southern Ireland, where frosts are not too severe. *D. antarctica* is a very exotic-looking tree with a fibrous trunk and large fern-like fronds, which can reach over 3m/10ft long. In Tasmania there are tree fern forests with specimens growing to more than 10m/33ft tall.

Perhaps the most beautiful of all temperate trees is the tulip tree, *Liriodendron tulipifera*. It is native to North America, where it grows from Nova Scotia to Florida. The tulip tree is a large species, growing to heights in excess of 40m/131ft. It has flowers that resemble greenish-orange tulips. Quite often a mature tree will be covered with a stunning spectacle of flowers.

*Below: Species such as oak and beech are slow to establish on new sites and move in only after pioneer species, such as birch, willow, pine and poplar, have improved soil conditions with their fallen leaves. Oak and beech are the predominant woodland species of Great Britain.*

# TROPICAL TREES

*Tropical trees are found in three main parts of the world: central Africa, Amazonia in South America, and South-east Asia. The total area they cover amounts to about 9 million square kilometres (3½ million square miles) and represents 7 per cent of the earth's land surface.*

In the rainforest, levels of rainfall, warmth and sunlight are constant, creating ideal conditions for tree growth throughout the year. Most tropical trees have evergreen leaves with pointed tips. These "drip tips" help the trees to keep their leaves dry, shedding excess water during tropical rainstorms. Tropical trees include the fastest growing trees in the world; 5m/16½ft of vertical growth per year is commonplace. Fast growth means a fast metabolism; consequently everything happens at a fast rate, including the advent of senility. Very few of the tropical trees live beyond 500 years of age, whereas many temperate trees are much older.

## Tropical diversity

The range of tropical species is amazing – there are over 2,000 tree species in Madagascar alone.

The reason for so many different species is not fully understood. However, the fact that today's tropical rainforests have existed for millions of years means that there has been plenty of time for new species to evolve. Evolution takes place primarily as a response to outside influence. It is possible there are so many tropical tree species because there are so many potential killers of trees in tropical forests. The climate is ideal for tree growth and for insects, fungi and viruses. New tree species may have evolved specifically to repel attackers.

Despite their great species diversity, most tropical rainforest trees look very similar to one another; they have tall, thin trunks supported by roots with prominent buttresses. The crowns of these trees are comparatively small and bear large, thick, evergreen leaves not dissimilar to those of laurel. Most tropical trees have thin bark because there is no need to provide protection against frost or water loss. Often, however, the wood of tropical trees is stained dark with chemicals for protection against fungal attack.

## Tropical pioneers

Such is the competition for space and light in a tropical rainforest that only those trees that can react quickly to changes in the density of the canopy survive. If a gap opens up in the canopy when a mature tree dies, light reaches the forest floor and there is a scramble by other plants to fill that gap. The first species to colonize gaps are herbaceous plants and climbers. These plants do their best to smother the ground to prevent another tree from filling the gap because they need the light to survive. Eventually a branchless, umbrella-like tree shoot with a thick, slippery trunk will emerge

*Below: Rainforests are characterized by layers of planting. At the top are the tallest trees, usually with large leaves to take any moisture and light, and buttress roots to anchor them into the ground. Below these are smaller trees with glorious flowers and luscious fruit to attract pollinators.*

1 Kapok
2 Palm tree
3 Brazil nut

from the ground. At the top of this trunk a huge canopy of leaves unfolds, desperate to capture as much light as possible. The thick, slippery trunk provides nothing for climbing plants to grip on to and the tree's leaves are held well out of reach of grasping tendrils. These pioneer trees can grow up to 10m/33ft tall in their first year, quickly filling the space left by the fallen tree.

## Tropical species

Outside the tropics most tropical trees are known for their products. Brazil nut, *Bertholletia excelsa*, is probably one of the best-known tropical trees because of the nuts it produces. It grows wild in Brazil and throughout Peru, Columbia, Venezuela and Ecuador. The Brazil nut is among the largest tropical trees, reaching heights in excess of 40m/131ft. It has thick, leathery, oval-shaped leaves up to 20cm/8in long. Brazil nuts flower in November, producing fruit pods at the end of thick branches the following June. Up to 25 individual nuts can be found in each large, spherical, woody fruit pod. Each tree can produce up to 300 fruit pods a year and thousands of tons of Brazil nuts are exported from South America each year. In economic terms, the Brazil nut is second only to rubber in importance to Brazil as an export cash crop.

The weeping fig, *Ficus benjamina*, originates from the tropical forests of South-east Asia and today is grown from India through to northern Australia. It is an attractive tree with narrow, leathery leaves, which can be up to 12cm/4¾in long. Mature weeping figs can have dramatic twisting branches. In temperate areas this species is grown as a conservatory or house plant. In the warmer tropical regions it produces small red figs in pairs along its twisting branches.

Palm trees have different leaf shapes from both Brazil nuts and weeping figs. Long, narrow and strap-like, the leaves branch out from the tree top. There are around 3,000 species of palm in the world, and the vast majority of them grow in the tropics.

*Above: Such is the diversity of the Amazon rainforest that over 500 different species have been found within a single hectare (2½ acres).*

# DESERT TREES

*There are few places on earth, other than the polar regions, where plants will not grow. Even in the harsh environment of the desert, plants – including trees – somehow manage to cling to life. Deserts are very inhospitable places for trees.*

Trees that survive in the desert have developed unique ways of coping with the day-to-day difficulties of survival. The main problems facing desert trees relate to water – or lack of it. Hot sun, drying winds and low, erratic rainfall make it difficult for tree roots to supply enough water to make up for that lost by transpiration from the leaves. Desert trees have adapted to the extremes of heat and aridity by using physical and behavioural mechanisms.

Plants that have adapted by altering their physical structure are called either xerophytes or phreatophytes. Xerophytes, such as cacti, usually have special means of storing and conserving water. They often have few or no leaves, which helps them to reduce transpiration. Phreatophytes are plants, such as the African acacias, that have adapted to parched conditions by growing extremely long roots, allowing them to acquire moisture from the water table.

Other plants have altered their behaviour to cope. They have to make the most of the times of greatest moisture and coolest temperatures, remaining dormant in dry periods and springing to life when water is available. Many germinate after heavy seasonal rain and then complete their reproductive cycle very quickly. These plants produce heat- and drought-resistant seeds that remain dormant in the soil until rain eventually arrives.

## The Joshua tree

The *Yucca brevifolia*, or Joshua tree, grows in the Mojave Desert of California, Nevada, Utah and Arizona. It has spiky, leathery, evergreen leaves at the tips of the branches, thus reducing the effects of transpiration. The leaves have a hard, waxy coating that also helps to reduce water loss. Originally considered a member of the agave family, the Joshua tree is now known to be the largest yucca in the world. It can grow up to 12m/40ft tall with a trunk diameter of 1m/3ft.

## Welwitschia

A dwarf species from Africa, *Welwitschia mirabilis* is one of the strangest trees on earth. It grows on the dry gravel plains of the Namib Desert in southern Angola and is a throwback to the prehistoric flora that existed on the supercontinent of Gondwanaland millions of years ago. Its shape and growing characteristics are so unusual that there is no comparable living plant. It is a unique species occupying its own genus.

The bulk of *Welwitschia's* "trunk" grows under the sand like a giant carrot. Its girth can be up to 1.5m/5ft

1 Baobab
2 Date palm
3 *Welwitschia mirabilis*

*Above: Mormon pioneers are said to have named this species the Joshua tree because it reminded them of the Old Testament prophet Joshua, with arms outstretched, waving them on towards the promised land.*

*Below: Tropical Africa is home to deserts and savannah. Trees that live in these habitats are exceptionally good at storing water. The* Welwitschia mirabilis *has a long tap root that can reach down to the water table. Succulents have few leaves and are best adapted to the desert. The baobab of the savannah can store vast amounts of water in its trunk.*

and its height (or in this case length) up to 4m/13ft, less than a third of which appears above ground. Its subterranean trunk is a water storage organ made of hard wood and covered with a cork-like bark. Broad, leathery leaves emerge from the part of the trunk that appears above ground. The leaves, which can reach 2m/6½ft long, sprawl across the desert floor. They have specially adapted pores to trap any moisture that condenses on the leaves during the night when the temperature falls. As rain falls about once in four years in the Namib Desert this method of moisture collection is vital. Recent carbon-dating has established that some of these trees are more than 2,000 years old.

## Other desert trees

The acacias and tamarisks, which grow in African deserts, are phreatophytes – they have developed incredibly long root systems to cope with the absence of surface water. These roots take water from the permanent water table, which may be anything up to 50m/164ft below the desert surface. Once mature, they have little trouble combating harsh desert conditions – the difficulty is in establishing themselves, as the roots have to first grow through great depths of bone-dry soil before they reach the water. Phreatophytes grow in places where the soil is occasionally wet, such as dried-up riverbeds, as these are the only spots where they can get started.

Perhaps the most successful desert plants are cacti. The giant saguaro cactus, *Carnegiea gigantea*, is the ultimate desert tree. It has no leaves at all but does have a thick green trunk, which is capable of photosynthesis and storing water. The giant saguaro can grow to heights in excess of 10m/33ft.

The Socotran desert-rose tree, *Adenium obesum*, grows in desert conditions on the Indian Ocean island of Socotra, Yemen. It has a swollen grey trunk, which looks like a sack of potatoes. This trunk has the ability to expand in size on the rare occasions that rain falls, enabling it to store huge quantities of water for the drought period to follow.

# MOUNTAIN TREES

*Mountains tend to be covered with conifers and most are members of the Pinaceae family – pines, spruces, hemlocks and firs. The higher the elevation, the slower the trees grow. The point beyond which no trees will survive is known as the tree line.*

In many ways mountains have the same climate as subarctic regions, having short summers, cold winters and a mean temperature that rarely rises above 10°C/50°F. Wind speeds tend to be greater at high altitudes. These drying winds and shallow soils, often frozen for long periods of time, mean that only those trees that are protected against water loss and frost damage will survive.

## Conifers and evergreens

A characteristic that conifers and broad-leaved evergreens share is leaves that are resistant to water loss and cold. Broad-leaved evergreens often have thick, leathery leaves with a waxy coating. Conifers further reduce water loss by having fine, rolled, needle-like leaves, that expose a small surface area to the elements.

Conifers and other evergreens are efficient at functioning in low light and temperature conditions. Once deciduous trees have lost their leaves in autumn they cannot produce food or grow until the next year's leaves grow – anything up to six months. Yet, during this time there are periods when the temperature and light is sufficient for photosynthesis to occur. Evergreens and conifers take advantage of this. Deciduous trees are also vulnerable when their young leaves are bursting from the bud in spring. These new leaves are sensitive to frost and can easily be damaged. Evergreens have tough leathery leaves that are never so vulnerable.

## Mountain characteristics

Trees become progressively shorter as they approach the tree line. The reason for their shortness is not cold but increasing wind – constant stem movement stunts a tree's growth. High winds can also damage trees and to avoid this some species have evolved a low-growing, almost sprawling habit.

Many mountain trees have adopted characteristics to cope with this harsh environment. They are conical or spire-shaped with branches and twigs that point downwards. This prevents snow from building up on the branches and breaking them. Instead it simply slides off the tree to the ground.

Mountain trees will also grow away from the direction of the prevailing wind, giving them a windswept appearance. The reason for this is that the waxy coating on the leaves or needles on the windward side gets worn away by the sandpaper effect of harsh winds carrying ice particles. Once the coating has gone the leaves and shoots are open to dehydration,

1 Sitka spruce
2 Brewer spruce

*Above: As trees approach the elevation beyond which they will not grow (known as the tree line), they become stunted and eventually prostrate.*

*Below: Most of the conifers found in mountainous regions are members of the Pinaceae family – pines, spruces, hemlocks and firs. Such trees have adapted to cope with the harsh and extreme conditions of the weather, from freezing snow to fierce winds, driving rain, and the blistering heat of the summer sun.*

and slowly die. The tree compensates for the lack of leaves and shoots on the windward side of the crown by producing more on the leeward side.

In exposed mountain regions, young trees can only grow in the shelter of other trees. This leads to clumps of trees scattered across the mountainsides. As trees die on the windward side and new ones grow on the leeward side, the whole clump moves slowly downwind. Research has shown that the average movement of these clumps is 2–7m/6½–23ft per century.

Often the branches at the bottom of mountain trees grow much better than those at the top, giving a skirted effect around the tree's base. This is because in the depths of winter these lower branches are protected from the ravages of the wind by snowdrifts.

## Mountain species

The dwarf mountain pine, *Pinus mugo*, is native to the mountains of central Europe, the Carpathians, the Balkans and the Italian Apennines. It is a low-growing, shrubby tree with twisting, snake-like stems and branches that form dense, impenetrable entanglements (known in Germany as *krummholz*).

Brewer's spruce, *Picea breweriana*, is a tree that originates from the Siskiyou Mountains of California and Oregon, where it grows at elevations of up to 2,100m/7,000ft. In Scotland the rowan or mountain ash, *Sorbus aucuparia*, will grow at altitudes in excess of 700m/2,300ft.

# COASTAL TREES

*Coastal conditions differ radically around the world. It is one of the most difficult environments for trees to grow in. Those that survive have adapted to the strong winds and salt-laden water by growing additional roots on their windward sides to improve anchorage, and their habit becomes low and squat.*

Only the toughest tree species can survive a combination of strong winds and salt spray. Exposure to ocean storms, with winds in excess of 160km/h or 100mph, is only part of the problem. Strong wind alone is something that many trees are able to withstand. However, if those winds are laden with huge quantities of sea salt, most trees will simply die.

## Salt damage

Trees can be damaged by salt in two ways: through direct contact with the foliage and by absorption from the soil through the roots. Direct and prolonged contact with salt will cause leaf-burn, branch die-back and defoliation. This in turn will reduce the ability of the tree to photosynthesize and produce its own food, so eventually it dies. Salt can also dramatically reduce the amount of seed and fruit produced.

The most common cause of tree death by salt is through its uptake from the soil. When salt-laden winds that have travelled across the ocean reach land they condense, producing rain or dense sea mists. The salt precipitation from these mists and rain soaks into the soil. The highest salt concentrations are deposited closest to the coast.

Salt causes the soil structure to deteriorate, leading to a decrease in soil fertility. Natural calcium in the soil is replaced by sodium chloride. This increases soil alkalinity, making it dramatically harder for trees to survive. Salt also makes the soil less permeable and reduces the moisture content, causing its root systems to dehydrate and die back. The moisture that is absorbed by the roots can literally poison the tree. It takes only half a per cent of a tree's living tissue to contain salt before the tree starts to die. This process is also what damages trees planted on roadsides, where the road is regularly covered with salt to clear it of ice.

## Mangroves

One genus of trees has adapted so well to life alongside the coast that its members can actually grow with their roots in salt water. Called mangroves, they are found throughout the tropics,

*Below: New Zealand is home to a diverse collection of trees. The kanuka and puriri trees and the nikau palms all thrive here. The kanuka tree is a pioneer tree, endemic to the area. The nikau palm is New Zealand's only native palm tree and thrives in coastal areas and warmer, inland regions.*

particularly in shallow, muddy estuarine and coastal situations. They have to cope not only with waterlogging but also with the high salinity of seawater.

The most notable feature of mangroves is their roots. Many species are anchored in the soft mud by prop roots, which grow from the trunk, or drop roots, which grow from the branches. Oxygen is piped from the roots above ground to those below the water line. This aeration is particularly important to mangroves because they need oxygen to carry out the process of ultra-filtration, which they use to exclude salt from the tree. Each root cell works like a mini desalination plant, screening out the salt and allowing only fresh water to flow into the root system and on through the rest of the tree.

Mangroves display several other adaptations to their situation. They have leathery, evergreen leaves, which are able to conserve the fresh water within them but keep out salt-laden water that lands on

*Above: Mangroves have a root system that copes with continual immersion in water.*

them. They also have wind-pollinated flowers, which are able to take full advantage of sea breezes, and spear-shaped seed pods, which can stab into the mud or float away from the mother tree, coming to rest elsewhere.

## Monterey cypress

At the Monterey Peninsula in San Francisco, clinging to life and the cliff edge, are two groves of Monterey Cypress, *Cupressus macrocarpa*.

The trees grow on the shore cliffs and, being undermined by the waves, occasionally fall into the sea. There are fewer than 300 trees left, ancestors of a species that covered great swathes of the temperate world at the start of the glacial cool-down a million years ago. Monterey cypress, along with other American giants, such as the Douglas fir and the giant redwood, retreated to the Pacific coast to escape the worst of the cold. When the climate warmed up 12,000 years ago and the glaciers withdrew, the trees moved back to the land they had occupied before the ice ages – all, that is, except the Monterey cypress, which remained on the Californian coast, where it has been growing in decreasing numbers since.

The trees that are left are stunted and gnarled, seldom reaching more than 15m/50ft tall. Collect seed from any of them and sow it anywhere else in the temperate world however and it grows into a magnificent giant. Wherever there is the need for shelter from the wind and salt spray off the sea, this is the tree to plant.

# TROPICAL ISLAND TREES

*Islands often contain a diversity of plant life that is very different to that of the nearest mainland. This is because evolution on islands occurs in isolation. Some islands, such as New Caledonia in the Pacific Ocean, still have a range of trees which evolved during the Jurassic period.*

The reason for the often unique plant life on individual islands lies in the earth's history and how each island was first formed. Islands are normally formed as a result of continental drift or volcanic activity on the seabed.

About 200 million years ago, most of the world's land was clumped together in a single supercontinent, known as Pangaea. Pangaea began to break up about 190 million years ago. First it split in two. The northern part, Laurasia, contained what are now North America, Europe and Asia, while the southern part, Gondwanaland, consisted of present-day South America, Africa, India, Antarctica and Australasia. Gradually Laurasia and Gondwanaland also broke up to form the continents we recognize today.

This fragmentation process created the major continents, and thousands of islands. When these islands broke away from the continents, they carried with them a collection of the flora and fauna that existed on the larger land-masses at that time. Over the following millions of years, plants and animals on these isolated fragments of land adapted to their new environments, and often evolved in different ways to those on the mainland.

In some cases, evolution has continued on the continents, while little has changed on some of the islands. The island of New Caledonia off the east coast of Australia is home to an amazing collection of ancient trees no longer found anywhere else on earth. So primeval is its landscape that it has been used as a backdrop for films on dinosaurs. In other cases the reverse has been true, with island life forms changing quite dramatically.

Not all of the world's islands were created by the break-up of the continents. Many were formed more recently by undersea volcanic activity and have never been attached to the continents at all. At first these islands had no plants of their own.

## The Hawaiian Islands

Archipelagos such as the Hawaiian Islands began as barren outcrops of rock. Hawaii's native trees are all descendants of the few plants whose seeds washed up on its shores or were carried there by birds.

Tarweed is a daisy-like plant from California. Fragments of this plant floated across the ocean to Hawaii millions of years ago. Tarweed gradually colonized the island and then began to evolve into new plants, filling the empty niches. Today Hawaii has 28 species whose ancestry can

*Below: Islands have unique eco-systems. The weather they receive, their landmass and the vegetation that thrives on them can differ dramatically to that of the nearest mainland.*

1 Coco-de-mer
2 Palm tree

Above: Coconut palms have large seeds that can float for hundreds of miles across the ocean.

be traced back to tarweed. One, *Dubautia reticulata*, is a tree that can grow to more than 10m/33ft tall.

## The Galapagos Islands

Like the Hawaiian Islands, the Galapagos Islands formed in volcanic activity after the break-up of the continents. Mangroves were among the first and most successful tree colonizers of the Galapagos Islands. Four species exist there today: the black mangrove, *Avicennia germinans*; the red mangrove, *Rhizophora mangle*; the button mangrove, *Conocarpus erecta*; and the white mangrove, *Laguncularia racemosa*. Mangroves are able to live in shallow seawater and grow on the shores of almost all the islands. They are a vital part of the coastal ecosystem, as fallen leaves and branches provide nutrients and shelter for a wide variety of sea creatures, and their tangled roots protect the coastline from erosion and storm damage. The Galapagos Islands' mangroves are thought to have established themselves from plants and seeds that floated from the Far East across the Pacific Ocean.

## The Virgin Islands

The warm, moist climate on the northern coasts of the Virgin Islands in the West Indies supports an amazing array of tree species. Growing wild here are West Indian locust, bay rum, sandbox, kapok and hog plum. To the south and east the climate becomes much drier, creating ideal growing conditions for the turpentine tree, acacia, white cedar and the poisonous manchineel tree.

## The Seychelles

More than 80 species of tree grow on the Seychelles in the Indian Ocean that grow nowhere else on earth. Among them is the record-breaking coco de mer palm with its 20kg/45lb nut. This is the largest seed of any plant in the world and it has been found washed ashore in places as far away as Africa, India and Indonesia. Before the islands were discovered the coco-de-mer seed was thought to have grown on the seabed, hence its name.

## New Caledonia

Situated off Australia's east coast, this island has been described as having "one of the richest and most beautiful flora in the world". The island is home to trees that are remnants of families that became extinct elsewhere millions of years ago, some as far back as the Jurassic period. *Araucaria columnaris* is a rocket-shaped relative of the monkey puzzle tree which grows wild in Chile and Argentina. Of the 19 living species of *Araucaria*, 13 are found in New Caledonia and nowhere else. The island also has unique members of the podocarp family, to which most of New Zealand conifers belong, and proteas, which only occur otherwise in South Africa.

New Caledonia is also home to some of the tallest tree ferns in the world, many of them over 30m/98ft tall. The island's rarest tree is a small evergreen called *Xeronema moorei*; this unique tree grows in isolated pockets high in the mountains and is found nowhere else on earth.

# URBAN TREES

*Trees have become a vital part of urban areas around the world. From plane tree avenues of Europe, the leafy avenues of downtown Manhattan to the cherry-covered walkways of Tokyo, they bring beauty and environmental benefits right to the heart of our cities.*

Trees have been planted in large numbers in our towns and cities ever since the 18th century. Before that time, urban trees were the privilege of royal palaces, cathedrals, churches, monasteries and universities. Some of the earliest town plantings were in specially landscaped town gardens, squares and crescents, such as Berkeley Square in London, which was planted with London plane trees in 1789. These trees still exist today, tall spreading giants bringing shade and cool in summer.

Quite often these early plantings only took place in the more affluent areas of towns and were for the private enjoyment of those who lived there. The poorer residential and industrial areas were left largely devoid of trees.

*Below: The London plane tree,* Platanus acerifolia, *is popular in urban settings throughout the temperate world. It grows quickly, is hardy and tolerates the pollution of modern cities.*

It wasn't until the Victorian era that municipal parks were laid out for the benefit of all town dwellers. At this time, the idea of parks as the "green lungs" of towns and cities developed, improving citizens' health as well as giving them opportunities to walk, meet and relax. Public parks began to appear in North America and all over the British Empire, and trees were seen as an integral part of them. Today some of the finest tree collections in the world are found in city parks.

Urban street planting also became prevalent during this time, although the planting of trees along roadsides between towns had been going on for centuries. Plane and poplar trees were planted by the Roman troops to provide shade and shelter for their legions as they marched back and forth across southern Europe. This tradition was repeated by Napoleon for his armies and many Napoleonic roadside trees can still be seen today in France, Germany and Spain.

The environmental benefits of trees in towns and cities were recognized towards the end of the Victorian era by Ebenezer Howard, whose book *Garden Cities of Tomorrow* inspired the early landscaping of suburbs and new towns that were being built outside the cities to house rapidly increasing populations. These new towns were built on "green field" sites and the inclusion of street trees, park trees and areas of woodland between housing were drawn into landscape plans long before the houses were even built. Howard's ideas quickly spread and were used by town planners across Europe and the Americas.

## Urban trees today

Trees have become an integral part of cities around the world. In terms of planning, they have become almost as important a feature of the urban landscape as the buildings themselves. Trees have a higher priority in our towns and cities now than at any time previously.

Above: Urban trees provide shade in summer and shelter in winter.

The architectural value of trees and the health benefits they offer are now well-recognized and some cities have instigated massive tree-planting campaigns. By 2010, one million more trees will be planted in the centre of London.

## Benefits of urban trees

Trees reduce air pollution. They help to trap particle pollutants such as dust, ash and smoke, which can damage human lungs, and they absorb carbon dioxide and other dangerous gases, releasing vital oxygen in their place. In a year, 0.4ha/1 acre of trees in a city park absorbs the same amount of carbon dioxide as is produced by 41,850km/26,000 miles of car driving.

Urban trees conserve water and reduce flooding. They lessen surface runoff from storms as their roots increase soil permeability. Reduced overloading of drainage systems, the main cause of localized flooding, occurs in towns with a high tree population.

Trees modify local climates as they help to cool the "heat island" effect in inner cities caused by the storage of thermal energy in concrete, steel and tarmac. They also provide a more pleasant living and working environment. They reduce wind speed around high-rise buildings, increase humidity in dry climates and offer cooling shade on hot, sunny days.

Without trees, towns and cities are sterile landscapes. Trees add natural character; they provide colour, flowers, fragrance, and beautiful shapes and textures. They screen unsightly buildings and soften the outline of masonry, mortar and glass.

## Trees for urban environments

One of the finest large trees for planting in towns and cities is the London plane, *Platanus acerifolia*. Most trees suffer in urban areas as their bark's "breathing pores", known as lenticels, get clogged with soot and grime. The London plane frequently sheds its old bark, revealing fresh, clean bark beneath.

The maidenhair tree, *Ginkgo biloba*, native to China, is also tolerant of air pollution. Its slow growth and narrow habit make it an ideal tree for street planting. Other trees suitable for the urban environment include laburnum, *Laburnum* x *watereri* 'Vossii'; black locust, *Robinia pseudoacacia*; hawthorn, *Crataegus laevigata* 'Paul's Scarlet'; Indian bean tree, *Catalpa bignonioides*; and cherry, *Prunus* species.

# HEDGEROW TREES

*Hedges are live fences or barriers consisting of trees and shrubs, which are used to contain livestock, act as boundary markers or provide shelter. Most hedges are thought to have been created in the last 200 years, as a result of Enclosure Acts passed in Britain and other European countries.*

There are three ways of producing a hedge. The first and most obvious is by planting. Second, hedges may be the remnants of woodland that was cleared for farming, leaving only narrow strips to stand as field boundaries. Finally, hedges can develop naturally by seeding into uncultivated land at the edges of fields, where banks, fences, or ditches exist or once existed.

## Early hedges

The Romans are known to have planted hedges right across their empire. These early hedges were established to contain livestock and also as defensive, impenetrable barriers against attack. At the very least they would slow down an invading force.

The Anglo-Saxons introduced the concept of villages and marked out many of today's parish boundaries. Natural markers such as old trees, large rocks, cliffs and streams were used to denote boundaries; on open ground where no such features existed, marker trees were planted, which over the years seeded and developed into hedges. Words such as "*hegeraewe*", "*hege*", "hazel-rows", "willow rows", "*hagaporn* (hawthorn) rows" and "hedge-sparrow" regularly occur in charters of this time.

## Fields and Boundaries

In Britain in the 18th and 19th centuries, various Acts of Parliament brought the ancient open field system to an end, dividing the countryside into smaller, more manageable fields – normally around 4ha/10 acres in size. The new fields were enclosed by hedges, except in areas where stone was readily available to create dry-stone wall boundaries. The most popular tree for planting was hawthorn, *Crataegus monogyna*.

Below: A familiar landscape of the British countryside with fields bounded by hedgerows.

*Above: Trees are a natural feature of hedgerows and field boundaries.*

## The best hedgerow trees

Hedges differ in content from region to region and reflect the local soil, topographical and climatic conditions, and their original purpose. A hedge planted to contain livestock may contain different tree species from one planted to create shelter. The local natural tree population should feature within the hedge, and sometimes the region's economic past will have a bearing on the content too: for example in some regions blackthorn, *Prunus spinosa*, was planted in hedgerows because the berry was used to make dye for wool.

The best trees, most commonly used for stockproof hedging, are those that naturally develop a thick, impenetrable framework of branches, ideally protected by an armoury of thorns. They should be tough, able to withstand constant browsing by livestock, and respond readily to pruning. The tree's natural response to browsing and pruning should thicken and strengthen the hedge.

Hawthorn, *Crataegus monogyna*, and blackthorn offer all the qualities described above and are widely planted in hedgerows across Europe. They are invaluable to wildlife, particularly insects and birds, as they produce flowers for pollen in spring and edible fruits in late summer and autumn. Their thick, dense foliage also provides ideal shelter and nesting habitat for birds. Other trees regularly found in hedgerows include beech, *Fagus sylvatica*, which provides an effective screen right through winter by retaining its dead leaves; hazel, *Corylus avellana*, which was once valued for supplying the wood needed to make sheep hurdles; holly, *Ilex aquifolium*, which produces spiny dense evergreen foliage; and elm, *Ulmus* spp., the leaves of which provide a browse for livestock.

# WETLAND TREES

*All trees require water to survive and grow. Some, such as palm trees, can manage with very little moisture while others require a constant supply of fresh water and are therefore commonly found growing alongside rivers and streams, and on flood plains.*

The banks of rivers and streams provide ideal growing conditions for trees. In a forest or woodland situation the river itself provides a gap in the overhead canopy, allowing the trees plenty of natural light. Waterways are also rich in minerals and nutrients flushed from the soil further upstream. Flooding and constant fluctuations in water levels ensure regular deposits of minerals and nutrients to keep riverside soils fertile. The riverside soils are also normally deep and moist.

There are times, however, when this environment may become inhospitable to trees. During prolonged flooding the soil will become waterlogged, preventing oxygen getting to the roots. Very bad floods may cause soil erosion: the bank may be undermined or in extreme cases even washed away. Parts of the tree's root system may become exposed or broken off, and the force of the water may even wash whole trees away.

## How wetland trees survive

Wetland trees survive flooding by absorbing air through the lenticels (pores) in their bark and trunk. This air is piped down to the roots immersed in water, where it is stored in large air pockets called aerenchyma. Some trees can survive being submerged in water for over six months of the year. Willows and alders are particularly good at doing this.

The root systems of wetland trees tend to be much larger than those of other trees. Not only does this help to stabilize the riverbank and reduce soil erosion, it also helps the tree cope with physical adaptations that may be required in this environment. Often, tree trunks are not vertical but grow or lean out across the water – this helps them exploit the extra light in that direction. To counteract the strain this

*Below: Where willows thrive, there will be a large supply of water nearby.*

*Above: Willow trees are a good indicator of the presence of water, and are frequently to be found along riverbanks.*

leaning growth habit puts on the tree, and the root system in particular, long "anchor roots" develop, growing deep into the soil on the opposite side of the tree. Having evolved the leaning habit, such species grow this way even when they are planted away from rivers and in full light.

### Wetland tree dispersal

Some wetland trees take advantage of their proximity to flowing water and flooding to aid their dispersal and population advancement. Crack willow, *Salix fragilis*, is particularly adept at this technique. Its twigs readily break off and float downstream, eventually being washed up on a new bank where they grow roots and develop into new trees. Another tree that increases its population in this way is the common alder, *Alnus glutinosa*. It produces cone-like seed capsules with inbuilt air pockets to aid buoyancy. They fall from the tree into the water and float downstream, releasing seeds as they go. Eventually the seeds float to shore, where many germinate in the damp, muddy conditions.

### Drying out waterlogged soil

Trees such as willow, alder and poplar are often planted in wetland areas to help dry out the soil and improve drainage (a specific need if housing is nearby). Depending on its species and size, during the growing season one tree may suck more than 1,000 litres/ 220 gallons of water from the soil each day. If trees are removed from an area that is prone to waterlogging, the water table will rise and anaerobic conditions will be produced, thereby preventing other plants from colonizing the area. In extreme cases water run-off and soil erosion may occur. Planting trees in wetland water-catchment areas substantially reduces the risk of flash flooding. Waterlogged soil is very likely to have low nitrogen levels, which means it has low fertility. Trees such as alders not only help to dry out the soil but can also improve its fertility.

# ENDANGERED TREES

*Trees are one of the most successful groups of plants on earth, but despite their proliferation, some trees are increasingly under threat. Ten per cent of the world's tree species are currently threatened with extinction. Across the world more than 40 hectares (100 acres) of forest are felled every minute.*

One third of the land on earth is covered by trees. But that figure is set to decrease. As the human population expands, so ever larger areas of the world are changed to meet people's needs. One of the first things to be decimated is forest; the timber is used for industry and the land for housing.

## Ten per cent threatened with extinction

There are more than 80,000 species of tree in the world. Around 8,750 of them are threatened with extinction. Almost 1,000 of those are critically endangered and some species are literally down to just one or two trees.

The threats to tree species are many and varied. They include felling of woodlands and forests for timber and fuel, agricultural development, expansion of human settlements, uncontrolled forest fires and the introduction of invasive alien tree species. Across the world we are losing at least 40ha/100 acres of forest every minute. At the same time we are planting only 4 ha/10 acres.

## Can we live in a world without trees?

The simple answer is no. Trees are essential to all life and incredibly important to the planet as a whole. They provide services of incalculable value to humans, including climate control, production of oxygen, pollution control and flood prevention. They also prevent soil erosion and provide food, medicine, shelter and timber. Forests are also extremely important from an ecological point of view – tropical forests contain almost 90 per cent of the world's land-based plant species, for example.

## Trees in danger

The monkey puzzle tree, *Araucaria araucana*, has become one of the most familiar trees in the temperate world. As an ornamental species, it is grown in virtually every botanical garden in Europe. Yet, in its native homeland, high in the Andes Mountains of Chile and southern Argentina, the monkey puzzle is threatened with extinction. Thanks to its tall, straight trunk, its timber is highly sought after and the land it once stood on claimed for new uses. Monkey puzzle forests have been felled on a massive scale and it is thought there are now more monkey puzzle trees growing in Britain than there are in South America.

Another native of Chile and Argentina, the alerce tree, *Fitzroya cupressoides*, is a magnificent slow-growing conifer. Its Latin name was given in honour of Captain Robert Fitzroy, who captained HMS *Beagle* on Charles Darwin's epic voyage around the world in the 1830s. Even back then alerce was being felled for timber. Today it is one of the world's rarest

*Below: Destruction of woodland and unsustainable forestry have contributed to the rarity of some of the world's trees.*

conifers, with only 15 per cent of the original trees remaining. Although international trade in alerce timber is banned, illegal felling still continues.

Wilmott's whitebeam, *Sorbus wilmottiana*, grows in only one place in the entire world and that is the Avon Gorge, which passes through the west of the city of Bristol in England. This beautiful little tree, which produces clusters of attractive creamy white flowers in June and bunches of red berries in September, is critically endangered. There are only about 20 trees now remaining in the wild.

The Australian wollemi pine, *Wollemia nobilis*, was thought to be extinct until 40 survivors were found in a remote canyon in the Blue Mountains of New South Wales in 1994. A distant relative of the monkey puzzle tree, the wollemi pine has existed unchanged for almost 200 million years.

The Pacific yew, *Taxus brevifolia*, hit the headlines in the 1990s when it was found to contain a toxin called taxol, which, when administered to humans, helped in the treatment of breast, ovarian and lung cancer. The greatest concentrations of taxol were found to exist in the tree's bark and for a while wholesale bark stripping took place, threatening the survival of what was already a rare species of tree. Bark from ten trees is needed to produce enough taxol to treat a single patient but steps have now been taken to protect the tree in the wild. Pacific yew plantations have been established and this, together with the recent chemical synthesis of taxol, has taken the pressure off the species in the wild.

*Below:* Fitzroya cupressoides *is one of the world's rarest conifers.*

*Above: Lawson cypress has become an endangered species since felling for timber and disease have taken their toll.*

Madagascar has some of the world's most extraordinary flora, including six different species of baobab, three of which are found nowhere else in the world. *Adansonia grandidieri*, named after Grandidier, is the grandest of them all. It is also the rarest. Although recognized by botanists the world over as a tree that must be conserved, numbers continue to dwindle. The problem here is not logging but human overpopulation. As Madagascar's people continue to increase in number more and more of the island's wilderness is turned into agricultural land.

## Hope for the future

The problem with man's exploitation of trees is that it is very often done in an unsustainable way. Areas of forest are felled and cleared, often with little regard to replanting. Once the tree cover is removed, the animals, insects and birds that populated the area move away or die and soil erosion occurs, making it very difficult for trees to recolonize the felled area.

In some parts of the world, notably western Europe, sustainable silviculture is now practised with excellent results. Large areas of forest or woodland are never felled; trees are selectively thinned and removed one at a time, or in small clearings. These trees are then replaced by young seedlings that thrive naturally in the gaps once the light is allowed in. If this sustainable method of management could be adopted in other parts of the world then the future for some of the world's endangered trees might not be so bleak.

*Below: The monkey puzzle tree is now threatened with extinction in the wild.*

# TREES OF BRITAIN AND EUROPE

The range of climatic variation found within Europe makes it one of the most diverse continents. There are sun-baked lands surrounding the Mediterranean, exposed snow-covered mountains in Scandinavia, plentiful rainfall and mild temperatures in the United Kingdom and dry tundra permafrost in North-eastern Russia. These diverse and sometimes extreme conditions enable a rich and varied population of tree species to grow. Ironically, the majority of these trees are not native to Europe, as the effects of the last Ice Age, which ended 12,000 years ago, meant that many European species perished. Since then, through the travels of European plant hunters, Europe has become home to a vast collection of species obtained from temperate regions elsewhere. Trees from South and North America, South Africa, Australasia and Asia are all now to be found thriving in Europe.

The following pages provide a comprehensive profile of the trees that thrive in Britain and Europe. The associated descriptions will clarify how each tree can be identified at all times of the year, even in winter. Other information contained in fact boxes provides general points of interest about each tree and covers such things as map and country of origin, average size, shape and how each tree is pollinated.

Iceland
Norwegian Sea
Finland
Norway
Sweden
Estonia
Russian Federation
Latvia
North Sea
Denmark
Lithuania
Republic of Ireland
United Kingdom
Belarus
Luxembourg
Netherlands
Poland
Belgium
Germany
Ukraine
Czech Republic
Atlantic Ocean
Slovakia
Moldova
France
Austria
Switzerland
Hungary
Romania
Slovenia
Croatia
Bosnia & Herzegovina
Yugoslavia
Bulgaria
Portugal
Spain
Corsica
Italy
Turkey
Albania
Macedonia
Sardinia
Balearic Islands
Mediterranean Sea
Greece
Sicily
Cyprus
Malta
Crete
Black Sea

# HOW TO IDENTIFY A TREE

*Looking at and identifying trees can be an immensely enjoyable and fascinating pastime, but, unless you know what to look for, it can be confusing. The following information should help to reduce the confusion and provide a clear route to tree identification.*

Whether growing in a woodland or forest, lining the hedgerows, bringing green to our city streets or standing in defiant isolation on some windswept hillside, trees form an integral part of the landscape. They are the most diverse group of plants on the planet, providing variation in shape, size, colour and texture, and in the detail of their leaves, flowers, fruit and bark.

## What to look for

There are many clues to a tree's identity, primarily built around seven main features. These features will generally not all be visible at the same time, flowers and berries are normally only present during certain seasons, for example, but some features are constant. The colour and texture of bark changes little throughout the lives of most trees.

**Shape and size** – Is the tree tall and spire-like or low and wide-spreading?
**Evergreen or deciduous** – Are there leaves on the tree all year round or do they fall in autumn?
**Leaves** – Are they long and needle-like or broad and flat?
**Flowers** – Are flowers (or flower buds) present? If so, what colour and shape are they?
**Fruit** – Does the tree have any fruit, berries, seeds, nuts or cones on it, and if so what are they like?
**Bark** – Does the bark have distinctive colouring or patterning?
**Buds** – In winter, buds can be a tremendous help in identifying temperate trees. What colour and shape are they, and how are they positioned on the twig?

By working through these features step by step, it should be possible to identify any tree.

There are other points to consider that relate to the tree's location and the environment surrounding the tree, which may yield some clues. The acidity of the soil will dictate what species will grow successfully, for example. Some trees, such as red oak, *Quercus rubra*, will only grow well on acidic soil (low pH), while others, such as whitebeam, *Sorbus aria*, prefer chalky, alkaline soil (high pH). The position of wild trees should also be taken into account. Some trees grow well alongside, or even in, water, for instance. Willow and alder enjoy damp conditions and grow naturally next to rivers; hawthorn on the other hand does not. Some trees, such as beech, will grow well in dense shade, others, such as the Judas tree, will only thrive in full sun.

*Above: In winter the buds and bark are important clues to identification.*

*Below: It is quite often possible to identify a tree from a distance by its overall shape.*

It is generally easier to identify trees in the wild than in a park or arboretum. This is simply because the pool of species is likely to be greater in a park or arboretum than in a natural setting. Most hedgerows in Europe will contain fewer than ten tree species, for example, and the majority of those will be common native species. At the other extreme, an arboretum may contain up to 4,000 trees, brought together from various habitats in different countries all over the world.

## Shape and size

Some trees have such a distinctive shape that it becomes almost unnecessary to continue down the identification trail, other than to confirm the initial assumption. The Lombardy poplar, *Populus nigra* 'Italica', is particularly distinctive with its remarkable narrow shape and upright habit. This shape is known as "fastigiate". Another very distinctive tree is the monkey puzzle, *Araucaria araucana*. No other tree has such sharply toothed evergreen foliage and stiff branching. Once a tree has been identified, stand well back from it and try to commit its overall shape to memory. Then look for other trees with similar shape and confirm their identity. After a while you will find as you walk or drive around the countryside that certain species become instantly recognizable.

Spherical

Coniferous spreading

Deciduous spreading

Ovoid

Conical

Weeping

Columnar

*Above: Tree shape or form is the first step in identifying trees.*

## Evergreen or deciduous

In winter, in temperate regions, this is a fairly obvious feature to substantiate; at other times of year, or in the tropics, it may require a little more detective work. Most evergreen leaves fall into two categories. They will either be long, thin and needle-like, which will suggest that they belong to a conifer, or they will be thick and leathery, quite often with a shiny surface. In most temperate countries the latter are few and far between, making identification relatively easy. A non-conifer evergreen in Britain is almost certain to be either holly, *Ilex aquifolium*, or holm oak, *Quercus ilex*, for example. In hot areas of Europe, you may well need to look at the leaves more closely and take other features of the tree into account before its identity becomes clear.

## Leaves

For most trees, the leaves are probably the most important aid to identification. There are many different leaf shapes but almost all of them fall into the following six categories. Leaves may be "entire", which means that they are undivided and have no serrations around the edge, such as those of magnolia. They may be "serrated" with sharp serrations around the edge, as with the leaves of sweet chestnut, or be "lobed", curving in towards the centre of the leaf and then back out again, as in oak. They may be "palmate", which means

*Below: Palm trees are clearly identifiable by their frond-like leaves and single trunk.*

*Below: Deciduous trees are easier to identify when in full leaf in summer.*

*Below: Evergreen trees can be identified by their cones or flowers.*

*Above: There are hundreds of different leaf shapes, colours and arrangements to help identify trees. Needles too are quite distinctive, and, like leaves, are arranged differently on different trees.*

hand- or palm-like – sycamore and maple leaves are palmate. On some leaves the indentations may go right down to the petiole (leaf-stalk) as with horse chestnut; then the leaf is called a "compound palmate" leaf. Sometimes the leaf is sub-divided into smaller leaflets. The leaf is then called "pinnate". Temperate trees with pinnate leaves include ash and rowan. Many tropical trees have pinnate leaves.

## Flowers

Most trees produce flowers in spring, although some, such as the Indian bean tree, *Catalpa bignonioides*, wait for summer. Relatively few temperate trees flower in autumn or winter although some tropical trees flower all year round. Tree identification can be much easier when flowers are evident. Cherries are instantly recognizable by their flowers, as are magnolias. The difficulty arises when individual species or varieties of cherry or magnolia are required. Here again the flower can help. What colour is it? Does it have double or single petals? How long are the flower stalks? Close examination of flowers will always enable trees to be separated. The way that flowers are held on the tree is also important. Where do they appear, on the ends of twigs or in the leaf axils? Are they individual or do they appear in clusters? If they are held in clusters, what are those clusters like?

## Fruit

Late summer to autumn is the best time to identify trees by their fruit. Some fruit or seeds are instantly recognizable – acorns will immediately identify an oak tree and conkers a horse chestnut. Fallen fruit are particularly useful indicators for tropical trees, which may be too tall for flowers or leaves to be visible. Some fruit are particularly distinctive, such as those of spindle trees, *Euonymus* species. The casing is normally bright pink and opens very much like a parasol to reveal orange seeds, which hang on tiny threads.

*Below: Fruit and seeds appear in a variety of forms to attract a wide range of pollinators.*

## Bark

Some trees are probably better known by their bark than any other feature. Silver birch, for example, has striking silvery white bark, while the Tibetan cherry, *Prunus serrula*, has bark that is polished, peeling and mahogany red. Other trees may not have such striking bark colour but bark may still be a useful feature to aid identification. For instance, beech has smooth light grey bark, cherry has distinctive horizontal

*Below: Indian bean tree flower clusters help identify the tree in spring.*

*Below: Simpoh air has distinctive yellow flowers that form in racemes.*

*Below: Buds can help identify a tree.*

*Above: Breadfruit are instantly recognizable in their native West Indies.*

*Above: The cones of the Likiang spruce age to become almost purple in colour.*

*Above: The sweet chestnut tree can be identified by its distinctive fruit in autumn.*

banding and plane trees have buff-coloured bark which is constantly "flaking" to reveal fresh, light fawn bark beneath.

## Buds

Winter is the time when tree identification can be most difficult. However, close inspection of the buds can be of considerable help. Ash has very distinctive black buds, for example, while those of magnolias tend to be large and covered with a dense coating of light-grey hairs. Horse chestnut buds are large and sticky, sycamore buds lime-green in colour. The positioning of buds can also help identify a tree. They may be in pairs on opposite sides of the twig, or they may be alternately positioned with one on the left followed by one on the right. Some buds hug the twig, such as those of willow, while others, such as oak buds, appear in clusters.

## Equipment

When identifying trees in the field it is worth having one or two pieces of equipment with you. A good field guide is essential. It is worth getting one that fits inside your pocket and preferably has a waterproof cover. A pair of binoculars can be useful to get a closer look at leaves, flowers or buds, which may be at the top of the tree. A notepad and pencil will allow you to sketch relevant features and make notes on locations. Finally, sealable plastic bags are particularly useful. They enable you to collect specimen leaves, fruit or seeds and take them home for closer examination.

## How to use the encyclopedia

The trees in the encyclopedia are arranged according to the plant family to which they each belong and are arranged in order using the recognized Cronquist system for classification of flowering plants as developed by A. Cronquist in 1981. The coniferous trees precede the broadleaved trees.

Each main group of trees is divided into families, then genus and finally species. For each species the botanical name and common name is given. Under the common name of each tree, is first any other common name the tree is known by, followed by the Latin name. The Latin names are constantly under review so may not match older published material.

*Right: In winter the overall shape of a tree and its bark, twigs and buds will all help towards identification. Quite often, in managed woodlands, the task of identification will be made simpler by the fact that many trees of the same species will be planted together.*

# HOW TO USE THE DIRECTORY

*The directory will help you to identify the most popular, most common and best-known trees in Europe. The associated descriptions will clarify how each tree can be identified at all times of the year, even in winter when deciduous trees only have bare branches, bark and twigs to show.*

The following directory includes trees that thrive all over Europe. They are subdivided using the recognized Cronquist system for classification of flowering plants, as developed by Arthur Cronquist in 1981. Each main group of trees is divided into families, then genus and finally species. Each double page spread features four main trees from one or more genera, and up to four other species of note contained within a tinted text box. The introduction to each group of trees describes common characteristics for the whole group.

Each main entry (shown with artwork) discusses the primary characteristics of the tree together with the uses to which it is put, such as timber, medicine, shelter or hedging. There is a detailed description to aid identification plus an accurate watercolour illustration of the tree in leaf and also in winter profile, where appropriate. An additional tinted box on the page describes other trees of interest within the same family group. Colour maps of all the featured trees show at a glance the natural distribution of the trees.

**Other Common Names(s)**
Occasionally some trees have different common names in other countries to the UK. These are listed underneath its UK common name.

**Common Name**
This is the popular, non-scientific name of the tree entry.

**Botanical Name**
This is the internationally accepted botanical name for the tree entry. It is always Latin.

**Tree Introduction**
This provides a general introduction to the tree and may include information on usage, preferred conditions, and other general information of interest about each tree.

**Identification**
This description will enable the reader to properly identify the tree in any season. It gives information on leaf shape, size, colour and arrangement, type of flower, type of bark, number of buds and type of fruit.

## Sweet Chestnut

Spanish chestnut *Castanea sativa*

This fast-growing ornamental tree is native to warm, temperate regions around the Mediterranean and in south-western Asia. It has also been widely cultivated elsewhere, often introduced by the Romans, who valued the tree as a source of food for themselves and their animal stock.

**Identification**: As a young tree the bark is smooth and grey. Older trees develop spiral fissures, which immediately distinguish the tree from oak. The leaves are oblong, up to 20cm/8in long, sharply pointed at the tip and rounded at the base. The leaf margin is edged with coarse teeth, each tooth linking to a strong vein running back to the midrib. Each catkin may be up to 25cm/10in long. The fruit is a spiny greenish-yellow husk, up to 6cm/2½in across, with up to three edible brown nuts.

*Above right: The male and female flowers are borne on the same upright yellow catkin in summer, making it one of the last trees to come into flower.*

*Right: The chestnuts ripen in autumn.*

**Flower Illustration**
A watercolour illustration shows the colour and shape of the tree's flowers.

**Leaf and Fruit Illustration**
A watercolour illustration shows the colour, size, shape and arrangement of leaf and fruit.

**Summer Profile**
A watercolour illustration of the tree in full leaf is given for each entry.

**Winter Profile**
Where appropriate, there is a watercolour illustration showing the branch outline of the tree during the winter months.

**1 Map**
A map shows the area of natural distribution of the featured tree. The relevant area is shaded in yellow. The natural distribution of a tree shows where in the world the tree originated. It does not mean that this is the only place where the tree now grows. All the trees featured in this volume grow in Britain and Europe, even though most did not originate here.

**2 Distribution**
Describes the tree's natural distribution throughout the world.

**3 Height**
Describes the average height of the tree given optimal growing conditions.

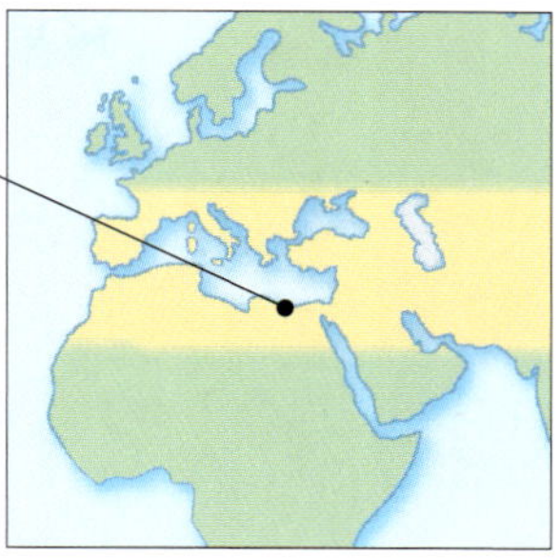

**Distribution**: Southern Europe, North Africa and south-west Asia.
**Height**: 30m/100ft
**Shape**: Broadly columnar
**Deciduous**
**Pollinated**: Insect
**Leaf shape**: Oblong

**4 Shape**
Describes the overall shape of the tree. For more information on tree shapes, turn to the section "How to Identify a Tree".

**5 Deciduous**
Trees are either deciduous or evergreen.

**6 Pollinated**
Describes the method of pollination.

**7 Leaf Shape**
Describes the shape of the leaf. For more information on leaf shapes, turn to the section "How to Identify a Tree".

## Other information

**1 Other species of note**
The trees featured in this tinted box are usually less well-known species of the genus. They are included because they have some outstanding features worthy of note, but may be less common in cultivation or in the wild. There most interesting features are noted, as well as any significant differences from other species in the genus. Occasionally the box may feature entries from more than one genus.

**2 Photographs**
Occasionally photographs are included within this box as an aid to identification. These show a specific feature of the tree or the whole tree profile. Occasionally an illustration may be included.

OTHER SPECIES OF NOTE

**Wheel Tree** *Trochodendron aralioides*
This attractive, evergreen, Japanese tree is the sole species in the only genus in the family Trochodendraceae. Its nearest relative is believed to be *Drimys winteri*. It has shiny dark green, narrow, elliptical, leathery leaves and aromatic bark. Its most interesting feature is its wheel-like bright green flowers, 2cm/¾in across, with no petals and exposed stamens radiating outwards from a central disc. They appear on upright slender stalks in early summer.

**Gutta Percha**
*Eucommia ulmoides*
This Chinese tree is the only member of the Eucommiaceae family. It is believed to be most closely related to the elms, and is the only temperate tree that produces rubber. If the leaf is gently torn in half, the two halves will still hang together, held by thin strands of sticky latex. In China, gutta percha has been cultivated for hundreds of years for medicinal purposes. It has never been found growing wild, so its origins are unknown.

**Sassafras** *Sassafras albidum*
This medium-sized deciduous tree produces a range of leaf shapes, from those with a large lobe on each side, like a fig leaf, to entire and oval. They turn yellow and orange in autumn. It has been cultivated in European gardens ever since its introduction in 1653.

**3 Species names**
The name by which the tree is most commonly known is presented first, followed by the Latin name and any other common name by which the tree is known.

**4 Entries**
The information given for each entry describes its main characteristics and the specific features it has that distinguish it from better known species.

# PRIMITIVE CONIFERS

*At the end of the Carboniferous period, 286 million years ago, the first conifers, or gymnosperms, began to evolve. These trees protected their seeds in cones and had a much more efficient reproductive system than earlier evolving trees. The following are direct descendants of the first gymnosperms.*

## Chinese Plum Yew

*Cephalotaxus fortunei*

This small, handsome evergreen tree, not dissimilar to yew, comprises a series of erect stems each clothed in distinctive whorls of branches. Chinese plum yew was introduced into Europe by the Scots Victorian plant hunter Robert Fortune in 1849, hence the botanical name *'fortunei'*. It grows naturally as a woodland species and, as such, prefers to be planted in dappled shade rather than full sun.

*Right: The leaves are glossy dark green above and a light matt green beneath.*

**Identification**: The bark of the Chinese plum yew is red-brown and flaking, rather like the English yew, *Taxus baccata*. The soft needle-like eaves are up to 7.5cm/3in long. Both male and female flowers are small and pale yellow, produced on separate trees in spring. The male flowers are borne in the leaf axils, the females at the shoot tips. On female trees these are followed by purple-brown oval fruits.

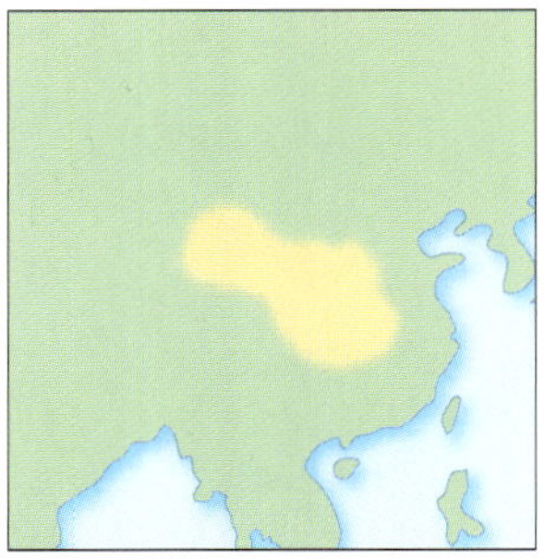

**Distribution**: Central China.
**Height**: 15m/50ft
**Shape**: Broadly spreading
**Evergreen**
**Pollinated**: Wind
**Leaf shape**: Linear

*Right: The needles each have two white bands of stomata.*

*Right: Each fruit is 2.5cm/1in long and contains a single seed.*

## Monkey Puzzle

Chile pine *Araucaria araucana*

This is a uniquely bizarre tree for its triangular, very sharp, pointed leaves and distinctive whorls of long branches. It was introduced into European cultivation in the late 18th century. It is widely admired for its architectural habit, but often looks misplaced. Even in its native Andean forest it is an impressive oddity. Female monkey puzzle trees produce cones 15cm/6in in length, which take more than two years to ripen and fall. The seed is edible.

**Identification**: When young, the tree has a slightly rounded conical outline, with foliage to ground level. As it matures, the crown broadens and the lower branches fall away. This reveals an impressive trunk with horizontal folds of grey bark, similar in appearance to elephant hide.

*Above: The mature tree's distinctive bark.*

*Right: The male cones are borne in clusters at the tips of shoots.*

*Left: The ovoid female cone contains up to 200 seeds.*

**Distribution**: Forms groves in the Andean forests of Chile and south-western Argentina.
**Height**: 50m/164ft
**Shape**: Broadly conical, becoming domed in maturity
**Evergreen**
**Pollinated**: Wind
**Leaf shape**: Linear to triangular

## Maidenhair Tree

*Ginkgo biloba*

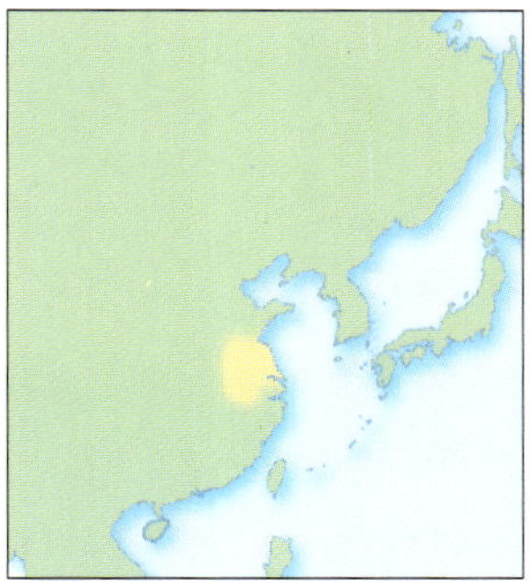

**Distribution**: Originating in China, thought to be from the provinces of Anhui and Jiangsu. It is widely cultivated throughout the Northern Hemisphere, including Japan and Europe.
**Height**: 40m/130ft
**Shape**: Broadly conical
**Deciduous**
**Pollinated**: Wind
**Leaf shape**: Fan

Fossil records show that *Gingko biloba* existed over 200 million years ago. It was introduced into general cultivation in 1754. It produces male and female flowers on separate trees. When ripe, the fruit has a rancid odour; the seed beneath this pungent flesh is edible if roasted. *Ginkgo* has an attractive outline.

**Identification**: This deciduous tree is unique in producing fan-shaped leaves, which resemble those of the maidenhair fern, *Adiantum*, hence its common name. The foliage is produced on characteristic short shoots, most apparent in winter. The bark is pale grey.

*Above left: The foliage turns golden yellow in autumn.*

*Left and above right: The fruit is orange-brown when ripe and has a single edible kernel.*

### OTHER SPECIES OF NOTE

**Manio** *Podocarpus salignus*
Native to Chile, this tree is commonly referred to as the willowleaf podocarp as its leaves are linear and sickle-shaped, resembling those of a willow. It has gently pendulous branches and graceful foliage, and can grow to 20m/65ft, though most specimens in Europe are far smaller.

**Californian Nutmeg** *Torreya californica*
At first glance the Californian nutmeg looks similar to the plum yews, *Cephalotaxus*. However it is easily distinguished by its sharply pointed, needle-like evergreen leaves. It originates from California, where it can grow to heights in excess of 30m/100ft. It has been widely planted in gardens and arboreta across northern Europe.

**Plum Yew** *Cephalotaxus harringtonia*
The origin of this tree is unknown but it has been in cultivation in Japan for centuries. It was introduced to Europe in 1829 and is sometimes referred to as the cow's tail pine. It is an attractive small evergreen, seldom attaining heights in excess of 10m/33ft. It has foliage that at first glance is similar to that of yew, but the fruit is purple-brown and nutmeg-shaped, and resembles a small domestic plum.

## Prince Albert's Yew

*Saxegothaea conspicua*

The genus *Saxegothaea* is monotypic, meaning that this tree is the only species in it. It is an evergreen tree that forms part of the temperate rainforests of southern Chile and adjacent Argentina. It is found growing in association with other forest species, such as *Nothofagus dombeyi*, *Drimys winteri* and *Podocarpus nubigena*, which are all prized for timber. It is cultivated throughout the warmer regions of the Northern Hemisphere, including Europe, as an ornamental tree. Both its generic and its common names commemorate Albert of Saxe-Coburg-Gotha, the consort of the British Queen Victoria.

**Distribution**: In lowland areas along the base of the west Andean slopes, from southern Chile into south-west Argentina.
**Height**: 15m/50ft
**Shape**: Broadly conical
**Evergreen**
**Pollinated**: Wind
**Leaf shape**: Linear

*Above: Needles of the Prince Albert's yew are slightly curved with a sharp tip and are up to 3cm/1¼in long.*

**Identification**: The tree grows to a height of 15m/50ft or more, developing a slender, conical crown in its native environment and a more bushy habit in cultivation. The foliage is similar in appearance to that of yews in the genus *Taxus*. The fruits are thick, round and composed of fleshy scales.

*Right: The leaf on the right shows the topside view and that on the left shows the underside colouring.*

# YEWS

*The genus* Taxus *is present in Europe, North America and Asia. All the species are mainly confined to the middle latitudes, though some grow in tropical highlands. Most are grown as ornamental trees and there are many cultivars. Yews are poisonous, and most species yield the alkaloid taxol, which is used in the treatment of cancer.*

## Japanese Yew

*Taxus cuspidata*

The Japanese yew grows predominantly in Japan but also occurs naturally in parts of China and eastern Russia. It is a slow-growing, very hardy species with a rather sprawling habit. It was introduced into Europe in 1855 by the Scottish plant hunter Robert Fortune. Today, there are more than 90 named forms of the Japanese yew and 130 hybrids – many crossed with the English yew, *T. baccata* – in common cultivation.

**Identification**: The bark is reddish brown, flaking in maturity. The leaves are 2.5cm/1in long, dark green above and yellow-green beneath with obvious yellow stomatic banding. They are arranged in two ranks and stand erect from the twig – often forming a narrow V-shaped trough. The fruit is similar to that of the English yew but is normally carried in greater profusion.

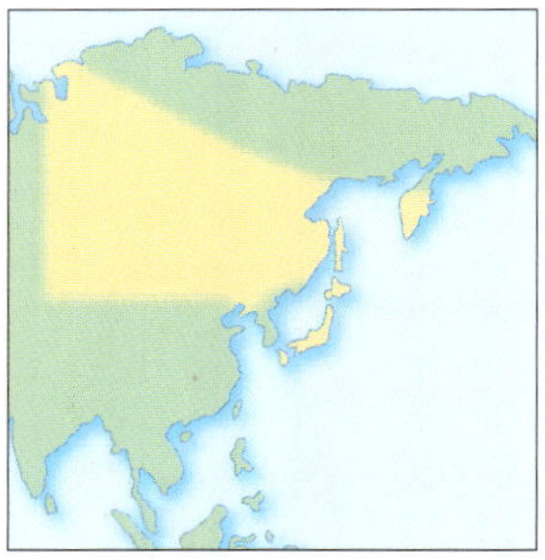

**Distribution**: Japan and north-east Asia.
**Height**: 15m/50ft
**Shape**: Broadly spreading
**Evergreen**
**Pollinated**: Wind
**Leaf shape**: Linear

*Left: The crown is broadly spreading with wide open branching, which gives a sprawling habit.*

*Above: The needle-like leaves are arranged in two erect ranks at each side of the twig forming a V-shaped trough.*

## English Yew

Common yew *Taxus baccata*

The English yew develops a very dense, evergreen canopy, which gives this tree a sombre feel that is heightened by its long association with burial grounds and churchyards. Yew wood is extremely durable and is valued in the production of furniture and highly decorative veneers used in cabinet-making. It was once commonly used for making bowstaves. A number of cultivars have been created, and one of the most striking is 'Standishii', which has an upright habit and golden-yellow foliage.

**Identification**: The tree develops a broad and loosely conical outline. Its leaves are glossy above with a central groove and the bark is rich brown with a purple hue. The male cones shed pollen with cloud-like abundance in spring. The fruit is fleshy, turning red at maturity, around an olive-green seed.

*Below: Yew berries are eaten by birds but the poisonous seeds inside are not digested.*

**Distribution**: Europe, including Britain, eastwards to northern Iran and the Atlas mountains of North Africa.
**Height**: 20m/65ft
**Shape**: Broadly conical
**Evergreen**
**Pollinated**: Wind
**Leaf shape**: Linear

*Right: Yew leaves are needle-like in appearance.*

## Himalayan Yew

*Taxus wallichiana*

The Himalayan yew grows naturally throughout the Northern Hemisphere including Europe and North America. It is prolific in India and the Himalayas. *T. wallichiana* is found growing wild from Afghanistan eastwards through the Himalayan region and north-eastern India into south-western China. The tree is particularly rare in cultivation, with few specimens having been planted outside botanical gardens and arboreta.

**Identification**: The bark is red-brown with purple patches, flaking in maturity. The leaves differ from those of the English yew in being narrower and longer, to 5cm/2in. Both male and female flowers are small and yellow-green, and are borne on separate trees in early spring. The male flowers produce copious amounts of yellow pollen. The ripe fruit, which is bright red and fleshy, contains a single seed.

**Distribution**: Asia.
**Height**: 20m/65ft
**Shape**: Broadly conical
**Evergreen**
**Pollinated**: Wind
**Leaf shape**: Linear

*Right: The leaves have a distinctive sickle-like curve and end in a sharp point.*

### OTHER SPECIES OF NOTE

**Pacific Yew** *Taxus brevifolia*
This small evergreen tree is native to western North America, from British Columbia to California, where it inhabits canyons and gullies, quite often alongside streams. The branches are slender and slightly pendulous and the winter buds are covered with golden scales. It has needle-like dark green leaves, 2.5cm/1in long. The anti-cancer drug taxol was originally isolated from the bark of this species in the 1960s.

**Canadian Yew**
*Taxus canadensis*
One of the hardiest of all yew species, this small evergreen tree is native to eastern North America from Newfoundland to Virginia. Unlike most yews the Canadian yew has both male and female flowers on the same tree. The leaves are more pointed than those of the English yew; otherwise it is very similar in appearance to its European cousin.

**Hybrid Yew** *Taxus* x *media*
This is a hybrid cross between the Japanese yew, *T. cuspidata*, and the English yew, *T. baccata*. It was bred by T. D. Hatfield at the Hunnewell Pinetum, Wellesley, Massachusetts, around 1900. It is a vigorous, medium-size tree with a spreading habit. Scores of cultivars have been developed from the hybrid, including 'Brownii', which is quite often used for hedging.

## Irish Yew

*Taxus baccata* 'Fastigiata'

Two female saplings of this distinctive erect form of common yew were originally found growing wild on the mountains above Florence Court in County Fermanagh, Northern Ireland, by a farmer called George Willis, in 1740. Since then, the cuttings taken from these two trees have resulted in thousands of Irish yews growing in churchyards, gardens and arboreta in western Europe.

**Identification**: The Irish yew is easily recognized by its sombre dark green foliage and upright pillar-like habit, which has made it a favourite for planting in classical or formal gardens throughout Europe. Although it can reach a height of nearly 20m/65ft it more commonly grows to 7–10m/22–33ft with a spread of up to 5m/16ft. Branches on old trees quite often break away from the vertical form, particularly when weighed down during and after heavy falls of wet snow. There is also a golden form of the Irish yew, *T. baccata* 'Fastigiata Aurea', which has bright yellow-green foliage.

**Distribution**: County Fermanagh, Northern Ireland.
**Height**: 19m/60ft
**Shape**: Narrowly columnar
**Evergreen**
**Pollinated**: Wind
**Leaf shape**: Linear

*Left and below: The needle-like leaves of Irish yew are slightly darker green than the English yew and are arranged radially on the twig.*

# FALSE CEDARS AND CYPRESS

*Trees belonging to the genus* Chamaecyparis, *or false cypress, have a number of obvious characteristics in common. All are evergreen, and their leaves are arranged in flattened sprays and have a pungent aroma when crushed. The habitats from which these species originate are generally wet and they all produce very durable timber.*

## Leyland Cypress

x *Cupressocyparis leylandii*

This fast-growing conifer is a hybrid between two American species: Monterey cypress, *Cupressus macrocarpa*, and Alaska cedar, *Chamaecyparis nootkatensis*. The hybrid cross has never occurred naturally in the USA, because the natural ranges of the two parents do not overlap. Instead it originated in 1888 at Leighton Hall, Powys, Wales, where the two parents were planted close to each other in a garden. Since then, Leyland cypress has become very popular for hedging and screening because it is extremely fast-growing, quite often exceeding 2m/6ft in one year. However, it is suitable for planting as hedging close to buildings only if it is regularly trimmed.

*Below: The shoot tip and fruit.*

**Identification**: The bark is red-brown, developing shallow fissures as it matures. The leaves are small and scale-like with pointed tips, dark green above, lighter green beneath and borne in flattened sprays. Male and female flowers are found on the same tree. The male flowers are yellow, the female green; both appear in early spring at the tips of the shoots. The fruit is a globular woody brown cone approximately 2cm/¾in diameter.

**Distribution**: Of garden origin as a hybrid in the United Kingdom. Widely planted throughout Britain and Europe.
**Height**: 30m/100ft
**Shape**: Narrowly columnar
**Evergreen**
**Pollinated**: Wind
**Leaf shape**: Scale-like

## Japanese Red Cedar

*Cryptomeria japonica*

**Distribution**: Found in Japan in Honshu, Shikuka and Kyushu. It is also found in Zhejiang and Fujian provinces in China.
**Height**: 30m/100ft
**Shape**: Broadly conical
**Evergreen**
**Pollinated**: Wind
**Leaf shape**: Linear

This stately, fast-growing conifer produces a large, straight trunk that tapers quickly from a broad base above the roots. Its reddish-brown bark is soft and fibrous and peels off to hang in long strips from the trunk. It was introduced into Europe in 1842 and is extensively grown in parks, gardens and arboreta across western Europe. Its timber is strong and pinkish-brown.

**Identification**: The crown of the tree is narrow when young, broadening with age. Often the heavy branches sweep downwards before rising again at the tips. The very aromatic foliage is a system of bright green branchlets covered with hard, forward-facing needles 1.5cm/⅗in long. The ovoid male flowers are yellowish-brown, clustered along the final 1cm/½in of each branchlet. They are bright yellow when ripe, and shed pollen in early spring. The female flowers are green rosettes, found on the same tree as the male flowers. The cones are globular, 2cm/¾in across, and held on upright, stiff stalks.

*Above: Branches occasionally touch the ground, causing layering.*

*Right: At the base of each needle is a long protruding keel, which runs down the branchlet.*

## Incense Cedar

*Calocedrus decurrens*

The incense cedar is grown widely in western Europe. The natural habit of the tree is unusual in that it develops a columnar, almost fastigiate, form. Shiny, mid-green leaves develop in flattened sprays produced on branches that are almost horizontal to the main stem. It has a very attractive, flaky bark, grey to reddish-brown. There are only two other species of tree in this genus, *C. macrolepis* from China and *C. formosana* from Taiwan.

**Identification**: The foliage of the incense cedar is dense, dark green and usually present to the base of the tree, with only a short exposed bole. The male and female flowers are produced on the same tree. Often, abundant quantities of oblong cones are produced and become pendulous with their own weight.

*Above: The red-brown bark of the incense cedar is similar to that of the giant redwood.*

*Far left: The yellow-brown cones have six overlapping scales.*

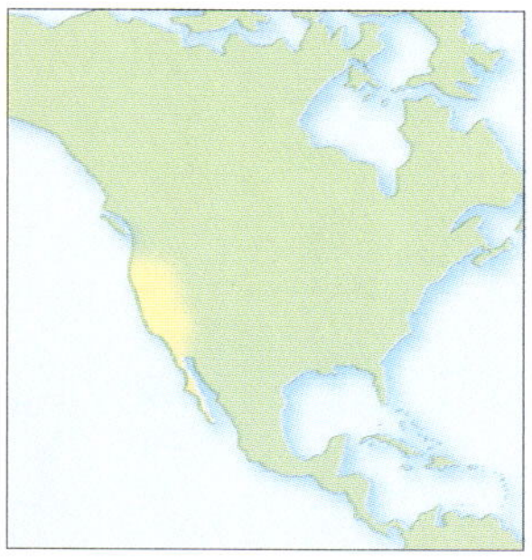

**Distribution**: Western North America, from mid-Oregon southwards to Baja California in northern Mexico.
**Height**: 40m/130ft
**Shape**: Narrowly columnar
**Evergreen**
**Pollinated**: Wind
**Leaf shape**: Linear scale-like

### OTHER SPECIES OF NOTE

**Hinoki Cypress** *Chamaecyparis obtusa*
This cypress is native to Japan, where it is cultivated as an ornamental tree and is highly prized for its beauty. It was first grown in Europe in the 1860s. It produces a valuable timber. The crown develops a fairly broad conical shape. The bark is soft, stringy and greyer than that of other *Chamaecyparis* species. Bright stomatal banding under the leaves gives an almost variegated appearance.

**Sawara Cypress** *Chamaecyparis pisifera*
This is a medium-size, highly ornamental Japanese evergreen tree, which has a conical shape and an open branching habit. It has attractive reddish-brown bark and light green foliage, which is produced in flattened sprays and, when crushed, emits a pungent, resinous scent. Small cones the size of peas are produced in dense clusters. It was introduced into Europe in 1861.

**Patagonian Cypress**
*Fitzroya cupressoides*
This beautiful large tree, with a cypress-like habit, is native to Chile and Argentina and was introduced into Europe in 1849. It produces pendulous sprays of blue-green foliage that are distinctly marked beneath each needle with two white stomatal bands. It is common in cultivation but seldom attains such large proportions as in the wild, where it may grow up to 50m/165ft.

## Lawson Cypress

Port Orford cedar, Oregon cedar
*Chamaecyparis lawsoniana*

The Lawson cypress develops into a conical tree up to 40m/130ft tall, with reddish-brown fibrous bark and scented foliage that has distinctive stomatal markings on the underside of the leaves. In the Pacific north-western USA, where this species originates, it remains a very important source of timber, with many uses from boat-building to cabinet-making. Many cultivars have been produced, which vary widely in form, foliage and colour.

**Identification**: The young tree has smooth, brown-green, shiny bark, and a pendulous dominant shoot, not seen in the mature tree. It produces globular cones on the foliage tips, which are at first fleshy with a bluish purple bloom and become woody and wrinkled.

*Below: The cones are globular, 7mm/⅓in in diameter and purple-brown. They remain on the tree long after the seed has been shed.*

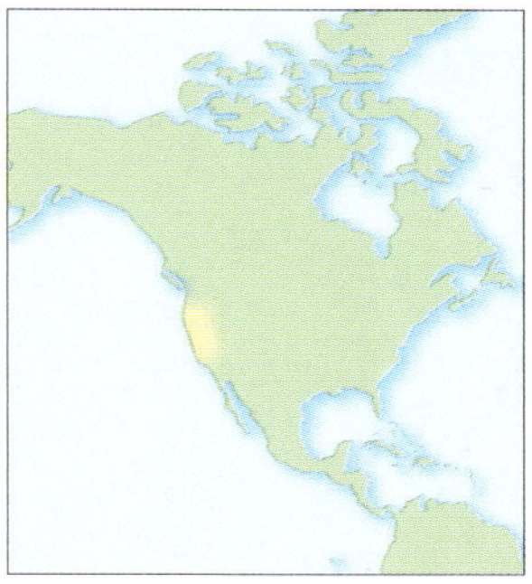

**Distribution**: North-western USA, from south-west Oregon to north-west California. Introduced into Europe in 1854 and widely planted in gardens across the continent.
**Height**: 40m/130ft
**Shape**: Narrowly conical
**Evergreen**
**Pollinated**: Wind
**Leaf shape**: Linear scale-like

*Below: The top side of the foliage is dark green to blue and when crushed smells of parsley.*

# FALSE CYPRESSES AND THEIR CULTIVARS

*The trees included below are commonly known as false cypresses. They have characteristics of the true cypresses, but belong to different genera. Like the true cypresses, they are grouped within the Cupressaceae family. However, true cypresses belong to the genus* Cupressus.

## Ellwood's Cypress

*Chamaecyparis lawsoniana* 'Ellwoodii'

This slow-growing, columnar cultivar of Lawson cypress was raised at Swanmore Park, Bishops Waltham, Hampshire, England, in the late 1920s. It takes its name from Mr Ellwood, the head gardener at Swanmore Park at the time. Originally thought of as a dwarf tree ideal for planting in rock gardens, it soon became clear that it had the potential to outgrow all but the largest rockeries.

**Identification**: Ellwood's cypress makes a distinctive, tightly packed, almost impenetrable column of strongly ascending branches. It has short, rather feathery, spray-like evergreen foliage, which is predominantly grey-green but turns a steely blue-green in winter. A feature of this tree is that the juvenile foliage is always present, giving a rather soft appearance. In maturity the top of the column becomes irregular, with multiple pinnacles of foliage.

**Distribution**: Of UK garden origin.
**Height**: 10m/33ft
**Shape**: Narrowly columnar
**Evergreen**
**Pollinated**: Wind
**Leaf shape**: Linear scale-like

*Above: Each small, irregularly shaped spherical cone contains hundreds of tiny seeds.*

*Left: A common and distinctive feature of this tree is the way it produces juvenile foliage throughout its life, which gives the whole tree a soft appearance even in maturity. The foliage is short, rather feathery and a metallic blue-green colour in winter.*

## Lawson Cypress 'Erecta Viridis'

*Chamaecyparis lawsoniana* 'Erecta Viridis'

This is one of the oldest of all Lawson cypress cultivars and still one of the most popular. It was raised at the Knaphill nursery of Anthony Waterer in Surrey, England, in the 1850s, and is believed to have arisen from only the second batch of seed of Lawson cypress ever imported into Europe from California. Originally called simply 'Erecta', its name was changed to 'Erecta Viridis' in 1867. Although relatively slow-growing, it is potentially a large tree and there are several old specimens around the British Isles that are 28m/80ft tall. It can be used as a hedging tree or as an isolated specimen to be grown in a lawn. It has a compact shape and dense foliage and grows well on free-draining soil.

**Identification**: 'Erecta Viridis' has perhaps the brightest green foliage of any Lawson cypress and maintains this vibrancy throughout the winter. It has a fastigiate shape, with ascending branches that sweep dramatically skywards at the tips. There will quite often be between two and six main stems growing from the base, which help to develop a rather billowing multiple top. In maturity its tight columnar form becomes more open and lax, as old branches fall outwards away from the main stem(s).

**Distribution**: Of UK garden origin.
**Height**: 28m/80ft
**Shape**: Columnar
**Evergreen**
**Pollinated**: Wind
**Leaf shape**: Linear scale-like

*Right: The foliage is bright, rich green and borne in large flattened sprays.*

## Taiwan Cypress

Formosan cypress *Chamaecyparis formosensis*

In its native habitat this is an impressive tree, growing to 60m/200ft, with a trunk of 20m/65ft or more in girth. Introduced into cultivation in the early part of the 20th century, it has yet to be recorded at the great size of its wild form. As with many of the false cypresses, its timber is noted for being durable and resistant to moisture. Since its introduction into Europe in 1910, it has been widely planted in arboreta and botanic gardens across the continent.

**Identification**: The tree has a reddish bark with regular, shallow fissuring. The leaves are flat and produced in broad sprays; the individual leaves are scale-like. The small oblong cones have brown wrinkled scales that are flat or slightly protuberant, and are carried above the foliage.

*Left: The lower branches are level and sweep up towards the tip.*

*Left: The scale-like needles are light yellow-green and when crushed have the aroma of seaweed.*

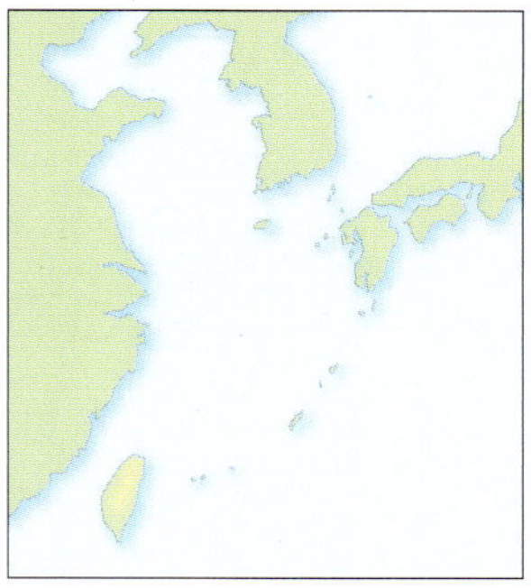

**Distribution**: Mountainous regions of central and northern Taiwan.
**Height**: 60m/200ft
**Shape**: Broadly conical
**Evergreen**
**Pollinated**: Wind
**Leaf shape**: Linear scale-like

## Fletcher's Cypress

*Chamaecyparis lawsoniana* 'Fletcheri'

This common, attractive cultivar is named after the Fletcher brothers, who ran a tree nursery at Chertsey in Surrey, England, in the early part of the 20th century. It was developed from a "sport" on a Lawson cypress, which was found growing near the nursery in 1911. The sport showed the interesting characteristic of having permanently semi-juvenile foliage. The Fletchers tested it in the nursery for reversion, and then named and distributed it in 1923. Today, one of the finest specimens is to be seen at Bedgebury Pinetum in Kent, England. It is a slow-growing tree that is often erroneously planted in rock gardens, where it soon becomes far too large.

**Identification**: Fletcher's cypress is normally a broad columnar multi-stemmed tree 10–15m/33–50ft tall, with regular ascending sprays of foliage on the outer side of vertical shoots. The feathery foliage is blue-grey-green, becoming bronze in winter. The shoots are dark pink-purple, turning purple-brown as they mature. In maturity the basal branches commonly brown and die.

**Distribution**: Of UK garden origin.
**Height**: 15m/50ft
**Shape**: Broadly columnar
**Evergreen**
**Pollinated**: Wind
**Leaf shape**: Linear scale-like

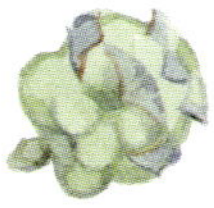

*Above: Small irregularly-shaped cones ripen in winter.*

*Below: The foliage is bronze tipped in winter.*

### OTHER SPECIES OF NOTE

***Chamaecyparis pisifera* 'Squarrosa'**
This widely planted cultivar of the Japanese Sawara cypress was raised in Japan in the 1840s and introduced into Belgium shortly afterwards, from where it was taken to Britain in 1861. It is a broad-crowned tree with bright red-brown bark and wide-spreading branches that are densely covered with sprays of soft, glaucous-coloured juvenile foliage.

***Chamaecyparis pisifera* 'Filifera'**
This is a small to medium-size tree of conical to domed habit that can be as broad as it is tall. It has spreading branches that carry long, pendulous, adult foliage, which resembles dark-green bootlaces. 'Filifera' was introduced into Europe from Japan in 1861 by the Scottish plant hunter Robert Fortune. There are several forms of this cultivar with golden foliage.

***Chamaecyparis lawsoniana* 'Intertexta'**
This cultivar of Lawson cypress was raised at the Lawson Nursery in Edinburgh, Scotland, in 1869, from seed collected by Andrew Murray in the upper Sacramento Valley, California, USA, in 1854. 'Intertexta' is a weeping cultivar with lax drooping tips to the branches and the main leader. It has dark green foliage, borne on the branches in flat, fan-shaped sprays. From a distance the outline of this cultivar is similar to that of the deodar, *Cedrus deodara*.

# TRUE CEDARS AND FALSE FIRS

*Although there are only four true cedars, they are without doubt the real stars of the coniferous world. Nothing can touch them for sheer majesty and dignity. Three originate from around the Mediterranean and the fourth grows a little further east in the Himalayas. The false firs, on the other hand, are all to be found in either North America or Asia.*

## Deodar

*Cedrus deodara*

In the days of the British Empire, huge deodars 75m/245ft tall were found, some of them over 900 years old. Sadly, most of these have been felled for their timber. The deodar has light brown, very durable timber, which is highly prized. It is widely planted all over Europe as an ornamental tree, rather than for its timber.

*Below: The barrel-shaped cones are 12cm/4¾in long, ripening to dull brown.*

**Identification**: In young trees the bark is smooth and dark grey, but on older trees it becomes cracked and covered with pink-grey fissures. The juvenile foliage is blue-grey, becoming dark green with age. The deodar is narrowly conical when young, and broadens as it matures. It is easily distinguished from other cedars by the drooping ends to its branches. The needles, which are 4cm/1½in long, are arranged in whorls around the shoot. The male flowers are very prominent and erect, purple ripening to yellow, up to 8cm/3in long; they shed copious amounts of pollen in late autumn.

**Distribution**: Western Himalayas, Kuram Valley to Kumaon and on to Afghanistan.
**Height**: 50m/165ft
**Shape**: Broadly conical
**Evergreen**
**Pollinated**: Wind
**Leaf shape**: Linear

*Right: The needles of the deodar are sparse.*

## Cedar of Lebanon

*Cedrus libani*

This large, stately tree is probably the best known of all the cedars. It has been revered for thousands of years, and in biblical times it was a symbol of fertility. King Solomon is believed to have built his temple using its timber. On Mount Lebanon it grows at altitudes up to 2,140m/7,020ft. Although numbers are decreasing in the wild, it is widely planted as an ornamental tree in parks, gardens and arboreta in Britain and Europe.

**Identification**: The young tree is narrow and conical; after about 40 years it broadens rapidly, with level branches so long they seem to defy gravity. The bark is a dull brown with even, shallow fissures. The 3cm/1¼in needles are grey-blue to dark green (depending on the provenance of the individual tree), growing in dense whorls on side shoots and singly on fast-growing main shoots. The erect, barrel-shaped cones are grey-green, maturing to purplish brown.

**Distribution**: Mount Lebanon, Syria and the Taurus Mountains in south-east Turkey.
**Height:** 40m/130ft
**Shape**: Spreading
**Evergreen**
**Pollinated**: Wind
**Leaf shape**: Linear

*Right: The cones are 12cm/4¾in long.*

*Left: Broad spreading habit.*

## Atlas Cedar

*Cedrus atlantica*

The natural range of this beautiful tree is in the Atlas Mountains, where it grows along the snowline at 1,500–2,200m/4,900–7,200ft above sea-level. It is sometimes considered to be a geographical sub-species of the cedar of Lebanon, *C. libani*, but is distinguished by its ascending branches. In its natural range the foliage varies from dark green to grey-blue. In cultivation it is the glaucous form, *C. atlantica* 'Glauca', that is most commonly planted. It grows in parks and gardens in western Europe.

**Identification**: The bark is slate grey, smooth at first becoming fissured into scaly plates in maturity. The needle-like leaves, up to 2.5cm/1in long, are borne in dense whorls on short side shoots and singly on the longer leading shoots. Both the male and female flowers are erect catkins; males are yellow, females green. The male flowers, up to 5cm/2in long, are conspicuous in autumn. Once the pollen has been released they fall to the ground. The fruits are barrel-shaped upright cones, to 7.5cm/3in long, borne in rows along the branches.

**Distribution**: Algeria and Morocco.
**Height**: 40m/130ft
**Shape**: Broadly conical
**Evergreen**
**Pollinated**: Wind
**Leaf shape**: Linear

*Left: Needles.*

*Left: After two to three years the cones ripen and release numerous winged seeds.*

## Douglas Fir

*Pseudotsuga menziesii*

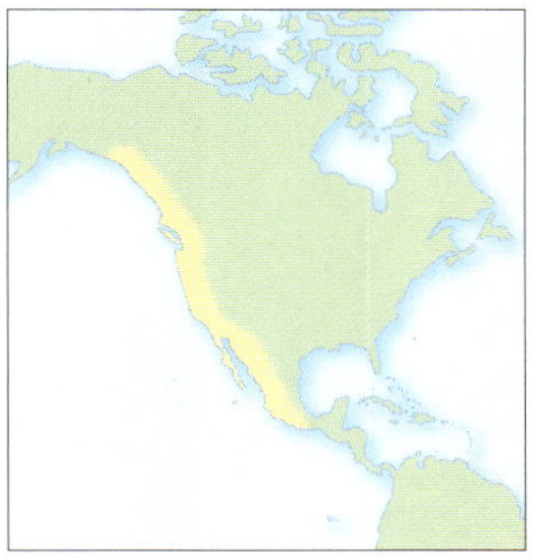

**Distribution**: North-west Pacific seaboard, from Mexico through USA to Canada, including Vancouver Island.
**Height**: 75m/245ft
**Shape**: Narrowly conical
**Evergreen**
**Pollinated**: Wind
**Leaf shape**: Linear

*Below: The cones have bracts that project from each scale.*

Douglas fir is one of the most commercially important timber-producing trees in the world. It has been planted throughout North America, Europe, Australia and New Zealand. It is a huge tree, attaining heights in excess of 75m/245ft. Quite often there is no branching for the first 33m/110ft. The bark is corky and deeply fissured in maturity; young trees have smooth, shiny grey-brown bark that is pock-marked with resin blisters.

**Identification**: When young this majestic tree is slender and regularly conical, with whorls of light ascending branches. In old age it becomes flat-topped, with heavy branches high up in the crown. The needles are up to 3cm/1¼in long and rounded at the tip. They are rich green, with distinctive white banding beneath, and arranged spirally on the shoot. When crushed the foliage emits a sweet citrus aroma. The male flowers are yellow and grow on the underside of the shoot. The female flowers are green, flushed pink to purple at the tip, and grow in separate clusters on the same tree. The fruit is a hanging cone, up to 10cm/4in long, green ripening to orange-brown, with distinctive three-pronged bracts.

### OTHER SPECIES OF NOTE

**Big-cone Fir**
*Pseudotsuga macrocarpa*

Native to south-west California, this rare tree has dull grey bark with wide, vertical orange fissures. Its crown is broadly conical with level branches. The needles are up to 5cm/2in long, stiff, pointed and widely spaced all around the shoot. The cylindrical cones grow up to 18cm/7in long; beneath each scale is a bract that only just protrudes. It was first grown in Britain in 1910.

**Japanese Fir**
*Pseudotsuga japonica*

This tree, which is found in south-east Japan, is rare in the wild and uncommon in cultivation. It has a flattened crown. The leaves are soft, light green, 2.5cm/1in long, blunt and notched at the tip. No fragrance is emitted when the foliage is crushed. The cone is the smallest of the genus at 5cm/2in long; its scales are few and smooth, with a spreading bract that is bent slightly outwards.

# TRUE CYPRESSES

*These trees are closely related to the genus* Chamaecyparis, *in that their leaves are scale-like and produced in sprays. Unlike the false cypresses, their foliage is not flattened. Their cones are composed of fewer scales – between six and eight – and are twice the diameter, but contain less seed. True cypresses are distributed throughout regions of North America, Europe and Asia.*

## Himalayan Cypress

Bhutan cypress *Cupressus torulosa*

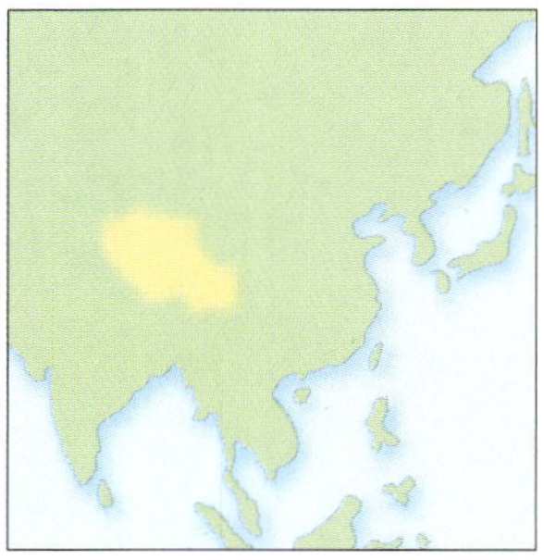

**Distribution**: Western Himalayas to south-west China; also central and north Vietnam.
**Height**: To 20m/65ft
**Shape**: Columnar with conical top
**Evergreen**
**Pollinated**: Wind
**Leaf shape**: Linear scale-like

Despite growing at elevations in excess of 3,300m/10,800ft in the Himalayas, this slow-growing species is not particularly hardy in cultivation in northern Europe. Consequently it is relatively uncommon and is mostly confined to arboreta, tree collections of botanic gardens and gardens that benefit from the warmth of the Atlantic Gulf Stream. Even here, patches of browning foliage in the canopy are commonplace. Where it does grow well it makes an attractive tree with weeping foliage. It was introduced into Europe in 1824.

**Identification**: The overall form of this tree is columnar, becoming conical towards the top, with slender branches that begin level, but then curve downwards towards the tips, where bright yellowish-green foliage hangs in dense irregular bunches. When crushed, the foliage smells of newly cut grass. The brown or greyish bark on mature trees peels off in long strips, revealing fresh red-brown bark beneath.

*Right: Clusters of small, 1cm/½in-long cones are conspicuous in autumn, when they ripen from blue-green to a dark red-brown.*

## Cedar of Goa

Mexican cypress, Portugal cedar *Cupressus lusitanica*

The origins of this tree are rather confusing. It is thought to have been cultivated in England since the late 17th century, having been introduced from Portugal, where it was once believed to be a native species. Its botanical name is derived from Lusitania, the Roman province established in the region occupied by Portugal. However, the tree has never been found growing wild there, or for that matter in the former Portuguese settlement of Goa in western India, despite its common name. It is now believed to have been introduced into Portugal from Mexico in the 16th century by Portuguese missionaries. More recently, it has been planted in parts of Africa as a forest tree. Its timber has been used for general construction, for posts and poles, although its durability is questionable.

**Identification**: On old trees, the rich, dark brown bark peels away in vertical fibrous strips. The leaves are small, scale-like, grey-green in colour and borne on pink to dark red shoots in short, broad sprays. In early spring the tips of each shoot are covered in numerous small male flowers of yellow to creamy-brown. These are followed by ovoid cones, 1.5cm/¾in across, which are glaucous green in the first year, ripening to dark, shiny purple-brown in the second year. Each cone scale has a hooked spine in the centre.

**Distribution**: Mexico and Guatemala.
**Height**: 30m/100ft
**Shape**: Narrowly conical
**Evergreen**
**Pollinated**: Wind
**Leaf shape**: Linear scale-like

*Left: The grey-green leaves are scale-like and borne on pink shoots in short, broad sprays.*

## Monterey Cypress

*Cupressus macrocarpa*

With an incredible ability to withstand exposure to salt-laden winds, this tree has become as common a sight in the exposed coastal habitats of Europe as in its native California. It can attain a height of 40m/130ft, but is often stunted by extreme conditions. In maturity it becomes flat-topped, with spreading horizontal branches. In cultivation it is best known for being the female parent of the leyland cypress, x *Cupressocyparis leylandii*. The strong, durable timber is often used for structural work.

*Right:* Cupressus macrocarpa *'Donald Gold' is a variety of Monterey cypress.*

**Identification**: Mature trees display a great variability in habit, from a dense crown of ascending branches to a more horizontal cedar-like form. Leaves are arranged in loose, circular sprays around the shoots. When crushed, the foliage releases an aromatic odour.

**Distribution**: USA: two sites near Monterey, California, at Cypress Point and Point Lobos.
**Height**: 40m/130ft
**Shape**: Columnar to spreading
**Evergreen**
**Pollinated**: Wind
**Leaf shape**: Linear scale-like

*Left: Cones are up to 4cm/1½in across.*

OTHER SPECIES OF NOTE

**Chinese Weeping Cypress** *Cupressus funebris*
Native to China, this species grows along the Yangtze Valley. It is planted an an ornamental tree in Europe. It is similar to the false cypresses in having foliage produced in flattened sprays. The habit is attractive and often pendulous. Immature plants have a fine, soft foliage that is retained for a number of years.

**Smooth Arizona Cypress** *Cupressus arizonica* var. *glabra*
This hardy, small to medium-size cypress grows in European arboreta, where it inhabits dry, rocky mountain slopes and has attractive, peeling red bark, which turns purple with age. It has a dense, conical habit with ascending branches that end in red shoots. The foliage is blue-grey speckled with spots of white resin. A form of this tree known as 'Pyramidalis', which has strikingly blue foliage, is common in cultivation.

**Kashmir Cypress** *Cupressus cashmeriana*
Curiously, this tree has never been found growing in the wild but it was introduced into Europe from Tibet in 1862. It is a graceful and beautiful tree with ascending branches, which are adorned with pendulous sprays of silver-blue foliage. In colder regions of northern Europe this tree will not survive outside, but it makes a splendid specimen for a large conservatory.

## Italian Cypress

*Cupressus sempervirens*

This is a fascinating tree because of its ability to retain a tight columnar form throughout its life. There is no other conifer, except the monkey puzzle, that has such a strong sense of architecture. It is the space between the plantings of these trees that characterizes the hills and roadside verges of Tuscany, in Italy, and the western Mediterranean. The foliage is a dull grey-green and, unusually for cypress, has no noticeable scent when crushed.

**Identification**: The Italian cypress can attain a height of 18m/60ft or more in its natural habitat. The bark is predominantly grey with some brown colouring. The cones are larger than in most trees in this genus, growing to 3.5cm/1⅓in, similar to those of the Monterey cypress. They are retained on the tree for many years.

**Distribution**: Predominantly Mediterranean, with a range extending northwards to Switzerland and east to northern Iran.
**Height**: 18m/60ft
**Shape**: Very narrowly columnar
**Evergreen**
**Pollinated**: Wind
**Leaf shape**: Linear scale-like

*Below: The cones are grey-brown, smooth and shiny.*

# JUNIPERS

*The junipers are similar to the true cypresses,* Cupressus, *in having two types of leaves on the same plant, both juvenile and scale-like. Unlike cypresses, their fruit is a fleshy cone in which the scales have fused together to give a berry-like appearance. There are at least 50 species, mostly occurring in the Northern Hemisphere, and in many dry regions junipers are the dominant forest cover.*

## Syrian Juniper

*Juniperus drupacea*

With the longest needles and largest fruit of the junipers, the Syrian juniper has a distinctive, fine outline, developing into a tall columnar tree. It thrives in Greece and Turkey in mountains and coastal forests. It usually reaches a height of about 15m/50ft, though the tallest recorded specimen, in Turkey, is about 40m/130ft. The Syrian juniper has a similar distribution to that of *J. excelsa*, the Grecian juniper. Between the two in distribution and extending further north is *J. foetidissima*, the stinking juniper.

**Identification**: The crown of this juniper is columnar to conical in shape. The trunk has orange-brown bark, which is shed in fine vertical strips. The leaves have two white bands of stomata on the inner surface and are shiny mid-green on the outer surface. They are arranged in groups of three.

*Above: The needles are very stiff, and can be up to 2.5cm/1in long. They are sharply pointed and are arranged in whorls of three. The cone is 2.5cm/1in across; blue-green at first, it changes colour to a blackish purple as it matures.*

**Distribution**: Range extends through southern Europe and northern Africa. Found on rocky slopes in forest or scrub, throughout Syria into Turkey and in parts of Greece.
**Height**: 15m/50ft
**Shape**: Columnar
**Evergreen**
**Pollinated**: Wind
**Leaf shape**: Linear scale-like

## Common Juniper

*Juniperus communis*

The common juniper is believed to be the most widespread tree in the world, growing naturally from Alaska, Greenland, Iceland and Siberia south through most of Europe, temperate Asia and North America. It is a hardy tree that tolerates intense cold, exposed coastal locations and high mountain ranges. As a tree it seldom attains heights in excess of 5m/16ft, and it often grows as a multi-stemmed shrub.

*Below: The needle-like leaves are grey-green, sharply pointed, carried in whorls of three with each needle up to 1cm/½in long. The juniper berries are easy to identify on the twigs.*

**Identification**: Depending on the degree of exposure of its location, the overall shape of common juniper may be thin and tree-like or wide-spreading and shrub-like. It has thin, dark red-brown bark that peels in vertical papery strips. The leaves are needle-like, sharply pointed and up to 1cm/½in long. They are grey-green and carried in whorls of three along the shoots. The yellow male and green female flowers are borne on separate trees in clusters within the leaf axils. The fruit is a green berry-like cone that takes between two and three years to ripen. When ripe it turns a glaucous purple-black.

*Right: The cones, which are commonly known as juniper berries, are used to give gin its characteristic flavour.*

**Distribution**: Europe, Asia and North America.
**Height**: 6m/20ft
**Shape**: Narrowly conical
**Evergreen**
**Pollinated**: Wind
**Leaf shape**: Needle-like

## Chinese Juniper

*Juniperus chinensis*

Above: *The mature foliage has dark green scale-like needles with paler banding.*

This is the conifer most commonly used for bonsai. It was introduced to Europe in 1804. Male and female flowers are borne on different trees. It has been in cultivation for centuries and has many cultivars, including 'Keteleeri', which makes a dense, regular and narrowly conical tree; the old and widely planted 'Pfitzeriana', known as the Pfitzer juniper; and 'Hetzii', a very vigorous hybrid, which has an upright and a spreading form.

*Right: When ripe, the cones are dark purple with a pale grey bloom.*

**Identification**: The Chinese juniper has an erect, narrow and conical growing habit. The bark is grey to reddish-brown, peeling off in strips. The leaves are dark green on the outer surface and have a broad green stripe on the inside, separated by two white stomatal bands. The cones are rather lumpy in appearance and measure approximately 6mm/¼in across.

**Distribution**: Widely distributed through north and east China. Also present in Inner Mongolia and around Japan, restricted to the coastal areas of Honshu, Kyushu and Shikoku.
**Height**: 25m/80ft
**Shape**: Narrowly conical
**Evergreen**
**Pollinated**: Wind
**Leaf shape**: Linear scale-like

*Right: Young needles are wedge-shaped.*

OTHER SPECIES OF NOTE

**Temple Juniper**
*Juniperus rigida*
Native to Japan, Korea and northern China, this slow-growing, small to medium-size tree is planted in the grounds of Japanese temples. It was introduced into Europe in 1861 but is relatively uncommon in cultivation. It has dull grey, peeling bark and glossy, bright green needles set in sparse whorls of three.

**Eastern Red Cedar**
*Juniperus virginiana*
Also known as pencil cedar, this North American species has a dense pyramidal to columnar habit and dense foliage that is sage-green above and grey-green beneath. It has red-brown bark that exfoliates in long strips. The fruit is berry-like, light green in spring and dark blue when mature in autumn. The wood has moth-repellent properties and is commonly used to make blanket boxes. It is common in Europe.

**Flaky Juniper** *Juniperus squamata*
Also known as the Himalayan juniper, this small bushy tree, with a natural range from Afghanistan to China, is uncommon in Europe. It has nodding tips to the shoots, dense foliage with awl-shaped, sharply pointed leaves in whorls of three, the upper surface of which are green-white, and purple-black fruits containing one seed. It was introduced into Europe in 1824.

## Drooping Juniper

*Juniperus recurva*

Drooping juniper is so-called because of its weeping foliage, which descends from upswept branch ends. This interesting, slow-growing tree was introduced into Europe by the Veitch Nursery of Chelsea, England, in 1861. Its natural range in the Himalayas overlaps with that of the flaky juniper, *J. squamata*. Although drooping juniper is relatively uncommon in cultivation, the variety *coxii* is widely cultivated, and differs from the species in having orange peeling bark and a much more open habit.

**Identification**: The bark is dark red-brown and peels in maturity into long vertical strips. The leaves are needle-like, forward pointing, 5–8mm/¼–⅓in long, grey-green with two distinct bands of blue-white stomata. They are borne on red-brown shoots, which, when shaken, emit a dry rustling sound. Both male and female flowers are green and are borne in small clusters at the tips of the shoots on the same tree. The small oval fruit is purplish-brown to black, resembling a black berry.

**Distribution**: From eastern Himalayas to Japan.
**Height**: 15m/50ft
**Shape**: Narrowly conical
**Evergreen**
**Pollinated**: Wind
**Leaf shape**: Linear scale-like

*Right: Small needle-like leaves point forward along the shoot.*

*Below: The foliage weeps or droops at the branch ends.*

## Canary Island Juniper

Canary cedar *Juniperus cedrus*

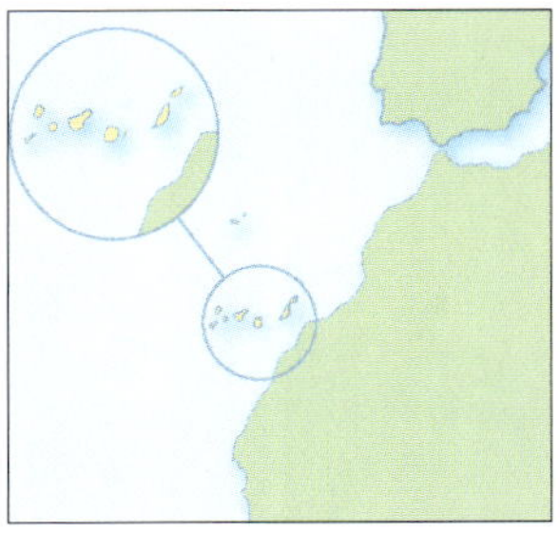

**Distribution**: Canary Islands and Madeira.
**Height**: 15m/50ft
**Shape**: Broadly conical
**Evergreen**
**Pollinated**: Wind
**Leaf shape**: Linear scale-like

The Canary Island juniper is native to the islands of La Palma, La Gomera, Tenerife, Gran Canaria and Madeira. It occurs in the high mountain regions of these islands alongside the Canary pine, *Pinus canariensis*. On Tenerife it can be found growing from old lava flows at 2,000m/6,560ft above sea level on Mount Teide. Its timber is highly prized, but in recent years the natural distribution of this tree has been dramatically reduced through deforestation and forest fires.

**Identification**: This wide and somewhat pendulous-branched, rather bushy tree has a low open crown and dark grey-green foliage arranged oppositely, or in whorls, upon silver-white pendulous shoots. The bark of the trunk and main branches is red-brown and the fruits are globular in shape, up to 1.5cm/½in in diameter, bright green at first, becoming dark red on ripening. This is a dioecious species, having both male and female individuals.

*Right top: The fruits of this juniper are rounded, bright green at first but ripening to a distinctive dark red colour.*

*Left and right: The foliage is a dark grey-green colour and borne upon silver-white pendulous shoots.*

## Greek Juniper

*Juniperus excelsa*

This hardy, shapely tree is native to mountainous regions from Greece and the southern Balkans to Pakistan. It occasionally occurs as pure forest and in Greece there are several ancient specimens that are over 500 years old. Greek juniper is normally a monoecious species, with both male and female flowers occurring on the same tree, although occasionally single sex species do occur. It was first introduced to cultivation in western Europe in 1806. A common garden cultivar of this species is the columnar 'Stricta', which retains its juvenile foliage.

*Below: The flattened foliage sprays carry small purple-brown fruits.*

**Identification**: Greek juniper has light brown bark, which in maturity, peels off in vertical strips, and long, slender, thread-like branches densely covered with sharply pointed grey-green foliage. The juvenile leaves are needle-like, 5mm/¼in long and borne in twos or threes on the shoots. The adult leaves are flattened and scale-like. The globular fruits, which ripen in the second year, are 1cm/½in in diameter, dark purple-brown covered with a blue-white bloom.

*Right: The mature leaves are scale-like.*

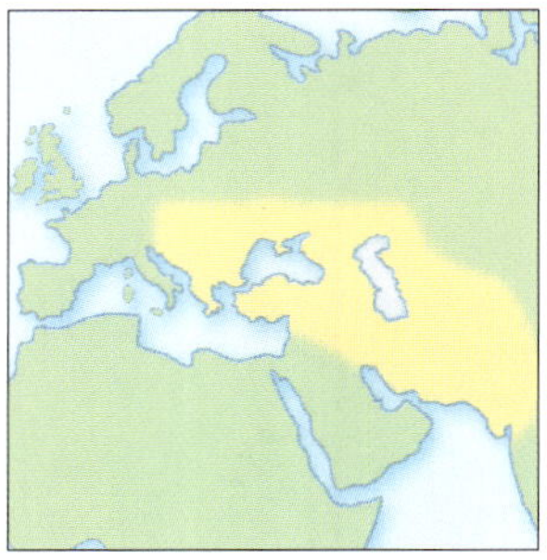

**Distribution**: South-east Europe to south-west Asia.
**Height**: 21m/70ft
**Shape**: Broadly conical
**Evergreen**
**Pollinated**: Wind
**Leaf shape**: Linear scale-like

OTHER SPECIES OF NOTE

***Juniperus* x *media***
This is a dwarf variable hybrid species between the Chinese juniper, *J. chinensis*, and the Common savin, *J. sabina*. It is the origin of around 85 named cultivars now common in cultivation in gardens throughout Europe. These include 'Pfitzeriana', 'Plumosa', 'Blue and Gold', and 'Mordigan Gold', which has bright golden summer foliage.

**Phoenician Juniper** *Juniperus phoenicea*
This small tree is native to southern Europe, North Africa and possibly the Canary Islands. It was introduced to Britain and northern Europe as early as 1683. It has a broadly conical habit and bright green foliage, which when crushed has an aroma reminiscent of paint.

**Creeping Juniper** *Juniperus procumbens*
As the name suggests, this is a low-growing dwarf juniper, which is much used for ground cover or as a rockery plant. In Japan it grows wild on the coast of Kyushu. It is a sturdy, dense little tree with stiff branches thickly covered with blue-green, sharply-pointed leaves.

**Common Savin** *Juniperus sabina*
This handsome dwarf juniper, widely distributed in the wild from the Pyrenees to the Caucasus, has been cultivated since ancient times in the Mediterranean region, and was introduced into England in 1548. It has resulted in many popular cultivars, including 'Blue Danube', 'Erecta' and 'Arcadia'. When crushed, the foliage of the true species emits a pungent, unpleasant odour.

## Prickly Cypress

Formosan juniper *Juniperus formosana*

Native to southern China, where it is widespread, and Taiwan, this beautiful small, graceful tree is somewhat tender and will not flourish in colder regions of Europe. There is a splendid specimen at Bedgebury Pinetum, Kent, England. It was introduced into Europe around 1844, although it was originally confused with *J. taxifolia* and was not correctly named until 1908. It has similar characteristics to those of the sharp cedar, *J. oxycedrus*, which is native to the Mediterranean, Iran and the Caucasus Mountains.

**Distribution**: South China and Taiwan.
**Height**: 12m/40ft
**Shape**: Broadly columnar
**Evergreen**
**Pollinated**: Wind
**Leaf shape**: Linear scale-like

**Identification**: This beautiful tree has fibrous brown bark, which peels in thin, narrow strips. It has a loose, elegant habit, open crown and sparsely set, very large, spiny forward-pointing leaves held on pendulous branchlets. The leaves are up to 1cm/½in long, awl-shaped, borne in whorls of three, blue-green in colour, with two bands of white stomata separated by the central midrib.

*Above: The globular fruits are olive-green in the first year, maturing to orange-red, or reddish-brown, in year two.*

## Shore Juniper

*Juniperus conferta*

This is a prostrate shrub rather than a tree, but it is widely used as low ground cover in areas close to the sea. It is native to coastal areas of Japan and Sakhalin, particularly in the sand dunes of Hakodate Bay in Hokkaido. It was identified by the Russian botanist Maximowicz in 1861 and introduced into Europe by the English plant hunter Ernest Wilson in 1914. It is closely related to the creeping juniper, *J. procumbens*.

**Distribution**: Japan.
**Height**: 2m/6ft
**Shape**: Prostrate
**Evergreen**
**Pollinated**: Wind
**Leaf shape**: Linear scale-like

**Identification**: This is a prostrate evergreen with short, angular, ascending branches and young shoots that are densely clothed with sharply pointed, glossy, pale green linear leaves, up to 5mm/¼in long, which have a distinctive band of white stomata on the upper surface. The globular fruits are 12mm/½in in diameter, purple-black and covered in a glaucous bloom. Inside each fruit there are three triangular seeds.

*Left: Shore juniper has short, pale green linear leaves which are sharply pointed. The tips turn yellow.*

# THUJAS AND HIBA

*The Arborvitae or* Thuja *species are similar to* Chamaecyparis *species, such as Lawson cypress, except that they have larger and broader scale-like leaves, which emit a more pleasant aromatic fragrance when crushed and have cones with overlapping scales. Both* Thuja *and* Thujopsis *are perfectly hardy in Europe and will thrive in almost any soil, providing it is well-drained.*

## Japanese Arborvitae

*Thuja standishii*

The name Arborvitae means 'tree of life' or 'tree of everlasting life' and refers to the evergreen foliage of this species. In fact, as with all evergreen trees, the foliage does die and is replaced, but gradually over a period of years rather than all at once each autumn. This tree is easily recognized by its lax yellow-green foliage and open crown, together with the citronella-like fragrance emitted by the foliage when crushed. In Japan it is one of the 'five sacred trees' from which Shinto shrines are built. It was introduced to Europe by the Scottish plant collector Robert Fortune in 1860.

**Identification**: The bark is deep red, smooth at first, peeling away into long strips and plates in maturity, from both the trunk and larger branches. Each spreading branch sweeps skywards towards the tip and is densely clothed in lax, nodding sprays of yellow-green foliage. The male flowers are dark red and the female flowers are green; both appear at the tips of the shoots on the same trees in spring.

The fruit is a very small cone, 1cm/½in long, covered with approximately ten scales, green at first, ripening to red-brown.

**Distribution**: Central Japan.
**Height**: 25m/80ft
**Shape**: Broadly conical
**Evergreen**
**Pollinated**: Wind
**Leaf shape**: Scale-like

*Right: The leaves are yellowish-green and carried in elegant drooping sprays.*

*Left: Male and female flowers are borne at the tips of the shoots in spring.*

## Korean Arborvitae

*Thuja koraiensis*

This beautiful small tree from north-east China and Korea, where it inhabits woodland in mountainous areas, was introduced into Europe by the English plant collector Ernest Wilson in 1917. It is a striking tree that deserves to be more widely planted, being largely confined at present to botanic gardens and arboreta. It has broad, flat foliage which, when crushed, emits a scent of almonds and marzipan.

*Right: Male and female flowers are green with black markings and borne at the shoot tips.*

*Right: The scale-like leaves are flat, broad and silver-white beneath.*

**Identification**: The bark is pinkish-brown, peeling away in thin scales. The scale-like leaves are flat, broad, grass green above and a remarkable silver-white colour beneath, bordered by pale green. Each spray of leaves is borne in a flattened plane on coppery-orange, stout shoots. Both male and female flowers are green tipped with black and appear at the ends of the shoots in spring. The fruit is an oblong, upright cone, 1cm/½in long, green-yellow maturing to orange-brown.

**Distribution**: North and central Korea.
**Height**: 10m/33ft
**Shape**: Narrowly conical
**Evergreen**
**Pollinated**: Wind
**Leaf shape**: Scale-like

## Western Red Cedar

Giant arborvitae *Thuja plicata*

This evergreen tree originates in the north-western Pacific coastline of America, where it is a major component of the moist, lowland coniferous forests. Individual living trees over 1,000 years old have been recorded, and its timber has been utilized for centuries. Native Americans burnt out the trunks to make canoes, and it is an economically important timber, being straight-grained, soft and easily worked. It has been widely used to make roofing shingles. The tree is grown for timber in Europe and New Zealand. Its many cultivars include a distinctive variegated form called 'Zebrina'.

**Identification**: This very tall, narrow, conical evergreen tree grows up to 50m/165ft. Individual specimens with low branching may layer to form a secondary ring of vigorous, upright trunks. The foliage is dark green and glossy above, with a sweetly aromatic scent when crushed. The bark is reddish-brown, forming plates in maturity. It is fibrous and ridged.

*Left: The shoots are coppery-brown with sprays of deep glossy green, scale-like needles that are flattened in one plane.*

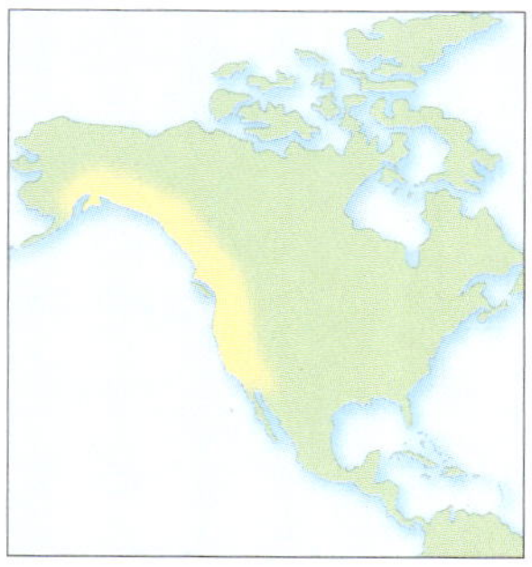

**Distribution**: USA: originating from the Pacific coastline of North America, it grows from southern Alaska, through British Columbia, southwards to Washington and Oregon to the giant coastal redwood forests of California.
**Height**: 50m/165ft
**Shape**: Narrowly conical
**Evergreen**
**Pollinated**: Wind
**Leaf shape**: Linear scale-like

OTHER SPECIES OF NOTE

**Chinese Thuja** *Thuja orientalis*
A species from Korea, northern and southern China, this differs from other *Thuja* in having foliage that is arranged in vertical sprays, giving it a distinct character as a specimen tree. Cultivars include 'Aurea Nana', a golden, slow-growing form, and 'Elegantissima', which has a less broadly pyramidal habit than the species.

**White Cedar** *Thuja occidentalis*
This narrowly conical, medium-size tree, to 20m/65ft, is native to eastern Canada and south-eastern USA, where it grows at high altitudes in forests on rocky outcrops. It has twisted sprays of mid-green foliage, which when crushed emit a scent of green apples. It is thought to be the first North American tree to have been introduced into Europe, in the mid-16th century.

**King William Pine** *Athrotaxis selaginiodes*
From a distance this tree looks like the Japanese cedar, *Cryptomeria japonica*, as it has similar foliage. Close up, the bark is like that of the American giant redwood, *Sequoiadendron giganteum*. However, the King William pine is in fact a Southern Hemisphere tree, originating from the temperate rainforests of Tasmania, where it grows alongside eucalyptus at altitudes in excess of 1,000m/3,300ft above sea level. It was introduced into Europe in 1857.

## Hiba

*Thujopsis dolabrata*

**Distribution**: Japan: From the southern islands of central Honshu, northwards to Shikoku and Kyhshu. A distinctive form, *T. dolabrata* var. *hondae* 'Makino', is unique to northern Honshu and southern Hokkaido.
**Height**: 20m/65ft
**Shape**: Broadly conical
**Evergreen**
**Pollinated**: Wind
**Leaf shape**: Linear scale-like

This monotypic genus consists of a single species of evergreen tree, which is distinguished from *Thuja* in having broad leaves with striking and distinctive white undersides. It was introduced to Britain in 1853. In cultivation it is slow in growth and for the first ten years often develops no further in form than that of a dense, evergreen shrub. As a mature tree, the habit of *Thujopsis* can vary from tall, upright and columnar to low-branching, multi-stemmed and broadly conical.

**Identification**: The foliage is yellow-green, glossy and hard. The white undersides of the needles are bordered by a dark green margin. The ovoid cones are 1.25cm/½in long, green becoming brown. The bark is red-brown to grey, peeling off in fine strips in maturity .

*Right: The scale-like, 6mm/¼in long needles form flattened sprays.*

# REDWOODS AND HEMLOCKS

*The conifers in this group are all members of the* Taxodiaceae and Pinaceae *families and are found in the Northern Hemisphere. The group includes some of the biggest, tallest and oldest trees in the world. The coast and giant redwood are evergreens originating in California, while the dawn redwood and swamp cypress are deciduous conifers. All have distinctive fibrous, reddish-brown bark.*

## Coast Redwood

*Sequoia sempervirens*

The *Sequoia* takes its name from a native American Cherokee called Sequoiah. It was introduced into Europe in 1840 and is widely grown in parks and arboreta in northern and western Europe. The trunk is quite often branchless for two-thirds of its height. It grows to 100m/330ft tall.

*Left: The cones are 2–3cm/¾–1¼in long.*

**Identification**: Young trees have a cone-like form with widely spaced, level, slender branches, up-curved at the tips. Old trees become columnar with flat tops and branches that sweep down. The leading shoots have small, pinkish-green needles arranged spirally. The needles on the main and side shoots are arranged in two flat rows, 1–2cm/½–¾in long, dark green above and speckled with two bands of white stomata on the underside. The male flowers are yellowish-brown; the female flowers are green, in separate clusters on the same tree.

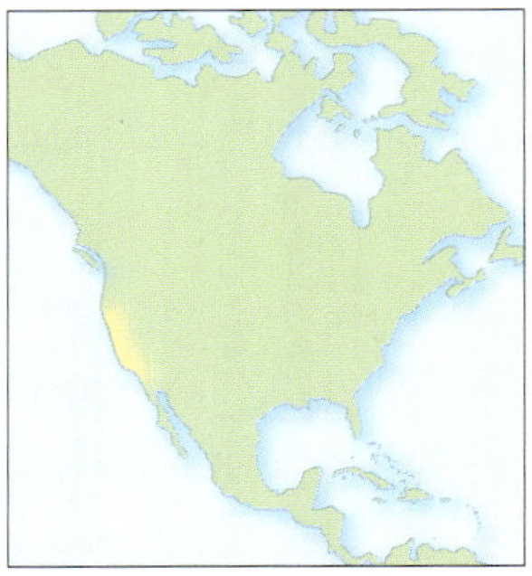

**Distribution**: USA: found in a narrow coastal band running for approximately 800km/500 miles from Monterey, California, to the Oregon border.
**Height**: 100m/330ft
**Shape**: Narrowly conical
**Evergreen**
**Pollinated**: Wind
**Leaf shape**: Linear

*Left: The fissured, reddish-brown, thick, spongy bark is fire resistant, protecting the tree from forest fire.*

## Giant Redwood

*Sequoiadendron giganteum*

**Distribution**: USA: restricted to 72 groves on the western slopes of the Sierra Nevada, California.
**Height**: 80m/260ft
**Shape**: Narrowly conical
**Evergreen**
**Pollinated**: Wind
**Leaf shape**: Linear

*Right: The trunk of a redwood may grow to more than 3m/10ft in diameter.*

The largest living organism in the world is a giant redwood called the 'General Sherman'. Some giant redwoods are up to 3,500 years old. They thrive in any soil, site or exposure with a moderate supply of moisture but do not grow well in heavy shade. The bark is red-brown, soft, thick and fibrous. It was introduced to Europe in 1853.

*Left: Needles and two-year-old brown cones.*

**Identification**: The crown of the tree is conical at first, becoming broad in old age. The leaves grow to 8mm/⅓in, and are sharp-pointed with spreading tips. They are matt grey-green at first, covered with stomata, and turn a dark, shiny green after three years. When crushed, the foliage emits a fragrance of aniseed. The male flowers are yellowish-white, held at the ends of minor shoots, and shed pollen in early spring. The female flowers are green, and develop into bunches of green ovoid cones, which ripen to brown in their second year.

OTHER SPECIES OF NOTE

**Southern Japanese Hemlock**
*Tsuga sieboldii*
In Europe this is a small to medium-size tree, but in its natural habitat in southern Japan it can reach heights in excess of 30m/100ft. It was introduced into Europe in 1853 by the German physician and plant collector Philipp Franz von Siebold. The foliage is similar to that of the American hemlocks, but it normally has multiple stems from ground level and dark grey bark with rectangular pink fissures.

**Himalayan Hemlock**
*Tsuga dumosa*
The Himalayan hemlock is found growing wild from north-west India to northern Burma and on into China, where it attains heights in excess of 50m/165ft. However, in cultivation elsewhere, including Europe, it seldom reaches 20m/65ft. Its bark is pink-brown and scaly, rather like that of the larch. The needles are relatively long for a hemlock, 3cm/1¼in, hard and rigid, green-blue above with two silver stomatal bands beneath.

## Western Hemlock

*Tsuga heterophylla*

This tall, elegant tree has weeping branches and soft, pendulous foliage. However, this softness is deceptive: western hemlock is as hardy as any other conifer. It has been widely planted for timber in forests throughout western Europe since its introduction in 1851. It is extremely shade-tolerant, out-growing its competitors in the thickest forest.

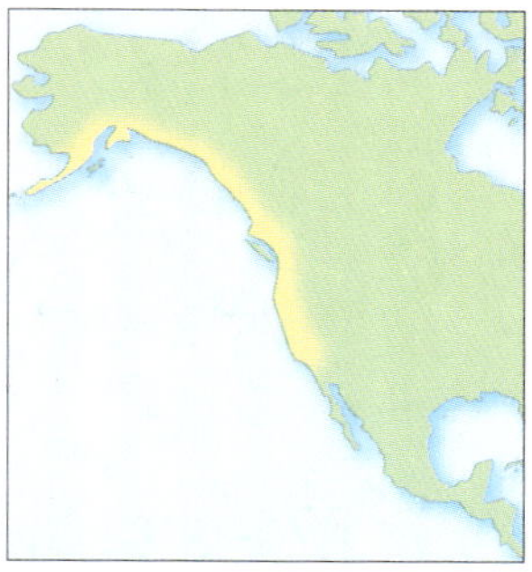

**Identification**: The tree has a narrow conical shape, with ascending branches that arch gently down towards the tip. The leading shoot is always lax. The bark is reddish-purple in young trees, becoming dark purple-brown with age. The needles are 2cm/¾in long, deep dark green above with two broad blue-white stomatal bands beneath. New growth is bright lime green in spring, contrasting dramatically with the sombre mature foliage. Both male and female flowers are red. Much pollen is shed in late spring. The cones are pendulous, egg-shaped, 2.5cm/1in long, with few scales, pale green ripening to deep brown.

**Distribution**: West coast of North America.
**Height**: 60m/200ft
**Shape**: Narrowly conical
**Evergreen**
**Pollinated**: Wind
**Leaf shape**: Linear

*Left and below: Small egg-shaped cones appear in late summer at the tips of branches.*

## Mountain Hemlock

*Tsuga mertensiana*

Native to the west coast of North America from Alaska to California, this handsome tree has a columnar crown of grey pendulous foliage. The mountain hemlock is sometimes mistaken for *Cedrus atlantica* 'Glauca', as it has thick blue-grey needles, which radiate all around the shoot. It thrives in Europe.

**Identification**: The bark is dark orange-brown, becoming vertically fissured into rectangular flakes in maturity. The branches are slightly drooping, with weeping branchlets hanging from them. The shoot is a shiny pale brown. The needles are similar to a cedar's, 2cm/¾in long, dark grey-green to blue-grey, and are borne radially all over the shoot. The cone is spruce-like, 7cm/2¾in long, cylindrical and buff-pink maturing to brown. The male flowers are violet-purple, borne on slender drooping stems. The female flowers are erect and have dark purple and yellow-green bracts. Although closely related to the western hemlock, it is seldom planted in European forests for timber production because it is a slower growing species and the timber is often extremely knotty and difficult to work. It was introduced into Europe in 1851.

*Above: The needles are blue-green and cedar-like. The cone resembles that of spruce.*

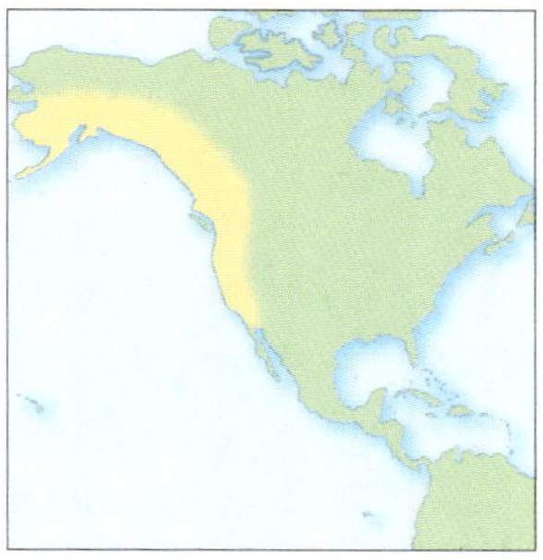

**Distribution**: West coast of North America, from Alaska to California.
**Height**: 30m/100ft
**Shape**: Columnar
**Evergreen**
**Pollinated**: Wind
**Leaf shape**: Linear

*Left: The needles have a definite bluish tinge and radiate out from the twigs. Viewed end-on, the shoots appear star-like.*

# SWAMP CYPRESS AND CHINESE FIR

*Swamp cypress and dawn redwood are widely planted as ornamental trees in parks and gardens throughout Europe – their main attraction being their soft foliage, which turns bright yellow or orange before falling in autumn. The remaining trees in this section are less common but can be found in botanic gardens and arboreta.*

## Japanese Umbrella Pine

*Sciadopitys verticillata*

The Japanese umbrella pine grows at elevations of 180–1,520m/590–5,000ft on rocky slopes and ridges. It has been widely planted in other parts of the world, including Europe, as an ornamental species, because of its unusual leaf formation and attractive, regular shape. In Japan its timber, which is pure white, springy and very durable, is used for making bathtubs, casks and boats.

**Identification**: This distinctive tree has widely spaced whorls of shiny green needles, which are deeply grooved and grow up to 12cm/4¾in long. Each whorl sits about 3.5cm/1⅓in apart from its neighbour on a buff-brown shoot. The male flowers are globular, yellow and green and bunched in clusters of 12. The green female flowers appear on the same tree, held at the end of each shoot.

*Below: The fruit is an egg-shaped cone up to 7.5cm/3in across.*

**Distribution**: Japan: the mountains of central and southern Honshu.
**Height**: 33m/110ft
**Shape**: Conical
**Evergreen**
**Pollinated**: Wind
**Leaf shape**: Linear

*Right and left: The arrangement of the needles within each whorl resembles the ribs of an umbrella – hence the common name.*

## Chinese Fir

*Cunninghamia lanceolata*

The Chinese fir is a handsome tree with a domed crown of short, drooping branches. At first glance there is a similarity with the monkey puzzle. It has prickly, lance-shape needles, which are glossy dark green. It grows in stands on Chinese mountainsides up to an elevation of 1,520m/5,000ft, enduring hot summers and high humidity. It thrives in Europe.

**Identification**: The bark is chestnut brown, with parallel, vertical shallow fissures running down the trunk. It has a columnar or conical shape with a domed top. The branches are widespread, giving a sparse appearance often concealed by the drooping foliage. The needles grow to 6cm/2½in long, spirally set on the shoots, but twisting to lie in just two planes on each side of the shoot. They are deep, glossy green on top, with two striking white bands beneath. The male flowers are yellow-brown, borne in clusters at the shoot tips, while the female flowers are yellow-green and held singly on terminal branchlets. The fruit is a rounded cone, 4cm/1½in across, bright green maturing to brown.

*Right: A male flower.*

*Right: Female flowers.*

**Distribution**: Central and southern China south of a west-east line from Sichuan, Hubei and Honan to Guangxi, Guangdong, Fujian and Hong Kong.
**Height**: 25m/80ft
**Shape**: Broadly columnar
**Evergreen**
**Pollinated**: Wind
**Leaf shape**: Lanceolate

*Left: The distinctive, glossy, dark green foliage of the Chinese fir makes it an excellent ornamental species.*

## Swamp Cypress

Bald cypress *Taxodium distichum*

Also known as the bald cypress because of its deciduous habit, this tree grows naturally in wet conditions and can tolerate having its roots submerged for several months. In these conditions it will produce aerial roots known as "knees" or "pneumatophores", which provide oxygen to the roots. It is an excellent tree for colour; the leaves turn from old gold to brick red in early to mid-autumn.

*Right: Autumn needles.*

*Left: Cones are borne on the same trees as male flowers.*

**Identification**: The crown is typically conical, although some trees develop a rather domed appearance in maturity, with heavy, low, upswept branches. The bark is a dull reddish-brown and is frequently fluted. The shoots are pale green, up to 10cm/4in long, with soft, flattened leaves 2cm/¾in long, arranged alternately along the shoot; they emerge late in the season. The male flowers, to 5–6cm/2–2½in, are prominent throughout the winter as three or four catkins held at the end of each shoot. These lengthen to 10–30cm/4–12in when pollen is shed in early spring. The female cones are on a short stalk, globular and light green until ripe.

*Right: The deciduous, needle-like foliage turns red in autumn.*

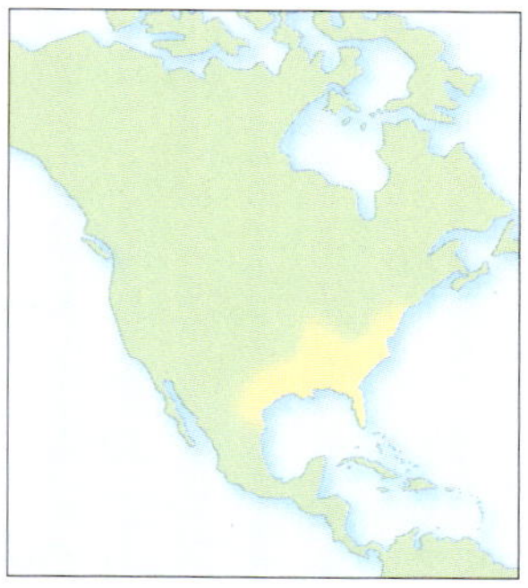

**Distribution**: South-eastern USA: Delaware to Texas and Missouri.
**Height**: 40m/130ft
**Shape**: Broadly conical
**Deciduous**
**Pollinated**: Wind
**Leaf shape**: Linear

OTHER SPECIES OF NOTE

**Pond Cypress** *Taxodium ascendens*
This broadly conical tree from the south-eastern USA reaches a height of 40m/130ft. It tolerates wet soil but also grows in drier situations. It has linear leaves 1cm/½in long, which are closely pressed around the upright, deciduous shoots. The bark is red-brown, thick and heavily fluted. The male flowers are yellow-green, and held in catkins up to 20cm/8in long. The female flowers are green and appear in clusters at the base of the male catkins. The fruit is a green globular cone, 3cm/1¼in across.

**Chinese Swamp Cypress**
*Glyptostrobus pensilis*
This small tree seldom reaches heights in excess of 10m/33ft. It originates from south-east China, where it grows wild in swamps and along riverbanks. However, it is now very rare in the wild. It has linear, scale-like leaves, 1.5cm/⅝in long, which are arranged spirally on deciduous side shoots. The bark is grey-brown and the flowers are insignificant. The fruit is an egg-shaped green cone up to 2.5cm/1in long. This fairly tender tree does not thrive in northern Europe.

## Dawn Redwood

*Metasequoia glyptostroboides*

Until this beautiful tree was discovered growing in east Sichuan by the Chinese botanist T. Kan in 1941, it had been seen only as a fossil and was deemed extinct. It was introduced to the West in 1948. Since then it has become a popular species for ornamental planting. It has bright orange-brown stringy bark and the trunk is quite often fluted.

**Identification**: The crown is conical in most trees, although some are broad with upswept branches. When grown in the open the crown is dense, but in shade it becomes sparse. The leaves are down-curved at the tips, 2cm/¾in long, bright green above with a pale band each side of the midrib beneath. The male flowers are ovoid, set on panicles up to 25cm/10in long. The female cones are green ripening to brown, 2cm/¾in across with stalks 2cm/¾in long.

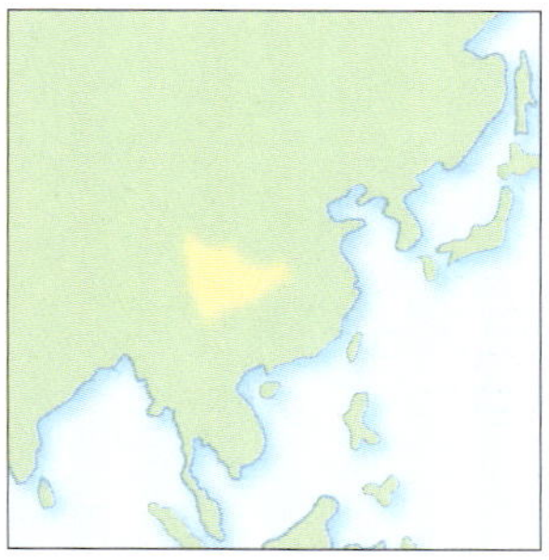

**Distribution** China: the Shui-sha valley, in the north-west part of Hubei and into Sichuan.
**Height**: 40m/130ft
**Shape**: Narrowly conical
**Deciduous**
**Pollinated**: Wind
**Leaf shape**: Linear

*Below: The leaves are positioned opposite each other on the shoot, which is bright green.*

# TRUE FIRS

*The diversity among European and Asian firs is quite remarkable. They include some of the tallest firs in the world and some of the smallest. They can be found growing wild from China to Spain. Several have been adopted as ornamental species and planted just about everywhere, from large arboreta to small town gardens. All are handsome trees, producing lush foliage, good symmetry and attractive cones.*

## European Silver Fir

*Abies alba*

**Distribution**: Pyrenees, France, Corsica, the Alps and the Black Forest south to the Balkans.
**Height**: 50m/165ft
**Shape**: Narrowly conical
**Evergreen**
**Pollinated**: Wind
**Leaf shape**: Linear

This species is long-lived for a conifer – some specimens are known to be over 300 years old. Although widely planted for timber, it is very susceptible to aphid damage, which can be fatal in close-grown plantation conditions. It is widely used as a Christmas tree in many parts of Europe. Prolific natural regeneration from seed is a characteristic of the species.

**Identification**: Young trees are symmetrical, with slightly ascending branches. In maturity the stubs of dead branches cover the trunk and the leading stem becomes heavily forked. In young trees the bark is smooth and dull grey. Older trees have a paler bark with shallow pink-brown fissures. The needles are 1–2cm/½–¾in long, shiny green above with noticeable linear grooves and white stomatal banding below. They have rounded tips and are flattened each side of the shoot. The cones are clustered on just a few branches at the top of the tree. They are red-brown, cylindrical, up to 15cm/6in long and disintegrate on the tree. In spring, new growth is an attractive bright lime-green but is quite often burnt by the sun melting late frost.

*Right: The long cones stand upright from the finger-like shoots.*

## Korean Fir

*Abies koreana*

This is an alpine species and the smallest of all firs. It originates from the volcanic island of Quelpeart, where it grows in vast forests on mountain slopes up to 1,000m/3,300ft. It has become a favourite for planting in gardens in Europe because of its manageable size and profusion of purple-blue cones on even the youngest trees.

**Identification**: The Korean fir forms a broad, tall shrub or small tree. It has dark olive-green to black bark, which is pock-marked with light freckle-like lenticels. The shoots are pale fawn and slightly hairy, and are covered in a profusion of short, stubby, dark green needles, 1–1.5cm/½–⅔in long, which curve upwards from the shoot, almost obscuring it from view. On the underside of the needles are two bright white stomatal bands. The male flowers are red-brown, normally covered in resin and clustered all around the side shoots. The female flowers are dark red to purple, ripening to attractive, dark blue-purple cones up to 7cm/2¾in long. These are normally covered with a sticky white resin.

*Above: The leaf buds.*

*Below: The male flowers.*

**Distribution**: South Korea.
**Height**: 15m/50ft
**Shape**: Broadly conical
**Evergreen**
**Pollinated**: Wind
**Leaf shape**: Linear

*Right: The cones stand upright from the shoots.*

OTHER SPECIES OF NOTE

**Noble Fir** *Abies procera*
As its name suggests, this tree has a stately, noble appearance with a long, straight stem, and large cones that stand proudly above the surrounding foliage. It has smooth, silver-grey bark with occasional blisters, and blue-green needles 3.5cm/1½in long. When crushed they emit a pungent smell. It grows in western Europe.

**Veitch's Silver Fir** *Abies veitchii*
This beautiful, fast-growing Japanese tree was discovered on Mount Fuji in 1860 and introduced into Europe in 1879. It has densely arranged, upcurved needles, 2.5cm/1in long, which are dark green above and silver beneath. The fruits are barrel-shaped, upright, purple-blue cones that stand proud on the topmost branches.

**Greek Fir** *Abies cephalonica*
Native to the island of Cephalonia and the mountains of southern Greece, where it reaches 30m/100ft tall, this handsome fir has rigid, sharply pointed, glossy dark green needles up to 2.5cm/1in long. The underside of each needle has two silver-grey bands of stomata. It is one of the earliest firs to start producing new growth in spring and as a result it is sometimes damaged by frost when planted in northern Europe.

**Grand Fir** *Abies grandis*
This tree has graceful, downward-sweeping branches with upturned tips, the lower boughs reaching the ground. The needles, up to 5cm/2in long, are glossy green on top, silvery below.

## Delavay Fir

*Abies delavayi*

This handsome medium-size tree is believed to have been introduced into Europe in 1918. However, there is some suggestion that the original trees may have been the Faber fir, *Abies fabri*. The Delavay fir is named after the French missionary Jean Marie Delavay, who discovered the tree in Yunnan in 1884. It is common in cultivation in botanic gardens and arboreta. Some of the largest specimens in Britain are found in west coast Scottish collections such as Benmore in Argyll.

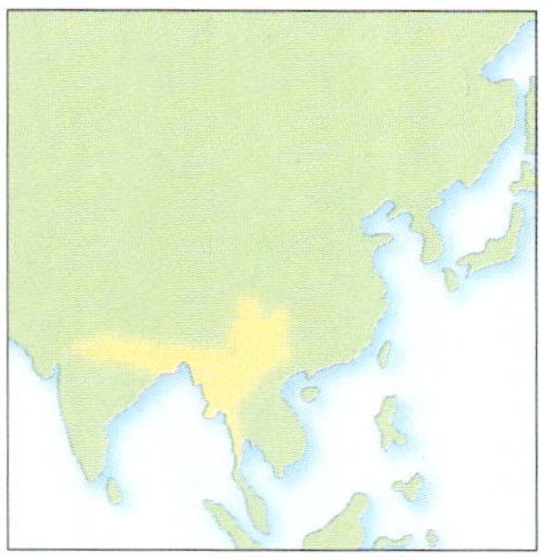

**Distribution**: Yunnan, China and Burma, extending into northern India.
**Height**: 20m/65ft
**Shape**: Broadly conical
**Evergreen**
**Pollinated**: Wind
**Leaf shape**: Linear

**Identification**: The crown is symmetrical but rather sparse and spiky, with strong ascending branches bearing short red-brown shoots. The bark is smooth, pale grey, becoming fissured with age. The needles, up to 4cm/1½in long, are borne all around the shoots. They are deep blue-green above with two bright silver stomata bands beneath. The fruit is a distinctive barrel-shaped upright cone, up to 10cm/4in long, purple-blue maturing to dark brown, with small bract-like curved spines emerging from each cone scale.

*Above: The needles are blue-green above with two silver-white stomata bands beneath. The cone is barrel-shaped.*

## Caucasian Fir

Nordmann fir, Crimean fir *Abies nordmanniana*

This slow-growing, densely branched tree is probably one of the most handsome and uniform-looking firs in the world. It makes a magnificent large specimen tree in parks, gardens and arboreta throughout Europe. It is also widely grown for sale as a Christmas tree, marketed as one of the 'non-drop' trees because of its ability to retain its needles after it has been cut. When it is crushed, the foliage emits a fruity, citrus aroma.

**Identification**: The bark is grey and smooth when the tree is young and becomes cracked into small, square plates in maturity. The leaves are linear, bluntly tipped, glossy dark green above with two white bands of stomata beneath, 2.5cm/1in long and carried all around the shoot but more densely on the top side. The male flowers are red and borne beneath the shoot, the female flowers are green and borne above the shoot, both on the same tree in spring. The fruit is a broad cylindrical cone up to 15cm/6in long, green maturing to a deep red-brown.

**Distribution**: Caucasus, north-east Turkey.
**Height**: 50m/165ft
**Shape**: Broadly conical
**Evergreen**
**Pollinated**: Wind
**Leaf shape**: Linear

*Right: The cone is broad, cylindrical and up to 15cm/6in long.*

*Left: Male flowers are red and borne beneath the shoot.*

*Left: Leaves are blunt and needle-like, glossy dark green above with two bands of white stomata beneath.*

## Himalayan Fir

*Abies spectabilis*

This magnificent, large, broad-shaped conifer, sometimes known as the east Himalayan fir, has a natural range from Afghanistan eastwards to Bhutan, where it grows at altitudes up to 4,000m/13,100ft above sea level, higher than any other Himalayan fir. It was introduced into Europe in 1822, and although it does well in mild, wet areas of Europe, it is susceptible to late spring frosts. Its species name, *spectabilis*, actually means 'beautiful'. It produces dark purple cones, which were at one time used to make a purple dye.

*Left: A single needle-like leaf.*

*Above: The fruit is an attractive dark purple cone which ripens brown and is borne on top of the shoot.*

**Identification**: The Himalayan fir has rough pink-grey bark that shreds into fine strips even when young. A distinctive feature of this tree is the way it produces low, heavy, horizontal branches, which frequently break off in maturity. The shoots are stout and reddish-brown, with a covering of hairs in the grooves between the leaf bases. The leaves are dense, and lie forward in two ranks, with the inner rank rising and the outer rank curving downwards beneath the shoot. They are up to 6cm/2¼in long, dark green above and silvery white beneath.

**Distribution**: Himalayas.
**Height**: 20m/65ft
**Shape**: Broadly columnar
**Evergreen**
**Pollinated**: Wind
**Leaf shape**: Linear

*Right: When young, the cones exude a sticky aromatic resin which crystallizes around the cone and on the surrounding leaves.*

## Algerian Fir

*Abies numidica*

This handsome fir is native to Mount Babor, Algeria, where it grows in association with the Atlas cedar, *Cedrus atlantica*, at approximately 1,500m/5,000ft above sea level. It was discovered in the mid-19th century and introduced into cultivation in Europe in 1861, where it was planted in the gardens of the Natural History Museum in Paris and named by head gardener E. A. Carrière. The Algerian fir is more lime-tolerant than most firs.

**Identification**: The overall shape of this tree is conical with an abruptly pointed top. It has dense foliage on numerous downward-sweeping branches, which then curve upwards at their tips. The bark is purplish-grey and smooth at first, becoming fissured and flaking in maturity. The leaves are rigid, to 2.5cm/1in long, with a blunt point. They are dark green above with two conspicuous bands of white stomata on the underside. The fruit is a cylindrical upright cone up to 18cm/7in long. It is pale green at first, ripening to brown, and ending with a short "peak" at the tip.

*Left: The Algerian fir produces a distinctive green cylindrical upright cone which ends in a peaked tip and matures to brown.*

**Distribution**: North-east Algeria.
**Height**: 25m/80ft
**Shape**: Narrowly conical
**Evergreen**
**Pollinated**: Wind
**Leaf shape**: Linear

*Below: The female flower.*

## Cheng Fir

*Abies chengii*

This uncommon, vigorous tree, found mainly in arboreta and botanic gardens, was first described by the English arboriculturalist Keith Rushforth in 1987. It is believed to have originated in north-west Yunnan in China and to have been introduced into Europe by the Scottish plant hunter George Forrest in 1931. It is closely related to the Forrest fir, *A. forrestii*, and some botanists claim it is a hybrid of this tree. It differs from that species in having longer needles and much more pointed buds.

*Right: The 4cm/1.5in–long needle-like leaves are dark green above, sage green beneath and surround cylindrical upright cones.*

**Identification**: The cheng fir has smooth grey bark when young, becoming fissured towards the base in maturity. It has shiny, mahogany-red shoots, pointed pale-brown buds and glossy, dark green needle-like leaves with sage green undersides. They are up to 4cm/1½in long, and are borne on the shoots in such a way as to create a 'V' down the centre. The fruit is an upright cylindrical cone, up to 10cm/4in tall, purple-brown in colour and disintegrating on the tree once ripe.

**Distribution**: Probably native to Yunnan, China.
**Height**: 15m/50ft
**Shape**: Broadly conical
**Evergreen**
**Pollinated**: Wind
**Leaf shape**: Linear

OTHER SPECIES OF NOTE

**Momi Fir** *Abies firma*
Also known as Japanese fir, this tree was introduced into Europe in 1853. It is a wide-spreading, large tree, to 30m/100ft tall. It has pink-grey bark and distinctive yellow-green, stiff leathery needles, each up to 5cm/2in long and rounded at the tips. It has a rather open appearance with well-spaced, long branches. The yellow-brown cones are 12cm/4½in long.

**Faber Fir** *Abies fabri*
This attractive, large tree, up to 40m/130ft tall, is native to western Sichuan, China. It was introduced into Europe in 1901. It has attractive, dark orange-brown scaly bark, deep green needles, 2.5cm/1in long, which are bright silver-white beneath, and bluish-black cones with protruding bracts.

**Sicilian Fir** *Abies nebrodensis*
The Sicilian fir is native to the slopes of Monte Scalone in northern Sicily where, due to deforestation, it is now very rare. It is similar to the European silver fir, *A. alba*, and it has been suggested that this tree is an island form of that species. However, it differs in having orange shredding bark and shorter needles, which are densely arranged on the top of the branchlets.

**Spanish Fir** *Abies pinsapo*
Also known as the hedgehog fir because of its distinctive blunt, rigid, dark green needles, which stick out from all sides of the branchlets, this Spanish, medium-size tree, 20m/65ft tall, is common in cultivation throughout Europe. It is one of the best firs for planting on chalk.

## Cilician Fir

*Abies cilicica*

This attractive, slow-growing large tree is native to the Taurus Mountains of southern Turkey, from where its range extends southwards to Mount Lebanon. It is closely related to the Caucasian fir, *A. nordmanniana*, but can be distinguished by its thinner, narrower leaves and crown. It is a relatively rare species, both in the wild and in cultivation, but is quite often found growing in botanic gardens and arboreta. It was introduced to western Europe, including Britain, in 1855.

**Distribution**: Lebanon, north-west Syria and southern Turkey.
**Height**: 30m/100ft
**Shape**: Narrowly conical
**Evergreen**
**Pollinated**: Wind
**Leaf shape**: Linear

**Identification**: The bark is dark grey, smooth at first, becoming distinctively wrinkled and cracked into rings, which radiate out from the knots of old branches, in maturity; the overall appearance is not dissimilar to that of an elephant's leg. Cilician fir foliage is variable, even on the same tree, with some shoots displaying very short leaves 1cm/½in long and others densely clothed with long leaves measuring 2.5cm/1in. All the needles point forward on the shoot and are shiny dark green above and pale green beneath. The tree has upright cones, 20cm/8in long, which disintegrate on the tree once ripe, leaving behind an upright woody spike.

*Below: The needles all point forward on the shoot and are dark shiny green above and pale green beneath.*

# SPRUCES

*The spruces,* Picea, *are a group of hardy evergreen conifers that grow throughout most of the colder regions of the Northern Hemisphere. They differ in one significant way from firs,* Abies, *which allows for quick genus identification: on all spruces there is a peg-like stump at the base of every needle. When the needles fall this peg remains, creating a rough texture to the shoot. Firs have smooth shoots.*

## Oriental Spruce

Caucasian spruce *Picea orientalis*

This graceful tree is one of the best spruces for cultivation, being more lime-tolerant and better able to withstand drier conditions than most other spruces. It was introduced into Europe in 1839 and is common in parks, large gardens and arboreta. There are trees in the wild reputed to be around 400 years old, something obviously yet to be achieved in cultivation. The oriental spruce is easily recognized by its dense, blunt-tipped short needles and conical form.

*Right: The female flowers are red.*

*Right: Both male and female flowers are borne on the same tree; the male flowers are bright red opening to yellow.*

**Identification**: The bark is pinkish-brown, relatively smooth at first, flaking into small irregular plates in maturity. The needle-like leaves are up to 1cm/½in long, bluntly pointed at the tips, shiny dark green above, paler beneath and pointing forward on buff-coloured shoots. The male flowers are bright red at first, opening yellow, and the female flowers are red. Both appear on the same tree, in separate clusters, in spring. The fruit is a cylindrical drooping cone, up to 10cm/4in long, purple ripening to brown, and covered with patches of sticky aromatic resin.

*Above: Short, shiny dark green needle-like leaves are borne on pale buff-coloured shoots.*

**Distribution**: Caucasus, north-east Turkey.
**Height**: 50m/165ft
**Shape**: Narrowly conical
**Evergreen**
**Pollinated**: Wind
**Leaf shape**: Linear

## Sitka Spruce

Coast spruce *Picea sitchensis*

**Distribution**: USA: narrow coastal strip from Kodiak Island, Alaska, to Mendocino County, California.
**Height**: 50m/165ft
**Shape**: Narrowly conical
**Evergreen**
**Pollinated**: Wind
**Leaf shape**: Linear

The largest of the North American spruces, this is a major species of north-west American forests. Valued for its timber, Sitka spruce has been widely planted across the Northern Hemisphere (including Europe) in forestry plantations. The timber is pinkish-brown, and is strong for its light weight. Originally used for aircraft frames, it is now the main species used in paper manufacture.

*Far left: Male flower.*

*Left: Female flower.*

*Right: Cones are pale buff, 10cm/4in long with thin, papery scales and are pendulous in habit.*

**Identification**: The overall shape is an open, narrow cone, with widely spaced, slender, ascending branches. Sitka spruce can easily grow more than 1m/3ft a year when young. The bark in young trees is a deep purple-brown. Older trees have large, curving cracks, which develop into plates of lifting bark. The needles are stiff with a sharp point, blue-green above with two white stomatal bands beneath, and up to 3cm/1¼in long. They are arranged all around the pale, buff-coloured shoot. The male flowers are reddish and occur in small quantities on each tree, shedding pollen in late spring. The female flowers are greenish-red and present on only the topmost shoots.

## Norway Spruce

*Picea abies*

Norway spruce is the Christmas tree of Europe. It has a regular, symmetrical form with horizontal branching at low levels, which gradually becomes upswept towards the top of the tree. It grows naturally throughout northern Europe (except in the United Kingdom) up to altitudes of around 1,500m/4,900ft. Elsewhere it has been widely cultivated for its timber. Norway spruce is traditionally used to make the bellies of violins and other stringed instruments.

*Left: The cones hang down.*

**Identification**: The bark in young trees is a deep coppery-pink; on older trees it becomes a dark purple, with shallow round or oval plates that lift away from the trunk. The needles are a rich dark green with a faint sheen and are up to 2cm/¾in long. When crushed they emit a citrus fragrance, which has become synonymous with Christmas. The male flowers are golden, shedding copious amounts of pollen in late spring. The female flowers are purple-red, and are frequently confined to the top of the tree. The pendulous cones are cylindrical, slightly curved and up to 15cm/6in long.

*Left: Male flowers occur in groups at the shoot tips.*

*Right: A female flower.*

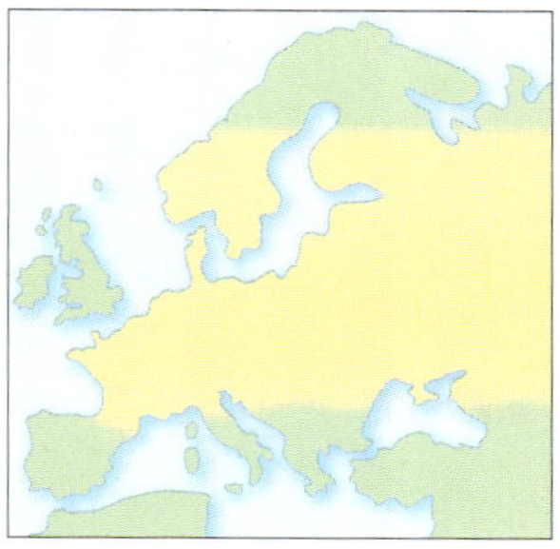

**Distribution**: Northern Europe (excluding the UK), from the Pyrenees to western Russia.
**Height**: 50m/165ft
**Shape**: Narrowly conical
**Evergreen**
**Pollinated**: Wind
**Leaf shape**: Linear

OTHER SPECIES OF NOTE

**Tiger-tail Spruce** *Picea polita*
This ornamental, pyramid-shaped, medium-size spruce originates from central and southern Japan. It is one of the most striking and attractive spruces. Consequently, it has been widely planted in parks and gardens throughout the world. It has long (up to 5cm/2in), rigid, spine-tipped needle-like leaves – probably the sharpest of any spruce – which are flattened and stand out at right angles to the chestnut-brown shoots. Its common name 'tiger-tail' comes from the fact that the foliage-covered branch ends are pendulous – looking very much like a tiger's tail.

## Serbian Spruce

*Picea omorika*

The Serbian spruce has a very small natural population and because of this is considered to be endangered in the wild. It is a beautiful, slender, spire-like tree with branches that sweep elegantly downwards, only to arch upwards at their tips. This habit means that it is able to resist damage by efficiently shedding snow rather than collecting it. It is also the spruce that is most resistant to atmospheric pollution.

**Identification**: The bark is orange-brown to copper, broken into irregular to square plates. The shoot is a similar colour to the bark and quite hairy. The needles are short with blunt tips, less than 2cm/¾in long, glossy dark green above and with two broad white stomatal bands underneath. The male flowers are crimson and held below the new shoots; the female flowers are also red but confined to the topmost branches. This tree's most distinctive characteristic is its spire-like form.

*Left: The needles are short with blunt tips.*

*Right: To help shed snow the branches sweep downwards.*

*Left: The cone is pendulous, held on a thick curved stalk, tear-shaped, 6cm/2½in long and purple-brown in colour.*

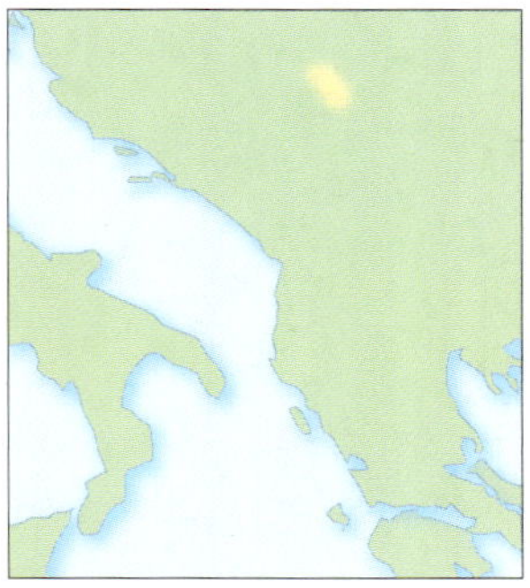

**Distribution**: Europe: confined to the Drina Valley in south-west Serbia.
**Height**: 30m/100ft
**Shape**: Very narrowly conical
**Evergreen**
**Pollinated**: Wind
**Leaf shape**: Linear

## Dragon Spruce

*Picea asperata*

The dragon spruce was discovered in China in 1903 and introduced into Europe by the English plant collector Ernest Wilson in 1910. Wilson described it as being similar in form to Norway spruce, *P. abies*. It is a slow-growing tree, which begins life narrowly conical in outline but becomes progressively broader as it matures. It is common in botanic gardens and arboreta across Europe.

*Below: The needle-like leaves are rigid and sharply pointed and borne on stout shoots.*

**Identification**: The bark is purplish at first, becoming purple-brown with loosely hung scales in maturity. Where the bark is exposed to direct sunlight it may fade to pale grey. The dragon spruce has rigid, sharply pointed, needle-like leaves, which are dark bluish-green and around 2cm/¾in long, borne on stout pink-buff shoots that are rough to the touch. The cones are cylindrical, shortly tapered to a rounded end and up to 15cm/6in long. Both male and female flowers are borne on the same tree and conspicuous in spring, when clouds of pollen are released.

*Left: Male and female flowers are borne on the same tree and release copious amounts of pollen in spring.*

*Right: The cones, which may be up to 15cm/6in long, hang in clusters from the branches.*

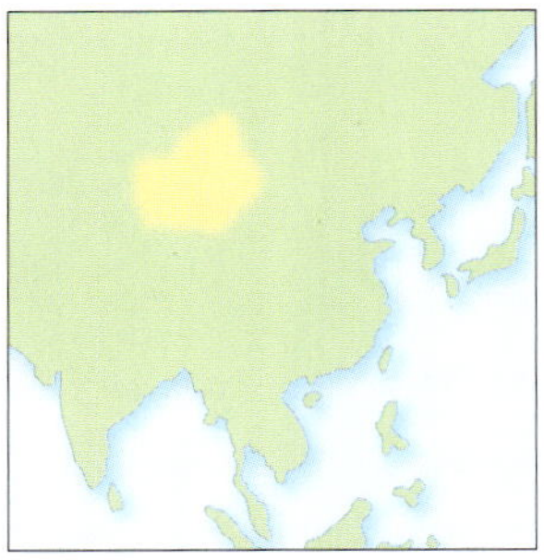

**Distribution**: North-west China.
**Height**: 20m/65ft
**Shape**: Broadly conical
**Evergreen**
**Pollinated**: Wind
**Leaf shape**: Linear

## Sargent Spruce

*Picea brachytyla*

This tree is one of the most ornamental of all spruces. It has two-toned, needle-like leaves, which are fresh, bright green on the top side and silver-white beneath. Each shoot is also white on the underside. It is a hardy species, growing in China at elevations in excess of 4,000m/13,000ft above sea level, and does not suffer damage from late spring frosts. It was discovered by the French missionaries Jean Marie Delavay and Paul Farges and was introduced into Europe by the English plant hunter Ernest Wilson in 1901.

*Right: The young shoot tips can be used to make a tea rich in vitamin C.*

*Right: Cones grow to a long elongated shape.*

**Identification**: The tree has long, gracefully ascending and spreading branches that arch downwards towards the tip. The bark on young trees is smooth, pinkish-grey and pockmarked with white resinous spots. In maturity it is pale grey and roughly fissured into irregular scales. The needle-like leaves are up to 2cm/¾in long and the purple-green cones are cylindrical, tapering to each end, up to 15cm/6in long, and held on a short stalk.

*Right: The seeds mature in October and November.*

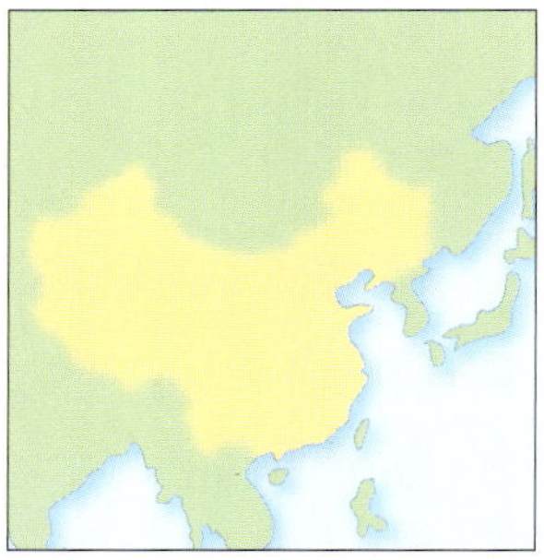

**Distribution**: China.
**Height**: 30m/100ft
**Shape**: Broadly conical
**Evergreen**
**Pollinated**: Wind
**Leaf shape**: Linear

OTHER SPECIES OF NOTE

**West Himalayan Spruce** *Picea smithiana*
The west Himalayan spruce (also known as the morinda spruce) has weeping foliage, which allows it to shed snow easily. It is native to Afghanistan, Kashmir and Nepal, where it grows at altitudes in excess of 3,500m/11,500ft. Its needles are slender, up to 4cm/1½in long and shiny dark green.

**Sikkim Spruce** *Picea spinulosa*
This rare, beautiful tree is native to Sikkim, Bhutan and the eastern Himalayas, where it may reach 60m/200ft in height. It was introduced into Europe as early as 1878. It is recognizable by its open green crown and spreading branches, draped with pendulous branchlets that have a covering of flat, green, needle-like leaves.

**Koster's Blue Spruce** *Picea pungens* 'Koster'
This cultivated form of the Colorado blue spruce, *P. pungens*, was raised in Holland in the late 19th century. It is a small to medium-size conical tree, suitable for garden use, with silvery-blue to almost white, stiff foliage.

**Creeping Spruce** *Picea abies* 'Nana'
'Nana' is a dwarf form, first raised around 1855 and of irregular conical habit, with sharply pointed, small, needle-like leaves that densely clothe its numerous branches.

## Likiang Spruce

*Picea likiangensis*

This is a beautiful, ornamental tree of medium to large size, which grows naturally from north-west Yunnan in China to south-east Tibet. The species displays considerable variation and was at one time split by taxonomists into several different species. The best-known form is the 'Yunnan' form, which was introduced into Europe around 1900.

**Identification**: Although this is a variable species, the following features are fairly constant. The bark is pale grey with a few shallow, vertical fissures. The needle-like leaves are blue-green above and distinctly banded with two white lines of stomata beneath. They are up to 2cm/¾in long and bluntly pointed. The cones are cylindrical, pale brown with red margins to each scale. Both male and female flowers are large, brilliant red and produced in profusion. In springtime they stand out on each branch like burning embers.

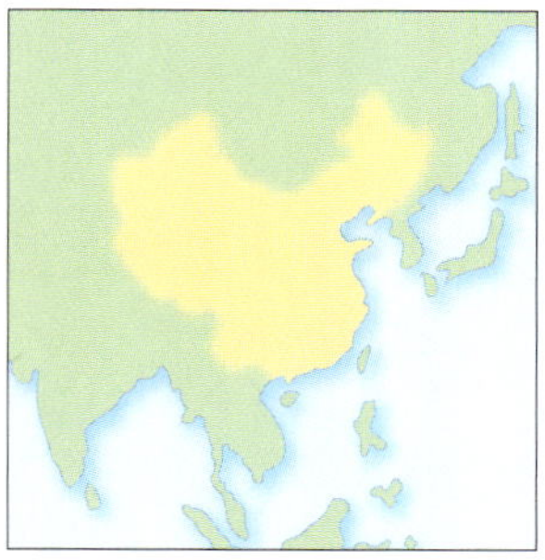

**Distribution**: China to Tibet.
**Height**: 30m/100ft
**Shape**: Broadly conical
**Evergreen**
**Pollinated**: Wind
**Leaf shape**: Linear

*Right: The cones are up to 13cm/5in long*

*Below: The distinctive feature of Likiang spruce is its flowers.*

## Siberian Spruce

*Picea obovata*

This hardy sub-Arctic tree has been in cultivation in Europe since the mid-19th century, although it did not reach Britain until 1908. In the wild in Scandinavia it hybridizes with Norway spruce, *P. abies*, which it resembles, except that it usually has shorter, matt, needle-like leaves and smaller cones. As with other Arctic species, Siberian spruce comes into leaf early in spring in response to increasing light levels and its young foliage is therefore at risk from frost damage.

**Identification**: In cultivation the Siberian spruce is sometimes a bushy, even shrubby, tree, but more often it is conical, with a broad base and ascending branches with pendulous foliage toward the tips. The bark is purple-grey flecked with white, becoming brown and flaking in maturity. The needle-like leaves are short (2cm/¾in), slender, pointed and forward-facing on buff to pale brown shoots. The cones are cylindrical, tapered at each end, golden brown and leathery and up to 10cm/4in long.

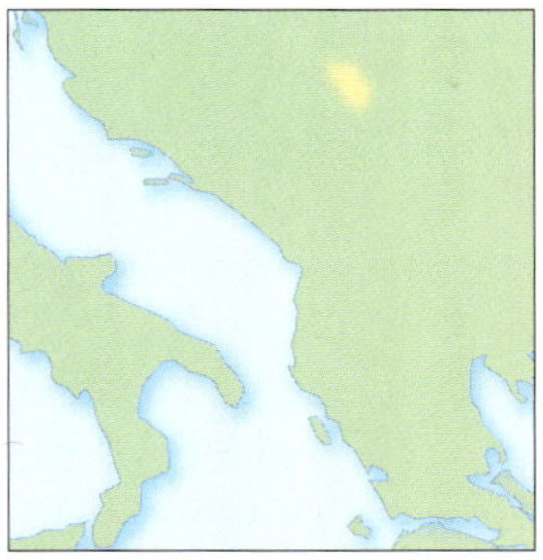

**Distribution**: From northern Scandinavia across Russia and Siberia to east Asia.
**Height**: 30m/100ft
**Shape**: Broadly conical
**Evergreen**
**Pollinated**: Wind
**Leaf shape**: Linear

*Far left to left: Short needle-like leaves face forward on the shoot. The cone is roughly cylindrical, tapering at both ends and up to 10cm/4in long.*

# DECIDUOUS LARCHES

*This small genus of fewer than a dozen species is confined to temperate regions of the Northern Hemisphere. Deciduous larches are fast-growing conifers – several species have been widely planted for forestry purposes. Larches are some of the most seasonally attractive of all conifers, as their needles turn gold and fall in autumn, to be renewed every spring with a flush of lime-green foliage.*

## European Larch

*Larix decidua*

This attractive, hardy tree grows naturally at altitudes up to 2,500m/8,200ft above sea level. It is a long-lived conifer, with some trees in the Alps recorded at over 700 years old. European larch has been widely planted throughout Europe and North America for both forestry and ornamental purposes.

**Identification**: On young trees the bark is pale grey and smooth; on old trees it is dark pink and heavily fissured. Whorls of upswept branches are well spaced. The 4cm/1½in needles are soft and bright green, becoming yellow before falling in autumn. They are carried singly on main shoots and in dense whorls on side shoots. The shoots are pendulous and straw-coloured. The male flowers are pink-yellow discs, normally on the undersides of shoots. The female flowers, appearing before the leaves in early spring, are purple-pink, upright and develop quickly into immature cones.

*Far right and left: Cones are 4.5cm/1¾in long.*

**Distribution**: From the southern Alps through Switzerland, Austria and Germany to the Carpathian Mountains of Slovakia and Romania.
**Height**: 40m/130ft
**Shape**: Narrowly conical
**Deciduous**
**Pollinated**: Wind
**Leaf shape**: Linear

## Japanese Larch

*Larix kaempferi*

In the wild, this larch is confined to the Japanese island of Honshu, where it grows at altitudes exceeding 2,750m/9,000ft. It was introduced to Britain in 1861 and makes up a large portion of the conifer forests of Wales where it is cultivated for timber. It is also grown as an ornamental species in Japanese temple gardens, where it is quite often trained as a bonsai.

*Above: Cones are borne intermittently along the shoot among the foliage.*

**Identification**: The form is broader than that of the European larch. Branches sweep upwards when young, becoming level or even slightly descending in older trees. The bark is reddish-brown and scaly. The young shoots are a distinctive purple-red, sometimes covered with a slight silver bloom. The needles are 5cm/2in long, and flatter and broader than those of European larch. They dull to grey-green in summer, then become orange before falling in autumn. The male flowers are yellow globules clustered on pendulous shoots. The female flowers, which occur all over the tree, have a pink centre and creamy-yellow margins. The bun-shaped cones are 3cm/1¼in long.

**Distribution**: Central Honshu, Japan.
**Height**: 30m/100ft
**Shape**: Broadly conical
**Deciduous**
**Pollinated**: Wind
**Leaf shape**: Linear

*Left: Japanese larch cones have scales that turn outwards. The needles are soft, bright green when they sprout in early spring, then fade to grey-green in summer and turn orange in autumn.*

## Golden Larch

*Pseudolarix amabilis*

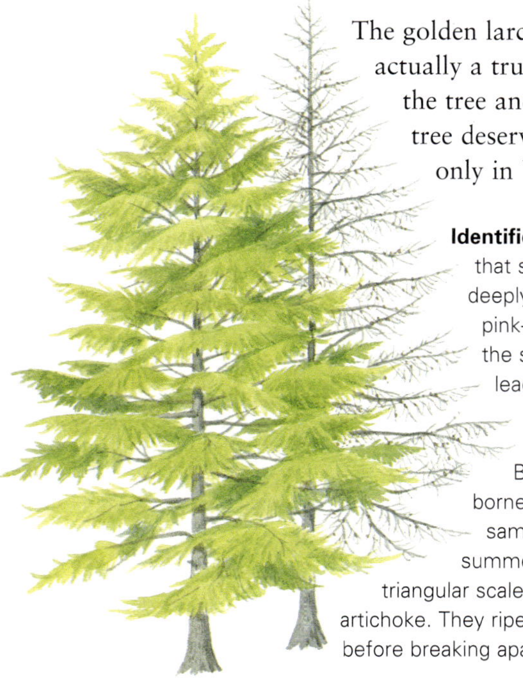

The golden larch is a slow-growing, deciduous conifer, but it is not actually a true larch. It differs in having cones that disintegrate on the tree and spur shoots that lengthen annually. This beautiful tree deserves to be more widely grown; it is normally found only in botanic gardens and arboreta across western Europe.

**Identification**: The overall shape is broadly conical with level branches that sometimes curve upwards at the tips. The bark is grey-brown, deeply fissured into thick, square plates. The shoot is pale yellow to pink-buff. The soft needles, 5cm/2in long, are set spirally around the shoots, occurring in dense whorls on side shoots and singly on leading shoots, where they face forwards. They are bright grass-green on emerging in spring and a glorious bright orange-gold in autumn just before falling. Both male and female flowers are yellow, borne in clusters at the ends of shoots on the same tree. Cones appear only after long, hot summers; they are 6cm/2½in long, with curious, triangular scales, reminiscent of a green globe artichoke. They ripen to golden brown in autumn before breaking apart on the tree.

*Left: Bright, new needles appear in spring. The cones ripen in autumn, then open to release their seeds.*

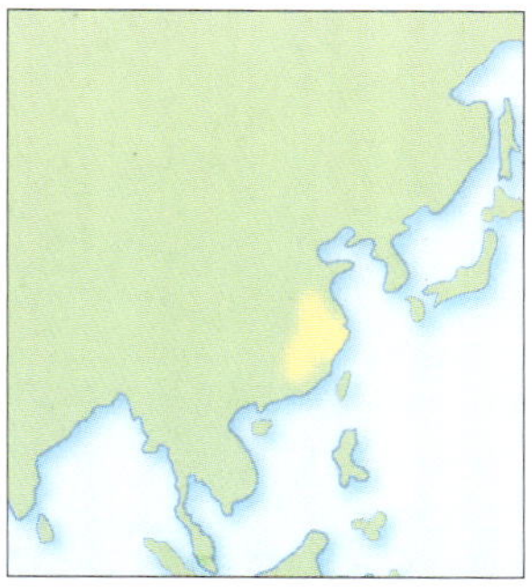

**Distribution**: Provinces of Zhejiang, Anhui and Guangxi in eastern China.
**Height**: 40m/130ft
**Shape**: Broadly conical
**Deciduous**
**Pollinated**: Wind
**Leaf shape**: Linear

OTHER SPECIES OF NOTE

**Dahurian Larch** *Larix gmelinii*
This broad, rather squat tree will, in good growing conditions, reach a height of 20m/65ft; elsewhere it may resemble a spreading bush. It grows in temperate regions of Europe where frost does not damage its early emerging new growth. The needles, which emerge bright green in mid-winter, are 4cm/1½in long and turn butter-yellow in autumn before falling. The cones are produced prolifically. They are ovoid, 2.5cm/1in long, shiny pale brown with broad scales that curve slightly outwards at the margins.

**Sikkim Larch** *Larix griffithiana*
This tree is native to eastern Nepal, Sikkim, Bhutan and Tibet. It also grows in the Alps. A narrow, conical tree up to 20m/65ft tall, it is rare both in the wild and in cultivation. The Sikkim larch has bark like a Corsican pine – grey-purple fissured into rough, scaly ridges. A graceful tree, it has pendulous red-brown shoots, bright green needles 5cm/2in long, and large erect cylindrical cones, 7–10cm/2¾–4in long.

**Siberian Larch** *Larix russica*
Sometimes referred to as *L. sibirica*, this tough, hardy, 30m/100ft tree originates from an area that extends from northern Russia east to the Yenisei River in Siberia. Its overall shape varies according to the location in which it is planted; in northern Europe it is narrow and conical with upswept branches. Further south it becomes broadly spreading with horizontal branching. This tree comes into leaf in late winter and consequently may suffer from frost damage.

## Dunkeld Larch

Hybrid larch *Larix* x *eurolepis* (*L.* x *marschlinsii*)

This vigorous, beautiful tree is a hybrid between the European larch, *L. decidua*, and the Japanese larch, *L. kaempferi*. Both species were planted in close proximity to each other in the grounds of Dunkeld House, Scotland, in 1885. Seed collected from the Japanese larch in 1897 produced seedlings, which were formally identified and described as hybrids in 1919. Since then the Dunkeld larch has been widely planted as a timber-producing forestry plantation species.

**Distribution**: Of garden origin but now cultivated throughout Europe.
**Height**: 35m/115ft
**Shape**: Broadly conical
**Deciduous**
**Pollinated**: Wind
**Leaf shape**: Linear

*Above: In early spring this tree produces pink-purple female flowers on bare branches just before the soft needle-like leaves appear.*

**Identification**: The Dunkeld larch has reddish-brown scaly bark, orange-brown to pink-buff shoots, and egg-shaped, upright cones, as tall as those of the European larch (4cm/1½in), but with reflexed scales like the cones of the Japanese larch. The needles are soft to the touch, up to 4cm/1½in long, borne singly on leading shoots and in whorls on side shoots. They are grey-green to blue-green in summer, becoming golden-brown before falling in autumn.

# TWO- AND THREE-NEEDLE PINES

*The pines of Europe, of the family Pinaceae, are as diverse as the landscapes they inhabit. From the sprawling shrub-like pines of the Alps to the stately giants of the Mediterranean coastline, there are pines for every location. Perhaps the most widespread and easily recognizable is the Scots pine,* Pinus sylvestris. *It has a natural range from Scotland to Siberia and occurs as far south as the Mediterranean.*

## Maritime Pine

*Pinus pinaster*

**Distribution**: Central and western Mediterranean regions including North Africa.
**Height**: 40m/130ft
**Shape**: Broadly columnar
**Evergreen**
**Pollinated**: Wind
**Leaf shape**: Linear

The maritime pine has been cultivated for forestry purposes since the 16th century because its timber is strong and straight-grained. It was first described in 1789 by the British botanist William Aiton. It thrives in coastal locations, in sandy soils, and is very tolerant of salt spray. In exposed conditions it becomes a twisted, irregularly-shaped tree, whereas in plantations, such as those along the southern coasts of Spain, Portugal and France, it will grow to a height of 40m/130ft with a clean, straight trunk.

**Identification**: Maritime pine bark is purple-brown and smooth at first, becoming ridged and deeply fissured in maturity. It has sharp, needle-like leaves, 20cm/8in long, which are grey-green to dark green and borne in pairs on stout, pale brown shoots. The yellow male and red female flowers are borne in separate clusters on the same tree in late spring to early summer. The cones are hard and woody with spines on each scale, up to 20cm/8in long. Borne in clusters, they persist on the branches for several years.

*Right: Sharp, slender needle-like leaves up to 20cm/8in long are borne in pairs on stout shoots.*

## Corsican Pine

*Pinus nigra* subsp. *laricio*

**Distribution**: Southern Italy and Corsica.
**Height**: 40m/130ft
**Shape**: Broadly columnar
**Evergreen**
**Pollinated**: Wind
**Leaf shape**: Linear

A large tree, this differs from the species, *P. nigra* (the Austrian pine) by having a more open crown with fewer, shorter branches, which are level rather than ascending. The Corsican pine is grown throughout Europe, including Britain, for its timber, which is strong and relatively free of knots.

**Identification**: The stiff, stout shoots are pale yellow-brown, with buds that are narrowly conical, sharply pointed and commonly covered with white resin. The needles, in pairs, are sparsely positioned on the shoot. They are pale grey-green, up to 18cm/7in long and twisted. The male flowers are golden yellow and abundant at the bases of shoots, shedding pollen from late spring to early summer. The female flowers are dull pink and positioned on the tips of growing shoots. The cone is ovoid to conical with a slight sweep and up to 8cm/3in long.

*Above: The bark is light grey to pink and is fissured from an early age.*

*Right: Needles are long, measuring up to 18cm/7in.*

*Left: Flowers appear from spring to early summer.*

OTHER SPECIES OF NOTE

**Lace-bark Pine** *Pinus bungeana*

This is a beautiful slow-growing, broad, widely branched pine, with smooth, grey-green bark, which the tree gradually sheds to reveal patches of new bark. This is pale yellow, but slowly turns olive-brown, red and purple when exposed to light. It has dark yellow-green, needle-like leaves, which are borne in threes and are up to 8cm/3in long. Lace-bark pine was introduced into Europe in 1846 from its native territory in central China. Since then it has been planted in gardens and arboreta, but perhaps not as widely as it should be, considering its beauty.

## Stone Pine

Umbrella pine *Pinus pinea*

Widely planted throughout the Mediterranean for its seeds, which are eaten as nuts, the tree is also known as the umbrella pine because of its flat-topped shape in maturity. With its long, horizontal branches and dense foliage, it is a tree of unusual and aesthetically pleasing habit, and has become a distinctive part of the Mediterranean landscape.

**Distribution**: Mediterranean from Portugal to Turkey.
**Height**: 20m/65ft
**Shape**: Broadly spreading
**Evergreen**
**Pollinated**: Wind
**Leaf shape**: Linear

**Identification**: The bark is grey, with orange fissure lines running vertically down the trunk. The tree has a short main trunk and has branching relatively low down. The shoots are pale green, smooth and curved, with a bright chestnut-red bud, fringed with white hairs, at each shoot tip. The needles are forward pointing, in pairs, grey-green in colour, stout and up to 12cm/4½in long. The cone has a flat base but is almost round, up to 10cm/4in across, glossy brown and smooth. It is relatively heavy and can weigh up to 375g/12oz. After forming, the cones remain tightly closed for three years before opening to reveal up to 100 edible seeds.

*Above: Female cone (left) and male cone (right).*

*Above: The needles are long and occur in pairs.*

*Right: The closed cone is almost egg-shaped.*

## Scots Pine

*Pinus sylvestris*

This is one of the temperate world's most prolific and popular trees, which most Americans would instantly recognize as their Christmas tree. The Scots pine flourishes on dry, sandy soils but will grow in wet conditions, although more slowly. It is a prolific seed producer and is able to colonize new territory quickly. It is well known as a pioneer species, establishing itself long before other trees begin to move in.

**Identification**: An old Scots pine has a distinctive low, broadly domed crown and large, level but snaking branches. The bark is one of its most distinctive features. It ranges from grey-green in a juvenile to a stunning orange-red in maturity. On branches this red bark peels and flakes away. On the main stem it becomes cracked and fissured with age. The paired needles are stiff, twisted, bluish green, set in an orange-brown basal sheath and up to 7cm/2¾in long. The male flowers are yellow and the females are red; they are held in separate clusters on young shoots in late spring and early summer. The cone is egg-shaped, up to 7cm/2¾in long and green, ripening to brown.

**Distribution**: From Scotland right across northern Europe to the Pacific coast and southwards to the Mediterranean and Turkey.
**Height**: 35m/115ft
**Shape**: Broadly spreading
**Evergreen**
**Pollinated**: Wind
**Leaf shape**: Linear

*Left: The trunk is often branchless.*

*Right: Cones may occur in pairs.*

## Bosnian Pine

*Pinus leucodermis*

A medium-sized, neat, distinctive tree, this grows particularly well on sunny slopes, on dry and shallow soils overlying chalk or limestone, where it can live for more than 1,000 years. The needles are a deep black-green, giving the whole tree a sombre, dark appearance that makes this pine instantly recognizable.

**Identification**: The overall shape is ovoid to conical, rather narrow, regular and dense. The branches ascend slightly from the trunk. The bark is greenish-grey and smooth when young, becoming finely fissured in maturity. The shoots are pale brown and slightly hairy, and have a glaucous bloom. The male flowers are yellow and the females purple-red. Both are held at the tips of the shoots on the same plant. Clouds of pollen are often seen blowing in the breeze around the tree in late spring.

*Left: Flat-bottomed cones, cobalt-blue or black when young, ripen to brown in the second year.*

*Above: The needles, which are in pairs, are prolific on the shoot; they are up to 9cm/3½in long and all point neatly forward at a 45° angle. They have a sharp point and are very rigid.*

*Left: The upright bundles of needles are a deep lustrous green colour. They grow on branches that are set tightly together.*

**Distribution**: Balkans, Bosnia-Herzegovina, Bulgaria, Albania into northern Greece and south-western Italy.
**Height**: 25m/80ft
**Shape**: Narrowly conical
**Evergreen**
**Pollinated**: Wind
**Leaf shape**: Linear

## Aleppo Pine

*Pinus halepensis*

The Aleppo pine is a medium-sized tree that has the ability to survive in hot, dry conditions, including desiccating winds, and in exposed coastal locations. Consequently, it has been widely planted in arid areas as part of afforestation schemes to help stabilize sandy soils. Due to this artificial distribution its original natural range is unknown. However, its common name is derived from the second largest city in Syria. It was introduced into Britain in 1683. In the eastern Mediterranean the timber is used for fuel.

**Identification**: The overall shape of mature trees is domed and rugged, with large, contorted branches. The bark on young trees is purple-brown with orange, shallow fissures. Older trees have dark red-brown bark and orange fissures. The needle-like leaves are borne in pairs; they are bright fresh green, slender, shiny and smooth, and up to 10cm/4in long. The cones are orange to red-brown, woody and up to 10cm/4in long; they are normally borne in whorls that face backwards along the shoot.

*Right: The cones are orange to red-brown at first becoming woody and up to 10cm (4in) long.*

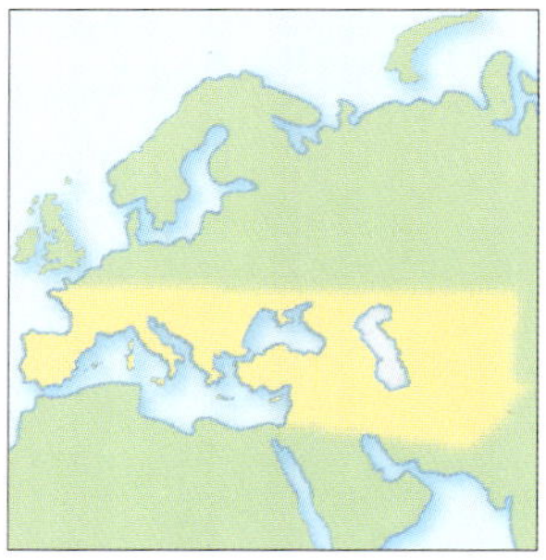

**Distribution**: Southern Europe/Mediterranean to south-west Asia and Afghanistan.
**Height**: 20m/65ft
**Shape**: Broadly conical
**Evergreen**
**Pollinated**: Wind
**Leaf shape**: Linear

## Canary Island Pine

*Pinus canariensis*

Distribution: Canary Islands.
**Height**: 30m/100ft
**Shape**: Broadly conical
**Evergreen**
**Pollinated**: Wind
**Leaf shape**: Linear

The Canary Island pine is restricted to the Canary archipelago, where it occurs naturally on all the islands except Lanzarote and Fuerteventura. It is one of the dominant tree species in the monteverde and montane cloud forests, which occur at high altitude, up to 2,500m/8,200ft above sea level, on the western and northern slopes of mountains such as Mount Teide, 3,718m/12,200ft, on Tenerife. Here, moisture-laden clouds, carried in on the winds, are "combed" for valuable moisture by the long needles of this pine. It is unlikely to survive outside in northern Europe.

**Identification**: The Canary Island pine has spreading branches and drooping branchlets. When young the overall form is conical, becoming broader as the tree matures. The bark is red-brown, thick, fissured and flaking. The long needle-like leaves are at first glaucous, becoming bright grass green as they mature. They are very slender and lax, up to 30cm/12in long and borne on new shoots in clusters of three. The cones are red-brown and may be up to 25cm/10in long.

*Left: Cones are cylindrical egg-shaped and have raised scales giving a rough appearance.*

OTHER SPECIES OF NOTE

**Dwarf Mountain Pine** *Pinus mugo*
This is a sprawling, small, but very hardy two-needled pine, which is native to the Alps and other mountainous regions of south-east Europe. In its native habitat it will crawl along the ground for great distances, but seldom attains heights in excess of 4m/13ft. It has been widely planted in Europe, both as an ornamental species in rock gardens and as a shelter species for protecting less hardy plants in exposed upland situations.

**Golden Scots Pine** *Pinus sylvestris* 'Aurea'
This is a popular garden cultivar of the Scots pine, which is native across Europe, from Spain to Siberia. As its name suggests, this particular cultivar has golden or yellow foliage, which is at its most vibrant in the winter months. The needle-like leaves are up to 7.5cm/3in long, borne in pairs on golden-brown shoots. It has been in cultivation since 1876 and may have been raised in Germany. It is slow-growing and seldom reaches heights in excess of 15m/50ft.

**Japanese Black Pine** *Pinus thunbergii*
This is a magnificent two-needled pine, which in maturity has a distinctive leaning top and numerous long, twisted branches. In its native Japan, where it is often found growing near the seashore, it may reach heights in excess of 40m/130ft, but in Europe it seldom reaches half that height. It was introduced into Europe in 1861 and is now widely grown in parks, large gardens and arboreta. It is also a favoured species for bonsai.

## Gerard's Pine

Chilgoza pine *Pinus gerardiana*

**Distribution**: North-west Himalayas and Afghanistan.
**Height**: 20m/65ft
**Shape**: Broadly conical
**Evergreen**
**Pollinated**: Wind
**Leaf shape**: Linear

This small to medium-sized rare ornamental tree inhabits dry valleys and mountainsides in the Himalayas and Afghanistan. It was discovered by Captain Gerard of the Bengal Native Infantry (hence its common name) in 1832, and was introduced into Europe seven years later. It is somewhat reminiscent of the lace-bark pine, *P. bungeana*: like that species it has exfoliating bark that creates a beautiful patchwork.

**Identification**: The overall appearance is of an open crown with few branches, but each one is clothed in dense, spiky foliage. The grey-pink bark flakes in papery scales to reveal patches of green, yellow and brown new bark beneath. The needle-like leaves are densely borne in spreading clusters of three all around the shoot; they are up to 10cm/4in long and dull green in colour. Gerard's pine has large, woody, oval-shaped cones, which may be up to 20cm/8in long and 10cm/4in wide.

*Below: Needle-like leaves are borne in threes.*

# FIVE-NEEDLE PINES

*Five-needled pines are those pines where the needles grow in clusters of five on the shoots. They are considered to include some of the most beautiful of all pines, quite often producing long soft foliage, which gives the trees an elegant feathery appearance. This group includes the ornamental Holford's pine.*

## Bhutan Pine

Himalayan white pine *Pinus wallichiana*

The soft, slender, pendulous appearance of this pine belies its resilience and ruggedness. In the Himalayas it is able to grow at altitudes in excess of 2,440–3,800m/8,000–12,470ft. It has also proved more resistant to air pollution than almost any other conifer. Although the tree is sometimes confused with *P. strobus*, its appearance has more in common with *P. ayacahuite*. The bark of young trees is grey with resin blistering. On older trees the bark becomes pinkish-orange and is lined with tiny fissures. It is common in cultivation in western Europe.

**Identification**: The crown is strongly whorled and relatively open. Young trees have a conical appearance; older trees become more broad and columnar. Lower branches descend gracefully from the trunk, curving upwards at their tips. Upper branches sweep skywards. The shoots are long, strong and pale grey with a purple bloom. The needles are light green, 18–20cm/7–8in long, in groups of five cupped in a red-brown basal sheath. They curve forward along the shoot, then droop at each side. The male flowers are pale yellow, ovoid, positioned at the bottom of new shoots; they shed their pollen in early summer. The female flowers also appear in early summer; they are dull purple and grow towards the tips of new shoots. The fruit is a long (up to 30cm/12in), drooping, banana-shaped, green cone covered in sticky white resin. It ripens to pale brown in its second year.

*Right: After hanging for more than a year on the tree, cones open to drop their seeds.*

**Distribution**: Himalayas, from Afghanistan to eastern Nepal and Bhutan.
**Height**: 40m/130ft
**Shape**: Broadly conical
**Evergreen**
**Pollinated**: Wind
**Leaf shape**: Linear

## Macedonian Pine

*Pinus peuce*

This is a distinctively dark-crowned pine, with dense, dark green foliage that is reminiscent of yew when viewed from a distance. The dense canopy extends downwards to the lower third of the tree, even on mature specimens. This pine was introduced into cultivation in 1864 by the Greek botanist Theodoros Orphanides. Today, as well as being used as an ornamental in gardens and arboreta across Europe, it is widely planted as a timber-producing species in forestry plantations.

**Identification**: The bark is dark purple, with large smooth areas between vertical fissures. New shoots are bright green, maturing to orange-brown. The dark green, stiff needle-like leaves are 10cm/4in long and borne in clusters of five. Male and female flowers are borne in separate clusters on the same trees in early summer. The leathery, cylindrical cones are resinous and up to 15cm/6in long.

*Above: The male flowers are yellow and the females red.*

*Left and right: The drooping cones are green at first, ripening to rich brown.*

**Distribution**: Balkans, Bulgaria and northern Greece.
**Height**: 30m/100ft
**Shape**: Narrowly conical
**Evergreen**
**Pollinated**: Wind
**Leaf shape**: Linear

## Armand's Pine

Chinese white pine *Pinus armandii*

This Asian five-needled pine is reasonably common in cultivation in Europe, particularly in botanic gardens and arboreta. It was discovered by the French Jesuit missionary Abbé Armand David in 1873 (hence its common name) and was introduced into Europe in 1895. Armand's pine is an attractive, medium-size tree with drooping foliage and large barrel-shaped cones.

**Identification**: The bark is a dull pink-grey, smooth at first but becoming deeply fissured into square or rectangular flaking plates. The crown is broadly conical with widely spaced whorls of horizontal branches, which give the tree an open, airy quality. Armand's pine has glaucous, blue-green, needle-like leaves that are long and lax, giving a "floppy" appearance. They are up to 15cm/6in long and gradually droop below the shoot as they mature. Each orange-brown cone is held by a thick 2.5cm/1in stalk; the cones are roughly cylindrical but broadest near the base, 15 x 7.5cm/6 x 3in with thick scales.

*Left: Long lax blue-green needles are borne in drooping clusters of five.*

*Right: The orange-brown cones may be up to 15cm/6in long.*

**Distribution**: Western China, south-east Tibet, Korea and Taiwan.
**Height**: 40m/130ft
**Shape**: Broadly conical
**Evergreen**
**Pollinated**: Wind
**Leaf shape**: Linear

OTHER SPECIES OF NOTE

**Mexican White Pine** *Pinus ayacahuite*
This beautiful tree attains heights of up to 35m/115ft. It is extremely hardy and grows high on mountain slopes. It has graceful, drooping foliage with blue-green, slender, lax needles up to 15cm/6in long, held in fives. The cone can grow up to 45cm/18in long and is normally covered with sticky white resin. It thrives in central Europe and the Alps.

**Weymouth Pine** *Pinus strobus*
Sometimes known as the eastern white pine, this North American species is one of the largest pines, reaching 75m/225ft in the wild. It was cultivated in Europe from around 1705, when trees were planted in Britain at the Badminton, Gloucestershire, estate of the Duchess of Beaufort and by Lord Weymouth (hence its name) at Longleat, Wiltshire. It is a five-needled pine, each needle being blue green, slender and up to 12cm/4¾in long.

## Arolla Pine

Swiss stone pine *Pinus cembra*

This is a dense, slow-growing, small to medium-size tree, and has been common in cultivation in Europe since at least the mid-18th century. It has a rather formal, neat appearance, which is favoured by some landscape architects. An unusual feature of this tree is that the cones seldom open on the tree to release the seeds inside. From two to three years after ripening, the cone begins to rot and either falls to the ground or is broken apart by birds and squirrels in search of the edible seeds inside.

**Identification**: The bark is grey-brown with vertical scales curling away from the trunk. The needle-like leaves are clustered in fives and densely borne on shoots covered in fine, rust-coloured hairs. The needles, up to 9cm/3½in long, are dark, shiny green on one side and bright bluish-white on the other. Male flowers are purple opening to yellow, female flowers red; both are borne on the same tree in late spring to early summer.

**Distribution**: Central Alps and Carpathians.
**Height**: 25m/80ft
**Shape**: Conical
**Evergreen**
**Pollinated**: Wind
**Leaf shape**: Linear

*Above: The cones are erect, up to 7.5cm (3in) long, deep blue at first.*

## Korean Pine

*Pinus koraiensis*

This tough, hardy Asian pine inhabits mountain slopes and river valleys throughout its extensive range. In cultivation it seldom exceeds 35m/115ft in height, but in the wild it is known to exceed 50m/165ft. It was introduced into Europe by the English nurseryman James Veitch in 1861. Since then it has become a popular ornamental tree in parks, gardens and arboreta and has spawned many cultivars.

**Identification**: The shape is loosely conical, with a feathery appearance. On young trees the bark is smooth and dark grey; on older trees it becomes pink-grey and thick, with curling, flaking scales. The needle-like leaves are densely clustered in groups of five and face forward on young shoots. Each needle is blunt-tipped, up to 12cm/4¾in long and has two, or sometimes three, sides: the outer side is a deep shining green and the inner surface is blue-white with distinct bands of stomata. The cones are oval-shaped, purple-brown and approximately 13cm/5in long.

*Above: The needles are grouped in fives, blunt-tipped and rather rough to the touch.*

*Left: The cones have short stalks, are oval-shaped and up to 13cm/5in long and up to 5cm/2in wide. They have one blunt end. Cones grow singly on the branch in groups of three.*

**Distribution**: North-east Asia, Japan and Korea.
**Height**: 50m/165ft
**Shape**: Broadly conical
**Evergreen**
**Pollinated**: Wind
**Leaf shape**: Linear

## Holford's Pine

*Pinus* x *holfordiana*

This beautiful tree is a hybrid between the Mexican white pine, *P. ayacahuite*, and the Bhutan pine, *P. wallichiana*, which initially developed by chance in Westonbirt Arboretum, Gloucestershire, England, where both parents were growing close together. In 1904 seed was collected from the Mexican white pine and propagated. The resulting progeny were planted in several locations in the arboretum and as they developed it became clear that they were different to the parent pine. The hybrid was officially recognized and named in 1933.

*Right: Cones are large – up to 30cm/12in long and quite often covered in a sticky white resin.*

*Right: The needles are blue-green, lax and hang from buff-coloured shoots.*

**Identification**: From a distance the tree has a graceful feathery appearance. The bark is grey and vertically fissured. The needle-like leaves, which are clustered in fives, are up to 18cm/7in long, slender, pointed, lax and blue-green. The cones are buff brown, covered with white sticky resin and occasionally up to 30cm/12in long. They are heavy and hang either singly or in twos and threes on the outside of the crown, creating a quite beautiful effect against the soft blue-green foliage. The tree is a vigorous hybrid.

*Right: Clusters of long lax needles give the tree a feathery appearance.*

**Distribution**: Of UK garden origin.
**Height**: 30m/100ft
**Shape**: Broadly conical
**Evergreen**
**Pollinated**: Wind
**Leaf shape**: Linear

OTHER SPECIES OF NOTE

***Pinus* x *schwerinii***
This elegant, large, five-needled tree is a hybrid between the Weymouth pine, *P. strobus*, and the Bhutan pine, *P. wallichiana*. It was found on the estate of Dr Graf von Schwerin near Berlin in 1905 and named in his honour in 1931. It differs from both parents in having densely hairy shoots and has shorter needles than the Bhutan pine.

**Cerro Potosi Pinyon** *Pinus culminicola*
This small five-needled pine takes its name from the mountain in north-east Mexico where it was first discovered. Since then it has been found in other mountainous locations in northern Mexico. It is a hardy, slow-growing, shrubby species with grey-green needles up to 5cm/2in long. It was introduced into Europe before 1979.

**Bristlecone Pine** *Pinus aristata*
This North American pine is closely related to the ancient pine, *P. longaeva*, which is also, confusingly, known as the bristlecone pine. However they are two different species. *P. aristata* is a small tree with thick, hairy, red-brown shoots. The needle-like leaves are in clusters of five, to 5cm/2in long and spotted with white resin, unlike those of *P. longaeva*. *P. aristata* also has longer bristle-like spines on the tip of each cone scale than *P. longaeva*. It was introduced into Europe in 1863.

## Dwarf Siberian Pine

Japanese stone pine *Pinus pumila*

This hardy, dwarf five-needled pine, which rarely grows to more than 6m/20ft in height, inhabits high, exposed rocky mountain slopes from eastern Siberia to the Pacific Ocean. It thrives in the Alps. In these habitats it is usually the dominant species, forming extensive forests of scrub. It is closely related to the Arolla pine, *P. cembra*, and it is often difficult to distinguish from dwarf forms of that species.

**Identification**: The bark is grey-brown, smooth at first becoming shallowly fissured in maturity. The shoots are green-brown and covered with grey down. The needle-like leaves, up to 10cm/4in long, are dark green and slightly glossy on the outside and bright blue-green with bands of stomata on the inner surface. The cones are purple when young, ripening to red-brown or golden yellow-brown.

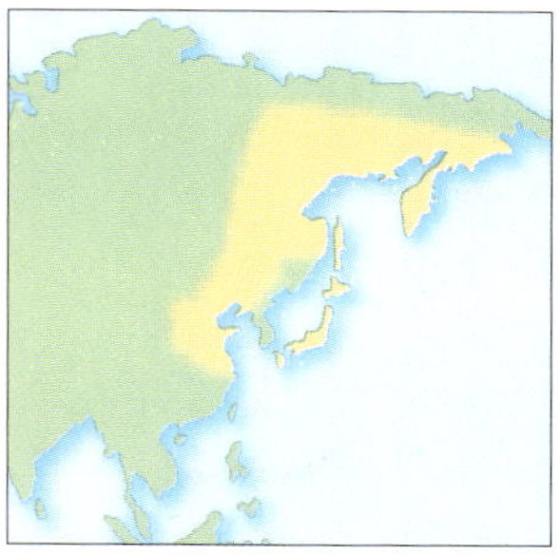

**Distribution**: Eastern Siberia south to north-eastern China and Japan.
**Height**: 6m/20ft
**Shape**: Broadly spreading
**Evergreen**
**Pollinated**: Wind
**Leaf shape**: Linear

*Left: The cones are small and egg-shaped, 5cm/2in long by 3cm/1¼in wide.*

*Left and above left: The leaves are densely borne along the branchlets in clusters of five.*

## Japanese White Pine

*Pinus parviflora*

This is a small to medium-size hardy tree that originally inhabited mountain slopes in its native Japan. It has been so widely planted, both in Japan and around the world, that it is now difficult to ascertain its exact natural distribution. It was introduced into Europe by English nurseryman James Veitch in 1861. This tree is much favoured for bonsai because of its broadly spreading, flat-topped shape. It also provided the motif for traditional "willow pattern" ceramic decoration.

**Identification**: The thick trunk supports a dense crown and old trees develop a picturesque shape as the upper branches lengthen to create a flat top. The bark is grey, scaly and in maturity covered with vertical fissures. The tree has twisted, needle-like leaves that are borne in clusters of five and emerge from bright orange leaf buds. Each needle is up to 6cm/2½in long and blue-green in colour. The male flowers are purple opening yellow, and the female flowers are red; they are both borne, in separate clusters, on the same tree in late spring.

**Distribution**: Japan.
**Height**: 25m/80ft
**Shape**: Broadly columnar
**Evergreen**
**Pollinated**: Wind
**Leaf shape**: Linear

*Right: The stiff, blue-green needles grow in brush-like tufts.*

*Left: The cones are small, to 7.5cm/3in long, and red-brown, with very few scales.*

# MAGNOLIAS

*The Magnoliaceae family contains 12 genera and just over 200 species, of which the majority are native to North America or Asia. They include some of the most beautiful of all flowering trees. Many cultivars have been developed from the true species. There are magnolias to suit all locations – some are giants, others little more than large shrubs. They were introduced to Europe in the 1700s.*

## Yulan Magnolia

Lily tree *Magnolia denudata*

**Distribution**: Central and eastern China.
**Height**: 15m/50ft
**Shape**: Broadly spreading
**Deciduous**
**Pollinated**: Insect
**Leaf shape**: Ovate

*Right: The beautiful white flowers are unfortunately prone to browning if subjected to frost.*

Sometimes known as the lily tree, this beautiful small to medium-size tree has been cultivated in Chinese Buddhist temple gardens since about AD600. Its pure white, cup-shaped, fragrant flowers were regarded as a symbol of purity in the Tang Dynasty period and it was planted in the grounds of the emperor's palace.

**Identification**: The Yulan magnolia is a rather low, rounded, thickly branched tree. It has thick, bright green leaves that are ovate, 15cm/6in long and 7.5cm/3in wide. After falling they decompose to leave a skeletal form. The pure white flowers are heavily scented with a citrus-lemon fragrance. They open from early to late spring: erect and goblet-shaped at first, they gradually spread to form a water-lily shape, as each thick petal curls outwards.

## Japanese Big-leaf Magnolia

White bark magnolia *Magnolia hypoleuca*

Sometimes known as *M. obovata*, this is one of the largest of all magnolias, not only in height but also in girth. In the forests of Hokkaido, Japan, girths in excess of 3m/10ft have been recorded. Japanese craftsmen prize the tree for its light but strong, easily worked timber, which is used for lacquerwork and for items such as sword sheaths and handles. In some parts of the country, the large leaves that earn this tree its name are used for wrapping food.

**Identification**: The bark in young trees is smooth and dark purple-brown. In maturity it becomes slate grey. The leaves are up to 45cm/18in long and 20cm/8in wide. They are thick and leathery, sage green above, with silvery blue downy undersides. They are held in whorls of five to eight, positioned at the end of each shoot. Large, cup-shaped, creamy pink flowers are produced in early summer. The flowers are up to 20cm/8in across and strongly scented. They have bright purple-red stamens, which create a distinctive central "eye" to each flower. Seeds are produced on a conspicuous red cylindrical pod, up to 20cm/8in long.

**Distribution**: Japan and the Pacific coast of Russia.
**Height**: 30m/100ft
**Shape**: Broadly columnar
**Deciduous**
**Pollinated**: Insect
**Leaf shape**: Obovate

*Left: The flowers of this tree, although large, are dwarfed by its massive leaves.*

## Bull Bay

*Magnolia grandiflora*

This magnificent evergreen flowering tree is more often than not grown as a wall shrub. However, given a warm, sheltered, sunny position it will develop into a broad-canopied, short-stemmed tree. The bull bay is common throughout warmer regions of Europe in parks, gardens and arboreta. The combination of glossy, dark green, leathery leaves and creamy white flowers makes it a very popular garden tree.

**Identification**: The bark is grey-brown, cracking into irregular small plates. The leaves, which grow up to 25cm/10in long and 10cm/4in across, are thick and rigid, glossy dark green above and either pale green or covered in copper-coloured hairs beneath. Flowers begin to appear when the tree is only around ten years old; they are wide-brimmed and cup-shaped, creamy white to pale lemon and deliciously scented. They can be up to 30cm/12in across and stand out splendidly against the dark foliage.

*Above: The spectacular flowers are like dinner plates, measuring up to 30cm/12in across.*

*Left: The flowers are produced in spring.*

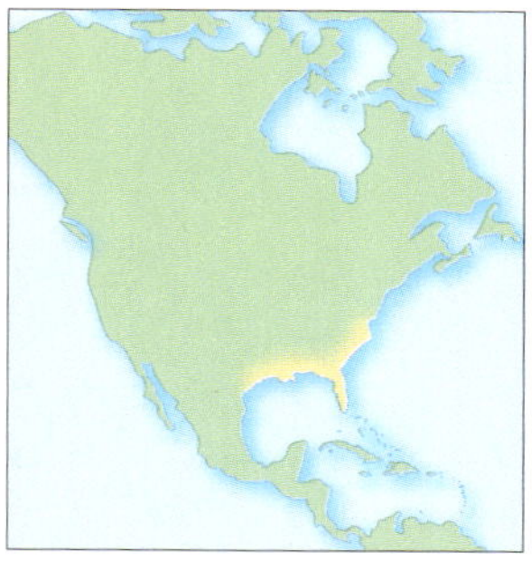

**Distribution**: North American south-east coastal strip, from North Carolina to Florida and west along the gulf to south-east Texas.
**Height**: 25m/80ft
**Shape**: Broadly conical
**Evergreen**
**Pollinated**: Insect
**Leaf shape**: Elliptic to ovate

*Right: The red seed pods first appear in midsummer.*

OTHER SPECIES OF NOTE

**Dawson's Magnolia** *Magnolia dawsoniana*
This beautiful, medium-size Chinese magnolia was discovered in western Sichuan in 1908 and was introduced into Europe in 1919. It is one of the earliest magnolias to flower, producing masses of pale pink, slightly fragrant, goblet-shaped flowers, with up to 12 lax tepals (petals) up to 13cm/5in long, in early spring.

**Sargent's Magnolia** *Magnolia sargentiana*
This large Chinese magnolia is named after Charles Sargent of the Arnold Arboretum, Boston, USA. It produces flowers like large pink water lilies in mid-spring, before the glossy, leathery, dark green leaves appear.

**Willow-leaf Magnolia** *Magnolia salicifolia*
This small and broadly conical Japanese magnolia was introduced into Europe in 1892. It has narrow, willow-like leaves (occasionally ovate), which emit a lemon fragrance when crushed. The fragrant flowers, up to 12.5cm/5in across, are white with yellow stamens, produced before the tree comes into leaf in mid-spring.

**Large-leaf Magnolia** *Magnolia macrophylla*
This south-eastern US tree has the largest leaves of any magnolia, up to 90cm/36in long and 30cm/12in wide, rather like tobacco leaves. Its large, creamy-yellow, fragrant flowers are borne at the ends of the shoots in early to mid-summer. It was discovered by the French explorer André Michaux in 1759.

## Umbrella Magnolia

*Magnolia tripetala*

This is a hardy small tree. It grows in woodland shade, quite often beside streams in valley bottoms. It has fragrant flowers and striking foliage, which make it a much planted ornamental tree. It was first identified in the 18th century and since then has been widely cultivated across the western world. It requires acid soil to grow well.

**Identification**: The bark is pale grey and smooth. The leaves can be up to 50cm/20in long and 25cm/10in wide, with a rich green upper surface and a sage green underside covered with soft down. They are borne in large whorls at the tips of the shoots, looking somewhat like umbrellas. The loose goblet-shaped flowers appear in late spring; they are 20cm/8in across, creamy-white and very fragrant, with up to 12 narrow, waxy, spreading tepals (petals). The fruit is a squat banana-shaped cone, up to 10cm/4in long, covered with crimson seeds.

**Distribution**: Eastern USA from Pennsylvania to Georgia.
**Height**: 12m/40ft
**Shape**: Broadly spreading
**Deciduous**
**Pollinated**: Insect
**Leaf shape**: Obovate to elliptic

*Right: The heavily veined leaves are huge and not dissimilar to tobacco leaves.*

## Hybrid Magnolia

*Magnolia* x *soulangiana*

This hybrid between *Magnolia denudata* and *M. liliiflora* has become the most widely planted ornamental magnolia in gardens throughout Europe. It was raised in the garden of Etienne Soulange-Bodin, at Fromont, near Paris, in the early 1800s, and first flowered in 1826. Since then, scores of forms of this hybrid have been named and introduced into cultivation, including 'Alba Superba', 'Rustica Rubra' and 'Brozzonii'.

**Identification**: This extremely ornamental small tree has wide-spreading multiple stems and grey, smooth bark. The flowers are large, goblet-shaped and normally creamy-white with pink-purple staining at the base of each tepal (petal). The colouring varies according to the different forms. The leaves, which taper to a narrow base and a pointed tip, are dark green above, paler green beneath and up to 20cm/8in long. Seeds are produced in irregular, cylindrical pink clusters in late summer to early autumn.

**Distribution**: Of garden origin, France.
**Height**: 9m/30ft
**Shape**: Broadly spreading
**Deciduous**
**Pollinated**: Insect
**Leaf shape**: Elliptic to obovate

*Below and left: The flowers appear from grey, downy buds, before the leaves, in spring.*

## Star Magnolia

*Magnolia stellata*

This beautiful, slow-growing, wide-spreading small tree or large shrub is considered by some authorities to be a variety of *M. kobus* and by others to be *M. tomentosa*. It is native to the mountains of Nagoya, Japan, from where it was introduced into Europe in around 1877. It is a perfect magnolia for planting in small gardens; it will start to flower at a young age and is more tolerant of limy soils than most other magnolias.

**Identification**: The bark is grey, smooth and aromatic when scratched. The flower buds, borne through winter, are covered with silver-grey down, and the star-like white flowers, which appear before the leaves in mid-spring, are fragrant. Some cultivars, such as 'Dawn', have flowers that are flushed with pink. The narrow leaves are deep green above and paler beneath.

**Distribution**: Japan.
**Height**: 3m/10ft
**Shape**: Broadly spreading
**Deciduous**
**Pollinated**: Insect
**Leaf shape**: Obovate

*Right: The narrow leaves are up to 10cm/4in long.*

*Left: The flowers have up to 18 narrow white tepals (petals), each up to 5cm/2in long.*

OTHER SPECIES OF NOTE

**Delavay's Magnolia** *Magnolia delavayi*
This broadly spreading, tender, Chinese evergreen, 10m/33ft tall, produces among the largest leaves of any tree grown in Europe: they are up to 35cm/14in long and 15cm/6in wide, dark glossy green and stiff to the touch. It produces cream, saucer-shape flowers in late summer to early autumn, which are up to 20cm/8in across and seldom last more than 48 hours before facing.

***Magnolia liliiflora***
This is an attractive, wide-spreading, deciduous small tree or large shrub. Slow-growing and shade tolerant, it is now considered to be native to China, but came to Europe from Japan in 1790. It has erect, tulip-shape flowers, which are purple-pink on the outside and creamy-white within. The flowering season is long, beginning in mid-spring and ending in early summer.

***Magnolia sinensis***
This elegant, small, deciduous tree, to 6m/20ft tall and wide, thrives on chalky soils. It is hardy, growing at elevations up to 2,750m/9,000ft above sea level, and has long been grown as an ornamental species in Europe. It has beautiful, pure white, nodding flowers, which are set off by bright crimson stamens. The flowers, which are strongly citrus-scented, do not open until after the leaves have appeared in early summer.

## Goddess Magnolia

*Magnolia sprengeri* var. *diva*

Almost all the trees of this variety growing in Europe are descended from a specimen at Caerhays Castle in Cornwall, England. It was raised from a seed collected in western China by Ernest Wilson in 1900 and is a large, vigorous tree with beautiful pink, goblet-shaped flowers. (All the other seeds Wilson sent in the same batch produced white flowers.)

**Identification**: The bark is grey and smooth. The leaves are up to 17.5cm/7in long. The flower buds are large, up to 7.5cm/3in long, and covered in silver-grey down. The erect flowers start to appear in early spring. As they mature the colour fades to light pink and the tepals become lax, spreading out to form an open flower the size of a small plate. The ripe seed is contained in an irregular bright red cylinder about 7.5cm/3in long.

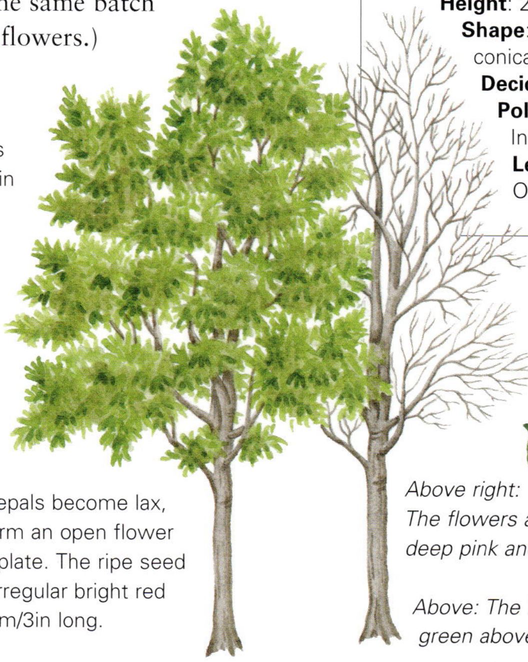

**Distribution**: China.
**Height**: 28m/90ft
**Shape**: Broadly conical
**Deciduous**
**Pollinated**: Insect
**Leaf shape**: Obovate

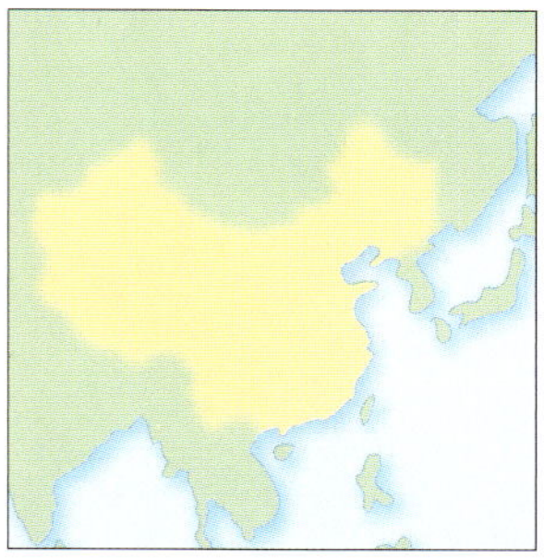

*Above right: The flowers are deep pink and goblet-shaped.*

*Above: The leaves are bright green above, pale beneath.*

## Campbell's Magnolia

*Magnolia campbellii*

This majestic tree is capable of attaining a height of up to 30m/100ft in less than 60 years. It is hardy, growing at up to 3,000m/9,850ft above sea level in the Himalayas. *M. campbellii* is grown widely for its dramatic flowers, which do not appear on seedlings for 20 years. They appear as early as mid-winter and are prone to frost damage.

*Below: The flowers are big, up to 30cm/12in across.*

**Identification**: The bark is smooth and grey, even in old age. The leaves are up to 30cm/12in long, with a pronounced point, medium green above and sometimes faintly hairy beneath. The flower buds are large, ovoid and covered in grey hairs: they stand out dramatically on the bare branches in late winter. The flowers are even more dramatic, beginning goblet-shape but opening to a lax cup-and-saucer shape, up to 30cm/12in across. Their colour can vary from deep pink to pale pinkish-white, and they have a slight fragrance. Each flower is held upright on a smooth green stalk. The fruit is a cylindrical, cone-like pod, up to 15cm/6in long, containing bright red seed.

*Right: Flowers often appear in profusion on both cultivated and wild trees.*

**Distribution**: Himalayas from Nepal to Assam and into south-west China.
**Height**: 30m/100ft
**Shape**: Broadly conical
**Deciduous**
**Pollinated**: Insect
**Leaf shape**: Obovate

# MAGNOLIAS AND TULIP TREES

*Tulip trees are closely related to magnolias. Both represent some of the oldest flowering trees on earth. Their flowers are usually large and showy single flowers that appear all over the tree, usually in spring. Also related to magnolias and tulip trees is the beautiful evergreen* Michelia doltsopa *which is native to China.*

## Chinese Tulip Tree

*Liriodendron chinense*

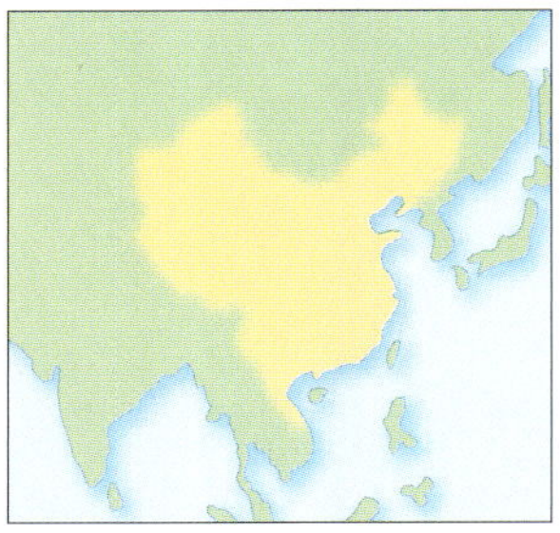

**Distribution**: China and North Vietnam.
**Height**: 20m/65ft
**Shape**: Broadly columnar
**Deciduous**
**Pollinated**: Insect
**Leaf shape**: Simple lobed

As its name suggests, this majestic medium-size tree is native to China (and Vietnam), where it grows in mixed woodland on mountain slopes up to 2,000m/6,560ft above sea level. It is one of only two species in the genus *Liriodendron*, the other being the American tulip tree, *L. tulipifera*. It was introduced into Europe by Ernest Wilson in 1901, and some of the original introductions are still growing in the grounds of the Royal Botanic Gardens at Kew, near London.

**Identification**: The bark is slate grey, smooth at first, becoming fissured in maturity. The leaves are dark green above and sage green below, to 15cm/6in long with two distinct lobes at the tip and two at the side. The flowers, resembling tulips in size and shape, are pale yellow-green on the outside and banded with orange-yellow inside. They appear singly at the ends of the shoots in mid-summer.

*Right: The leaves look as if their tips have been snipped off with scissors. They turn yellow in autumn.*

*Right: The flowers are produced on mature trees.*

## Tulip Tree

Yellow poplar *Liriodendron tulipifera*

This magnificent tree is one of the largest and fastest-growing deciduous trees. Fossil records show it lived in Europe before dying out in the last ice age. It stands out from the crowd for its size, its flowers, its unusual leaf shape and its ability to withstand atmospheric pollution. It is an adaptable tree, capable of growing in extreme climatic conditions.

**Identification**: The bark is grey-brown and smooth, becoming fissured with age. In maturity, tulip trees have clear, straight stems with broad crowns. The dark green leaves are up to 15cm/6in long and lobed on each side with a cut-off, indented leaf tip. The underside of the leaf is almost bluish-white. In autumn the leaves turn butter yellow before falling. The flowers are produced in summer once the tree reaches 12–15 years old. They are upright, 6cm/2½in long, tulip-shaped with nine petals: some are green, some are light green to yellowy-orange at the base. Inside each flower is a bright cluster of orange-yellow stamens. Unfortunately, because of the branchless stems of older trees, the flowers are very often positioned at the top of the tree, so it is difficult to admire their beauty.

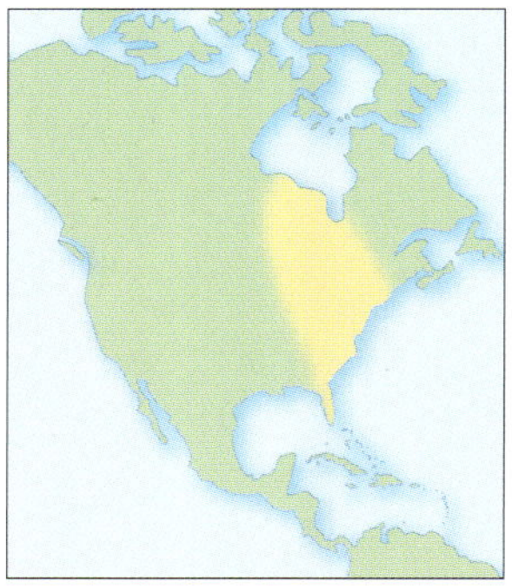

**Distribution**: Eastern North America from Ontario to New York in the north to Florida in the south.
**Height**: 50m/165ft
**Shape**: Broadly columnar
**Deciduous**
**Pollinated**: Bee
**Leaf shape**: Simple

*Left: As the flowers fade on the tree, the leaves change colour, giving the tulip tree a second flush of beauty.*

## Michelia

*Michelia doltsopa*

Without doubt this is one of the most beautiful of all evergreen trees and one that should be more widely planted in warmer regions of Europe. It is related to magnolias and was discovered in 1918 in western China by the Scottish plant collector George Forrest. He sent seeds to Caerhays Castle in Cornwall, England; these were propagated and the five resulting seedlings were planted in the grounds. The first seedling flowered in April 1933.

**Identification**: The leaves are dark green, glossy and leathery, 20cm/8in long, with rust-coloured down on the undersides. The flower buds form in autumn; they are like large magnolia flower buds, covered with fine, cinnamon-coloured hairs. They overwinter on the tree and begin to open early the following spring to reveal beautiful, multi-petalled flowers, soft lemon to white and reminiscent of water lily flowers. They have a strong, luscious fragrance that, on warm spring days, may be carried for quite some distance away from the tree.

**Distribution**: China, Tibet and eastern Himalayas.
**Height**: 20m/65ft
**Shape**: Broadly spreading
**Evergreen**
**Pollinated**: Insect
**Leaf shape**: Oval to oblong

*Above: The flowers are softly fragrant.*

*Left: Fragrant water-lily-like flowers are produced in spring, set among dark green leathery leaves.*

*Right: The flowers are cup shaped.*

### OTHER SPECIES OF NOTE

***Magnolia* x *loebneri***
'Leonard Messel'
This is a beautiful deciduous magnolia cultivar. It is a vigorous upright shrub or small tree that has lilac-pink flowers with 12 narrow lax tepals (petals). It is said to have resulted from a cross between *M. kobus* and *M. stellata* 'Rosea'. It thrives in Europe.

**Cucumber Tree**
*Magnolia acuminata*
Otherwise known as the mountain magnolia, because of its ability to grow at altitudes of up to 1,220m/4,000ft. This is a large, vigorous, deciduous tree that can grow to a height of 20m/70ft. It has elliptical, rich green leaves, 25cm/10in long, and blue-green to yellow-green cup-shaped flowers, which are reminiscent of those of its distant cousin, *Liriodendron tulipifera*. Its common name refers to the shape of the erect, cylindrical fruits, green when unripe, turning red, that contain the seed. It was introduced to Europe in the 16th century.

## Wilson's Magnolia

*Magnolia wilsonii*

One of the loveliest of all magnolias, this was named after the English plant collector Ernest Wilson, who discovered the species in western China in 1908. It is distinctive for having drooping flowerheads, which can be seen at their best when standing directly beneath the tree.

**Identification**: This large spreading shrub or small tree is primarily grown for its delightful flowers, which hang from arching stems. They are cup-shaped, up to 10cm/4in across and pure white with a brilliant crimson centre. The flower bud is covered in dense grey, soft down, which is obvious from mid-winter onwards. After flowering, a purple-pink, cylindrical-ovoid seed pod is produced, which is 5–7.5cm/2–3in long and contains scarlet-coated seed. The narrow leaves are almost spear-shaped, dull green above and covered in pale-brown down beneath. This magnolia is often found growing in shade.

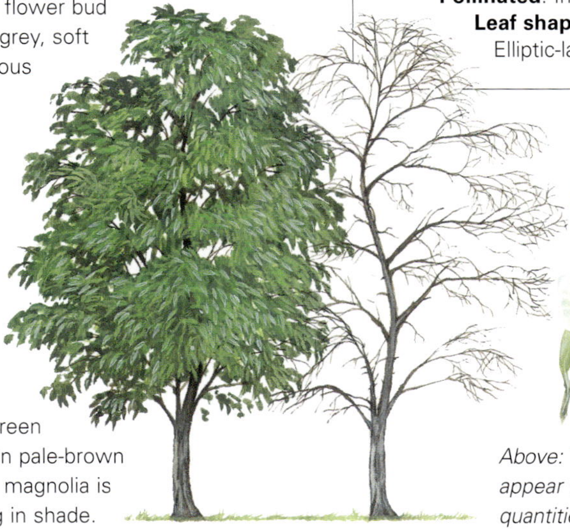

**Distribution**: Western Sichuan and Yunnan, China.
**Height**: 8m/26ft
**Shape**: Broadly spreading
**Deciduous**
**Pollinated**: Insect
**Leaf shape**: Elliptic-lanceolate

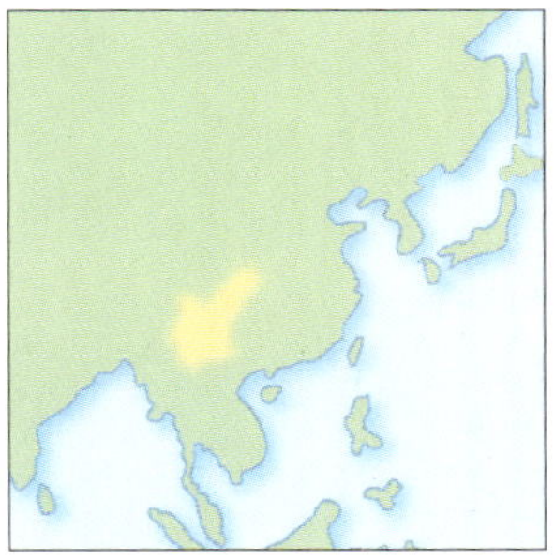

*Above: The flowers appear in great quantities in spring.*

# TREE RHODODENDRONS

*Commonly thought of as a genus of shrubs, there are in fact several rhododendrons that grow to tree-like proportions. Most of the species featured here attain heights in excess of 6m/20ft. In the wild they are an integral part of their native Himalayan and Chinese forests. All are widely cultivated as ornamental specimens in Europe. Some of the finest are to be found in Great Britain and Ireland.*

## Tree Rhododendron

*Rhododendron arboreum*

This was the first rhododendron to be introduced into Europe from the Himalayas, in around 1810. The first account of it came from the naturalist Thomas Hardwicke, then an army captain in India, who saw it flowering in 1796. Its natural range extends from Kashmir to Sikkim, where it grows at elevations in excess of 2,600m/8,500ft. It is now widely grown in gardens throughout northern and western Europe.

**Identification**: The tree is sometimes columnar and erect, with a rounded crown on a single stem, and sometimes multi-stemmed and broadly spreading. The red-brown bark flakes in maturity. The leaves are dark green above and covered with rust-coloured hairs beneath. The flowers are normally blood red, borne in dense clusters of up to 20 in early spring.

*Left: The flowers are bell-shaped and up to 5cm/2in long.*

*Right: The stiff, leathery leaves are up to 20cm/8in long.*

**Distribution**: Himalayas.
**Height**: 15m/50ft
**Shape**: Broadly columnar
**Evergreen**
**Pollinated**: Insect
**Leaf shape**: Oblong to lanceolate

## Rhododendron griffithianum

This majestic tree rhododendron, one of the finest for flower, originates from the Himalayas and can be found growing from Nepal to north-eastern India. It is not particularly common in the wild anywhere. It was introduced into Europe in 1850 by Sir Joseph Hooker of the Royal Botanic Gardens at Kew, England. In the wild it will grow up to 2,750m/9,000ft above sea level, but in Europe it thrives in only the mildest climates. It is the parent of numerous hybrids.

**Identification**: The bark, on both main stems and branches, is reddish-brown and peeling. The leaves are narrow and leathery, dark to mid-green, up to 30cm/12in long and held by a stalk that is up to 4cm/1½in long. *R. griffithianum* has possibly the most magnificent flowers of all rhododendrons. They are white, sometimes with the faintest blush of pink, and are speckled with green markings. They are bell-shaped with a wide spreading mouth, up to 7.5cm/3in long and up to 15cm/6in in diameter. They are borne in loose clusters of up to six in mid- to late spring and are deliciously fragrant.

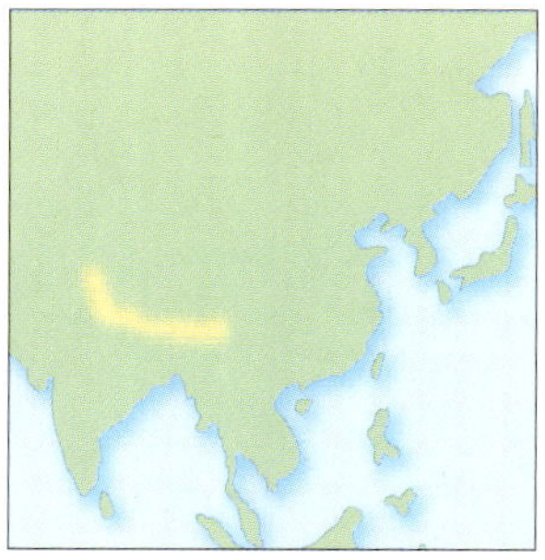

**Distribution**: Himalayas.
**Height**: 6m/20ft
**Shape**: Broadly columnar
**Evergreen**
**Pollinated**: Insect
**Leaf shape**: Narrowly oblong

*Left: The dramatic contrast between the milk-white flowers and the dark green leaves makes this species one of the most spectacular of all rhododendrons.*

## Rhododendron falconeri

This is a magnificent rhododendron, which is widely cultivated in parks and gardens throughout western and northern Europe. It is a moisture-loving species, needing more than 75cm/30in of rainfall a year to perform at its best. It is native to the Himalayas, from Nepal to Bhutan, and although it was first described by Joseph Hooker in Sikkim in 1849 it had actually been introduced into Europe 20 years earlier.

**Identification**: The leaves are large, up to 30cm/12in long and 15cm/6in wide, and very thick and wrinkled with pronounced veining. The upper surface of the leaf is glossy dark green; the lower surface has a thick felting of rust-coloured indumentum, and young shoots also have a temporary covering of brown down. The flowers are borne in huge, tightly packed, dome-shaped clusters. They are bell-shaped, cream-yellow with a distinctive dark purple blotch at the base. Each flower may be 5cm/2in long and roughly the same across. They open from large pointed flower buds in mid- to late spring.

**Distribution**: Himalayas.
**Height**: 10m/33ft
**Shape**: Broadly columnar
**Evergreen**
**Pollinated**: Insect
**Leaf shape**: Oval to obovate

*Left: The flowers emerge from large cone-shaped buds in mid- to late spring.*

*Left: The upper surface of the leaf is dark glossy green.*

OTHER SPECIES OF NOTE

***Rhododendron calophytum***
One of the largest rhododendrons, this species regularly attains heights in excess of 15m/50ft. It is a hardy Chinese rhododendron, native to western Sichuan. It was discovered by the French missionary Abbé Armand David in 1869. It has large trusses of white or blush-white, bell-shaped flowers, each with a dark crimson or maroon blotch at the base.

***Rhododendron sinogrande***
This rhododendron has some of the finest foliage. It has magnificent, glossy, dark green, leathery leaves, which may be up to 80cm/32in long and 30cm/12in wide. The lower surface of each is covered with silver-grey or buff-coloured indumentum. The flowers are creamy-white to soft yellow, with a crimson blotch at the base. It thrives in Europe.

***Rhododendron macabeanum***
This large rhododendron, from northern India, is a handsome small tree, which has been widely planted in European gardens since its introduction in the 1920s. It has large, shiny, dark green leaves, covered with silver-grey indumentum beneath. The bell-shape flowers are pale yellow, purple-blotched at the base.

***Rhododendron rex***
This is a beautiful Chinese tree rhododendron, with bell-shape, blush-white to rose colour flowers, with crimson blotches around the base. They are borne in large trusses of up to 20 flowers in mid- to late spring. It was introduced into Europe in the 1920s.

## Rhododendron barbatum

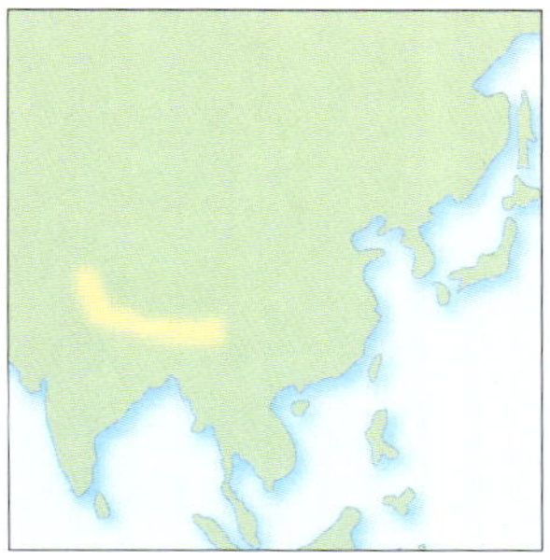

This is one of the most beautiful rhododendrons for bark colour. It has deep blood-red bark that peels to reveal patches of smooth, slate blue-grey, sometimes olive-green, bark beneath. It grows wild up to 3,700m/12,000ft above sea level in Nepal, Bhutan, southern Tibet and northern India. It was introduced into Europe around 1829 and has been a valued specimen for planting in woodland gardens in northern and western Europe ever since.

**Distribution**: Himalayas.
**Height**: 5m/16ft
**Shape**: Broadly spreading
**Evergreen**
**Pollinated**: Insect
**Leaf shape**: Oblong

**Identification**: This wide-spreading small tree or large shrub has leathery, oblong, evergreen leaves, heart-shaped at the base. They are borne in whorls at the tips of the branches and are matt dark green above, and covered with a pale felting beneath – even the leaf stalk is bristly. The flowers are bell-shaped and vivid scarlet to blood-red, and up to 20 flowers are densely packed in globular heads of about 10cm/4in across.

*Left: The oblong-shaped leaves are borne on bristly stalks.*

*Right: The vivid red flowers appear on the tree in early spring.*

# BAYS AND SPICE TREES

*The Lauraceae family contains more than 40 genera and 2,000 different species, most of which are tropical, originating from Asia and South America. Those that are hardier, and can survive in temperate regions of the world, tend to have several characteristics in common, including evergreen leaves and aromatic foliage or bark.*

## Spice Tree

Benjamin *Lindera benzoin*

**Distribution**: South-eastern USA.
**Height**: 5m/16ft
**Shape**: Columnar
**Deciduous**
**Pollinated**: Insect
**Leaf shape**: Obovate

This is a small to medium-size tree that is native to the eastern United States, from where it was introduced into Europe as early as 1683. Despite this early introduction the spice tree is relatively uncommon in cultivation, being confined to botanic gardens and arboreta across western and northern Europe. Its name is derived from the fact that when the leaves are crushed they emit a pungent, spicy odour.

**Identification**: *Lindera benzoin* has smooth grey bark, which becomes finely fissured in maturity. Its leaves, which are up to 12.5cm/5in long and 5cm/2in wide, are distinctly veined and taper to both the base and the tip. In autumn the leaves turn a clear bright yellow before falling. Both male and female flowers are small and yellow, or yellow-green, and are borne on separate trees in early to mid-spring. Although small, they are produced in profusion and are an attractive feature of the tree. On female trees the flowers are followed by red, oval berries.

*Right: The flowers are fragrant.*

*Right: Red berries appear in autumn.*

## Spice Bush

*Lindera praecox*

This beautiful small tree was introduced into Europe from Japan in 1891, but has not been widely cultivated since then, and today it is mainly found in tree collections and botanical gardens in northern and western areas. It is reasonably hardy and a specimen has been flowering in the Royal Botanic Gardens at Kew, near London, since the 1930s.

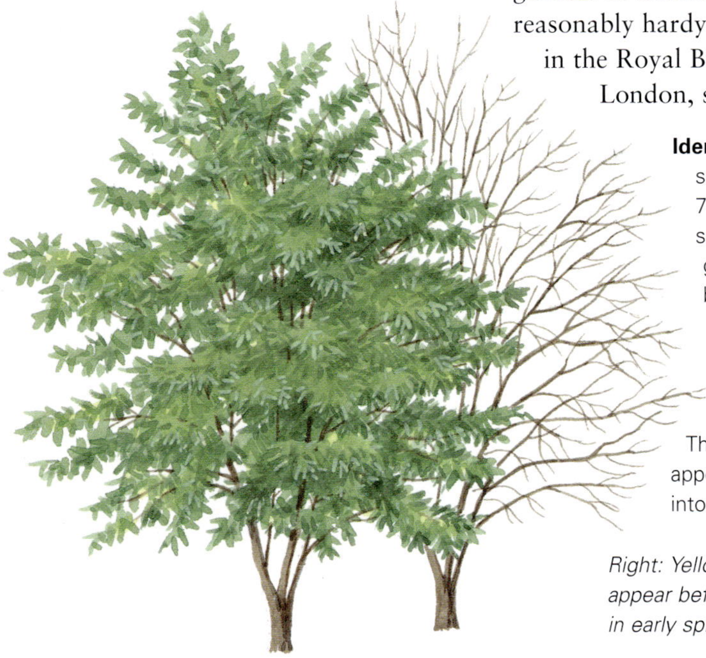

**Identification**: The bark is grey-brown and smooth. The leaves are thin, oval to ovate, 7.5cm/3in long, and translucent in bright sunlight. They are mid-green above and glaucous beneath. In autumn they turn bright butter-yellow before falling. The male and female flowers are borne on separate trees. They are yellow-green and appear in short clusters along the bare branches in early and mid-spring. The fruit is a red-brown, oval berry, which appears in late summer into autumn.

**Distribution**: Japan and Korea.
**Height**: 8m/26ft
**Shape**: Columnar
**Deciduous**
**Pollinated**: Insect
**Leaf shape**: Ovate to oval

*Right: Yellow-green flowers appear before the leaves in early spring.*

*Left: Butter-yellow leaves and red-brown berries appear in early autumn.*

## Bay Laurel

Sweet bay *Laurus nobilis*

This is the laurel used by the Greeks and Romans as a ceremonial symbol of victory; it was usually woven into crowns to be worn by champions. Fruiting sprays were also made into wreaths and given to acclaimed poets, hence the term "poet laureate". The title "bachelor", as awarded to those with university degrees, is derived from the French word *bachelier*, which means "laurel berry".

**Identification**: A dense small tree or shrub with aromatic leaves, which are widely used to flavour food. The bark is dark grey and smooth, even in old age. The leathery, glossy leaves are alternate, dark green above with a central lighter vein, and pale green beneath. They are 10cm/4in long, 4cm/1½in across and pointed at the tip. The male flowers, which appear in late winter, are greenish-yellow, 1cm/½in across, with many yellow stamens, positioned in the axils of the previous year's leaves.

*Above: The small, male flowers open in late winter. Bay leaves are commonly harvested for use in cooking.*

**Distribution**: Throughout Mediterranean regions.
**Height:** 15m/50ft
**Shape**: Broadly conical
**Evergreen**
**Pollinated**: Insect
**Leaf shape**: Elliptic

*Left: The fruit is a rounded berry, 1cm/½in across, green ripening to glossy black.*

OTHER SPECIES OF NOTE

**Canary Island Laurel** *Laurus azorica*
This handsome evergreen tree grows up to 15m/50ft in the wild and its trunk may grow to over 90cm/3ft in diameter. The young shoots emit a pleasant aroma, like that of bay laurel, *L. nobilis*, when brushed. It is an important component of the laurisilva forests of the humid northern and western slopes of the Canary Islands, and grows in parks, gardens and arboreta in the warmer regions of Europe.

***Lindera megaphylla***
This tender, small evergreen tree, native to southern China and Taiwan, was introduced into Europe in 1900. It has never become common and is confined to botanic gardens and arboreta. It has dark purple shoots and oblong, glossy, dark green leaves, which are pointed at the tip and rounded at the base, and when crushed emit a spicy aroma.

**Willow-leaved Bay** *Laurus nobilis* 'Angustifolia'
This form of the bay laurel, *L. nobilis*, is similar in all respects except for its long, narrow, pale green, pointed, wavy-edged, leathery leaves, which are reminiscent of willow leaves. It is perfectly hardy and common in cultivation across Europe.

## Japanese Spicebush

*Lindera obtusiloba*

This beautiful small Japanese tree, which is also native to China and Korea, was introduced into Europe in 1880 and although quite often cultivated in botanic gardens and arboreta its beauty suggests it should be more widely planted. It normally produces a compact rounded shape but on occasions may develop a more erect habit. It is a two-season tree, producing greenish-yellow flowers in early spring and butter-yellow leaves in autumn.

**Identification**: This large shrub or small tree has waxy blue-green leaves, which are irregularly lobed and turn butter-yellow in autumn. Small fragrant yellow flowers are produced in some profusion in early to mid-spring. The fruit is a shiny reddish round berry, ripening to black, that is evident against the backdrop of clear yellow autumn foliage.

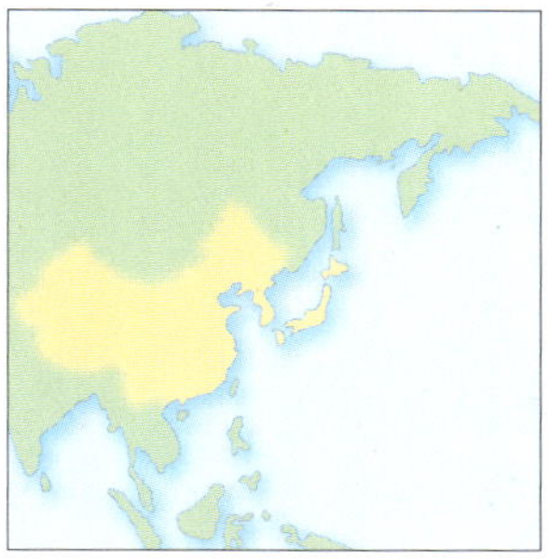

**Distribution**: China, Japan and Korea.
**Height**: 6m/20ft
**Shape**: Rounded
**Deciduous**
**Pollinated**: Insect
**Leaf shape**: Lobed

*Below left: The flowers are tiny, pale greenish-yellow and are produced in clusters in the leaf axils in spring.*

*Below right: The fruit is an egg-shaped berry, approximately 1cm/½in long.*

# MYRTLES AND WINTERWOODS

*Most of the trees represented here come from warmer regions of the Southern Hemisphere, such as Chile and Argentina, and as such are only cultivated in parts of Europe with a temperate mild climate. They include some of the most beautiful evergreen trees, such as the New Zealand Christmas tree, which produces intense red flowers, and the South American myrtle, which has bright cinnamon-orange bark.*

## Winter's Bark

*Drimys winteri*

This handsome Central and South American evergreen tree, with highly aromatic bark, is named after Captain William Winter, who sailed with Sir Francis Drake on his voyage round the world. Winter used the bark of the tree, which has a high vitamin C content, to treat sailors on his ship who were suffering from scurvy. It was cultivated in Europe in 1827 and is a valuable ornamental tree in parks and gardens throughout milder regions of Europe.

*Right: The white flowers with yellow centres are mildly fragrant.*

*Above: The elongated leaves have a peppery scent.*

**Identification**: The bark of Winter's bark is brown-grey, smooth and very aromatic when crushed or scratched. The tree has large, leathery, oblong to elliptic leaves, 20cm/8in long and 5cm/2in broad. They are a glossy dark green above and bluish-silver-green beneath, and when crushed are aromatic. The flowers are ivory white, up to 5cm/2in across and borne in branched clusters in late spring. These are followed by small, purple-black berries, also borne in branched clusters, in late summer and autumn.

**Distribution**: Mexico south to Chile and Argentina.
**Height**: 15m/50ft
**Shape**: Narrowly conical
**Evergreen**
**Pollinated**: Insect
**Leaf shape**: Oblong to elliptic

## New Zealand Christmas Tree

*Metrosideros robusta*

When in flower this is one of the world's most beautiful trees. In New Zealand it is known as Pohutukawa, which is Maori for "drenched with salt spray". As this name suggests, it is found growing around the coast of New Zealand. As well as being tolerant of sea air, it also tolerates wind and drought. It was introduced into Europe in 1840 and is found in cultivation in warmer coastal regions. Some of the finest European specimens grow on Tresco, part of the Isles of Scilly.

**Identification**: This is a broadly columnar tree that may become widely spreading in maturity. The bark is grey-brown and smooth. The leaves are elliptic, dark green on top with a light green midrib, and covered with a fine silver-grey down beneath. It has dazzling, bright crimson "bottlebrush" flowers, which are made up mainly of long red stamens. In large clusters, they smother the branches in summer, almost obscuring the dark evergreen foliage.

**Distribution**: New Zealand.
**Height**: 20m/65ft
**Shape**: Broadly columnar
**Evergreen**
**Pollinated**: Insect
**Leaf shape**: Elliptic

*Left: The bright crimson flowers appear in summer.*

*Right: Woody seed capsules are borne among the evergreen leaves.*

## Chilean Firebush

*Embothrium coccineum*

This beautiful South American small evergreen tree grows in exposed locations from the Pacific coast to high in the Andes Mountains. It was named from a specimen collected during Captain James Cook's second voyage to this region. It was introduced into Europe in 1846 by the plant collector William Lobb, and has been widely cultivated in parks, gardens and arboreta across milder regions of Europe.

*Left: The flowers, which appear in late spring to early summer, are reminiscent of a large honeysuckle.*

**Identification**: The bark is purple-brown and smooth at first, flaking in maturity. The soft leathery leaves, up to 15cm/6in long and 2.5cm/1in broad, are matt blue-green above and pale green beneath. The flowers, tubular at first, to 5cm/2in long, open into four narrow lobes that peel back to reveal a long, vivid red-orange style. They grow in clusters resembling glowing red embers along the branches. The fruit is a woody capsule, 2.5cm/1in long, which opens to release numerous winged seeds.

**Distribution**: Chile and Argentina.
**Height**: 9m/30ft
**Shape**: Broadly columnar
**Evergreen**
**Pollinated**: Insect
**Leaf shape**: Elliptic

OTHER SPECIES OF NOTE

**Wheel Tree** *Trochodendron aralioides*
This attractive, evergreen, Japanese tree is the sole species in the only genus in the family Trochodendraceae. Its nearest relative is believed to be *Drimys winteri*. It has shiny dark green, narrow, elliptical, leathery leaves and aromatic bark. Its most interesting feature is its wheel-like bright green flowers, 2cm/¾in across, with no petals and exposed stamens radiating outwards from a central disc. They appear on upright slender stalks in early summer.

**Gutta Percha**
*Eucommia ulmoides*
This Chinese tree is the only member of the Eucommiaceae family. It is believed to be most closely related to the elms, and is the only temperate tree that produces rubber. If the leaf is gently torn in half, the two halves will still hang together, held by thin strands of sticky latex. In China, gutta percha has been cultivated for hundreds of years for medicinal purposes. It has never been found growing wild, so its origins are unknown.

**Sassafras** *Sassafras albidum*
This eastern North American, medium-sized deciduous tree produces a range of leaf shapes, from those with a large lobe on each side, like a fig leaf, to entire and oval. They turn yellow and orange in autumn. It was introduced into Europe as early as 1633 and has been cultivated in European gardens ever since.

## Myrtle

*Myrtus luma*

This is one of the finest small trees for bark colour, and one that has been cultivated in warmer parts of Europe since its introduction from the temperate forests of Chile and Argentina by the English plant collector William Lobb in 1844. In some areas of Europe, such as southern Ireland, it has become naturalized, reproducing itself with ease. It may sometimes be found in gardens and arboreta labelled with its former botanical name, *Luma apiculata*.

**Distribution**: Chile and Argentina.
**Height**: 12m/40ft
**Shape**: Broadly columnar
**Evergreen**
**Pollinated**: Insect
**Leaf shape**: Broadly elliptic

**Identification**: *M. luma* has bright cinnamon-orange, soft-felted bark, which peels in patches to reveal cream-coloured fresh bark beneath. A curious feature of this tree is that the bark always feels cold to the touch, even when the tree is grown in full sun. The leaves are dark, dull green, oval to elliptic, up to 2.5cm/1in long and ending in a short point. When crushed they are pleasantly aromatic. *M. luma* has small, white, fragrant, four-petalled flowers, which are borne singly in late summer and early autumn.

*Above: Spherical red fruits ripen to purple in autumn.*

*Below: The flowers stand out against the dark leaves.*

# OLIVES AND PITTOSPORUMS

*One of the most widely recognized and cultivated trees is the olive. It has been grown for its fruits in warm regions of Europe for thousands of years. Due to recent increases in mean annual air temperatures and the popularity of domestic conservatories it is increasingly cultivated in cooler regions too. Along with the New Zealand pittosporums, the European future looks bright for these tender evergreens.*

## Common Olive

*Olea europaea*

The olive has been cultivated for its fruits and oil in the warm regions of Europe for thousands of years, to the extent that its true native distribution is now obscure. There are certainly specimens in excess of 1,000 years old growing in Italy and Greece. According to Greek mythology, the club wielded by the hero Hercules was made from the wood of a wild olive tree. Groves of olive trees have become an integral part of the Mediterranean landscape. Further north, in Britain for example, olive trees rarely produce fruit, but survive outside in warm, sheltered positions if well drained.

**Identification**: The olive has a distinctive, short grey, fissured trunk, which may become very thick and hollow in maturity, and a broadly spreading crown. It has grey-green, tough, leathery, opposite, evergreen leaves, which are silver-sage beneath. Small racemes of fragrant white cruciform flowers are produced in late summer. The green oval fruits, each containing a single seed or stone, ripen to a glossy purple-black.

**Distribution**: South-west Asia and the Mediterranean region.
**Height**: 9m/30ft
**Shape**: Broadly spreading
**Evergreen**
**Pollinated**: Insect
**Leaf shape**: Narrowly obovate

*Left: The leaves are an instant guide to identification.*
*Far left: Olive oil is extracted from the fruit.*

## Chinese Privet

*Ligustrum lucidum*

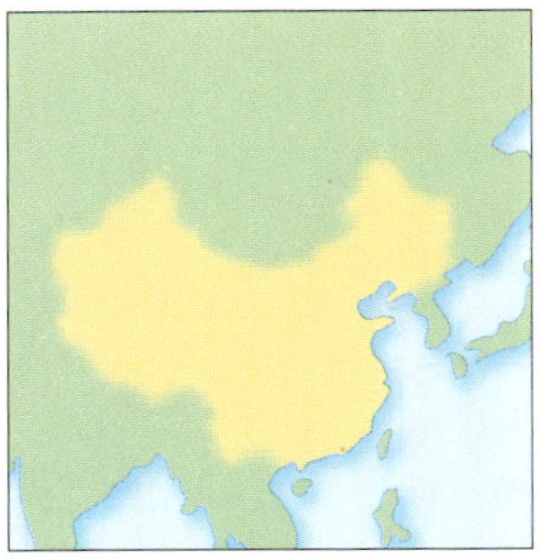

**Distribution**: China.
**Height**: 12m/40ft
**Shape**: Broadly spreading
**Evergreen**
**Pollinated**: Insect
**Leaf shape**: Ovate

This handsome, medium-sized tree is the finest of all the privets and from a distance, with its lush, lustrous evergreen leaves, it resembles a tropical rainforest tree. It grows naturally in the mountains of central and western China, particularly in hillside woodland above river valleys. It was introduced into Europe by the botanist Sir Joseph Banks in 1794. It is very hardy and well suited for urban planting as it tolerates air pollution and compacted dry soils. Chinese privet will also tolerate salt spray and is therefore suitable for coastal planting.

**Identification**: The bark is grey and smooth, even in maturity. The leaves, which taper to a fine point, are ovate, 10cm/4in long and 5cm/2in broad, bronze when first produced, becoming glossy dark green above and paler and matt beneath as they mature. The flowers are small, creamy-white, fragrant and profusely borne in large, upright, conical panicles, 20cm/8in long, which stand proud of the surrounding foliage, from mid-summer to early autumn. The fruit is a small blue-black berry.

*Right: Small blue-black berries are produced in autumn.*

OTHER SPECIES OF NOTE

**Karo** *Pittosporum crassifolium*
This small evergreen tree is native to the Kermadec islands in the Pacific Ocean and the North Island of New Zealand. One of the hardiest pittosporums in cultivation, it has thick, leathery, oval, lustrous leaves. These are grass-green above and covered with white pubescence beneath. Karo grows particularly well in coastal locations.

**Kohuhu** *Pittosporum tenuifolium*
As its common name may suggest, this is an evergreen New Zealand tree. Its name comes from the Maori language and roughly translates as "the black one". This is a reference to the colour of both the bark and the wood. The tree is native to the North and South Islands of New Zealand and has been cultivated in Europe for at least 150 years. It is a columnar tree, to 10m/30ft, with glossy, pale green, wavy-margined leaves, which are of variable shape (oblong to elliptic) and borne on black shoots. It makes an attractive hedging plant and is favoured for use in floral arrangements.

***Pittosporum turneri***
This New Zealand evergreen tree is one of the hardiest of all the pittosporums. It is native to only the North Island and has been widely cultivated in Europe since the late 1800s. It is a small tree, up to 6m/20ft tall, with an upright ascending trunk and branches which, when young, display a contorted, twisted habit. The leaves are pale lustrous green and 2.5–5cm/1–2in long.

## Phillyrea

*Phillyrea latifolia*

This elegant small evergreen tree, which is a member of the olive family (Oleaceae), is sometimes mistaken for holm oak, *Quercus ilex*. Like holm oak, phillyrea is native to southern Europe, especially around the Mediterranean coastline. It has been in cultivation in northern regions of Europe, such as Britain, since 1597. It is widely planted to provide shelter in exposed locations, especially along coastal strips, and is quite often used as a substitute for holm oak where space is limited.

**Identification**: In overall appearance, this tree resembles a small evergreen oak, or even an olive. Occasionally it is clipped to form a dense hedge. The bark is pewter-grey, smooth even in maturity, and relatively thin. Each long, spreading branch is covered with masses of small, glossy, dark green, opposite leaves. These are thick and leathery, elliptic to ovate, 5cm/2in long, have a bluntly toothed margin and are pale green on the underside. The flowers are small, creamy or yellowish-white, produced in clusters in the leaf axils in late spring.

**Distribution**: Southern Europe and Asia Minor.
**Height**: 10m/33ft
**Shape**: Broadly spreading
**Evergreen**
**Pollinated**: Insect
**Leaf shape**: Elliptic to ovate

*Below: The elongated leaf.*

*Right: In warm regions the flowers are followed by tiny, round, blue-black fruits.*

## Fringe Tree

*Chionanthus virginicus*

**Distribution**: Eastern United States.
**Height**: 10m/33ft
**Shape**: Rounded
**Deciduous**
**Pollinated**: Insect
**Leaf shape**: Elliptic

*Right: The fruit is a bloomy, egg-shaped, blue-black berry, up to 2.5cm/1in long.*

This attractive small tree inhabits moist woods and riverbanks in its natural range. It was introduced into Europe in 1736 and is cultivated in parks and gardens across the continent. It is sometimes known as old man's beard because of its feathery, beard-like flowers, which are borne in early summer. It is a member of the olive family and the connection can be seen in its olive-like fruits.

**Identification**: The bark is silver-grey and smooth becoming vertically fissured with age. The fringe tree has untoothed, lustrous leaves, which are elliptic, narrow, up to 20cm/8in long and 7.5cm/3in across, tapering to a short point. They are borne on short petioles (leaf stalks), to 2.5cm/1in long, and turn a clear butter yellow in autumn. The feathery flowers are creamy-white and slightly fragrant and are borne abundantly in conical erect panicles in summer.

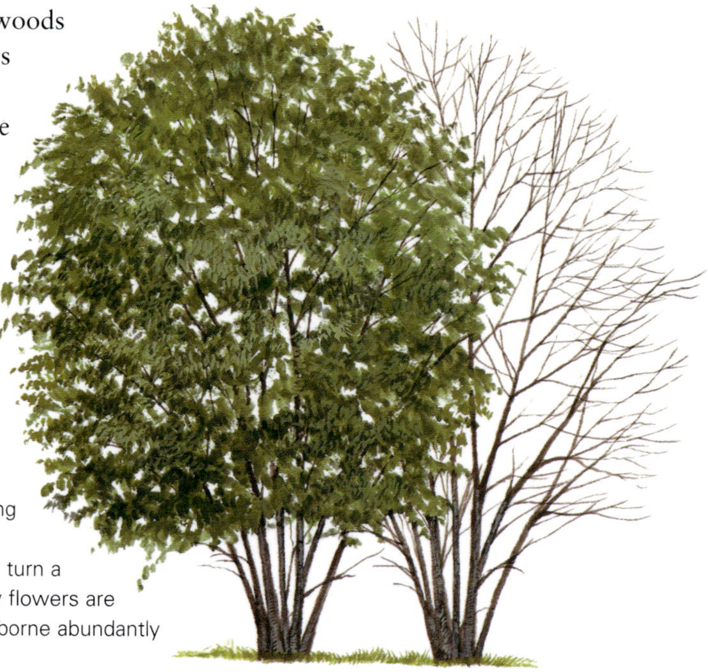

# ELMS

*The Ulmaceae family contains about 15 genera and 140 species of mainly deciduous trees. They thrive in all but the poorest of soils and are widespread throughout most temperate regions of the Northern Hemisphere, including Europe, North America and Asia – except, that is, where they have been affected by the fungus* Ophiostoma novo-ulmi, *which causes Dutch elm disease.*

## European Hybrid Elm

Dutch elm *Ulmus* x *hollandica*

This is an extremely variable, naturally occurring hybrid between two European elms, the wych elm, *U. glabra*, and the smooth-leaved elm, *U. minor*. It is found throughout western Europe, including Britain, and is prevalent in France, Belgium, Holland and Germany. This hybrid has resulted in several vigorous cultivars common in western Europe, including *U.* x *hollandica* 'Vegeta', *U.* x *hollandica* 'Major', and *U.* x *hollandica* 'Belgica'.

**Identification**: The bark is grey-brown, becoming regularly and vertically fissured in maturity. The leaves are up to 12.5cm/5in long and 7.5cm/3in broad, rough to the touch and unequal at the leaf base, where the petiole attaches the leaf to the twig. The flowers are small and red and are produced in obvious clusters on the shoots in early spring.

**Distribution**: Western Europe.
**Height**: 30m/100ft
**Shape**: Broadly columnar
**Deciduous**
**Pollinated**: Wind
**Leaf shape**: Ovate to elliptic

*Far left: The fruit is an oval-shaped, winged seed 2.5cm/1in long.*

*Left: The leaf.*

## European White Elm

Fluttering elm *Ulmus laevis*

**Distribution**: Central and eastern Europe to western Asia.
**Height**: 30m/100ft
**Shape**: Broadly spreading
**Deciduous**
**Pollinated**: Wind
**Leaf shape**: Obovate

This large, widely spreading elm is known as the fluttering elm because of its long-stalked flowers and fruit, which shiver in the slightest breeze. It is widely cultivated throughout eastern Europe, particularly in Russia, where it is planted alongside railway lines to protect them from snowdrifts. *Laevis* means "smooth" and refers to the smooth bark that this elm maintains into maturity. It has hard, durable timber, which is moisture-resistant even when in contact with water, and at one time it was favoured for use in the construction of waterwheels.

**Identification**: The bark is smooth, dull grey-brown and in old trees is covered in a shallow network of broad, smooth ridges. The leaves are double-toothed around the margin and distinctly oblique at the base. There may be up to 17 distinct veins on the long side of the leaf and only 14 on the short side. The winter leaf bud is a distinctive orange-brown and sharply pointed. Both flowers and fruit are held on long, pendulous stalks.

*Far right: After flowering pendulous bunches of oval-shaped individual paper-like seeds are produced in summer.*

OTHER SPECIES OF NOTE

**Cornish Elm** *Ulmus minor* var. *cornubiensis*
The Cornish elm is believed to be a smaller-leaved variety of the smooth-leaved elm, *U. minor* subsp. *minor*, and grows in isolated populations in north-west France and south-west England. It is a large tree with a dense, conical shape and sharply ascending branches. There is a suggestion that it survived the last Ice Age on land that is now submerged off the south coast of Cornwall. However, others believe it was introduced from France during the Anglo-Saxon period.

**Wheatley Elm** *Ulmus minor* 'Sarniensis'
This cultivar may also be a descendent of the smooth-leaved elm, *U. minor* subsp. *minor*. It has a narrower, denser crown and broader leaves than the Cornish elm. It occurs in all the Channel Islands, but Dutch elm disease has reduced it to mainly small, isolated clumps of sucker re-growth. A clone introduced from northern France has fared better and has produced narrow conical trees, ideal for roadside and coastal planting.

***Ulmus laciniata***
This east Asian elm is closely related to the European wych elm, *U. glabra*, and was introduced into Europe in 1905. It is a small tree in cultivation although in the wild it may reach 30m/100ft. Its large obovate leaves, up to 8in/20cm long, are rough to the touch and irregularly and sharply double-toothed. In Asia its fibrous inner bark (bast) is used to make rope and baskets.

## English Elm

*Ulmus minor* var. *vulgaris* (formerly *U. procera*)

This magnificent, large stately tree has for centuries been an inseparable part of the English landscape. Ironically, it is now believed that it is not truly native but was introduced from north-west Spain in early Neolithic times. Nevertheless, it was for centuries considered the quintessential English tree, but in the late 20th century it was severely affected by Dutch elm disease and populations of mature trees were virtually eradicated from western Europe.

**Distribution**: South-west Europe.
**Height**: 35m/115ft
**Shape**: Broadly columnar
**Deciduous**
**Pollinated**: Wind
**Leaf shape**: Ovate to oval

**Identification**: The bark is grey-brown and regularly fissured into rectangular plates. The leaves are 5–10cm/2–4in long, unequal at the base, pointed at the tip and coarsely toothed around the margin. Each has 10–12 pairs of veins. Small clusters of red flowers appear in early spring and are followed by oval-shaped, paper-thin winged seeds.

*Below: The leaves are dark green above, pale green beneath and rough to the touch.*

*Below right: The winged seeds are 2.5cm/1in long.*

## Siberian Elm

Dwarf elm *Ulmus pumila*

This hardy medium-size elm has a natural range that extends from Tibet in the south to eastern Siberia in the north. The further north it grows, the shrubbier the plant's habit becomes. It was introduced into cultivation in Europe in 1870; most specimens in cultivation are derived from taller southern specimens. It has some resistance to Dutch elm disease and grows better than any other elm in cold, dry, poor soil conditions. It has therefore become very popular for planting in the prairie lands of the mid-west United States.

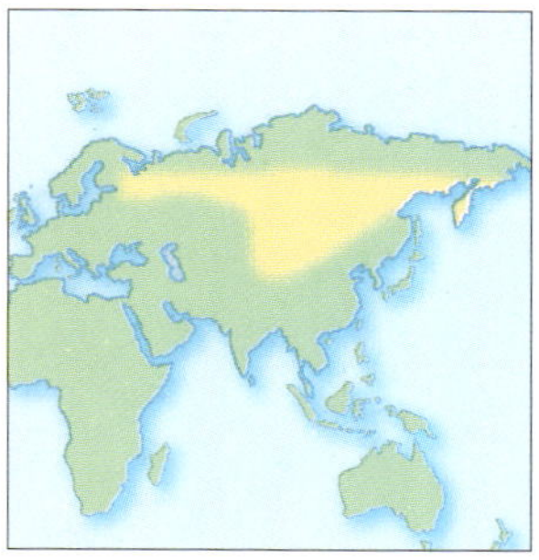

**Distribution**: Northern Asia.
**Height**: 20m/65ft
**Shape**: Broadly columnar
**Deciduous**
**Pollinated**: Wind
**Leaf shape**: Elliptic to ovate

**Identification**: The bark is grey-brown and smooth at first, becoming roughly ridged with long vertical fissures. The leaves are elliptic to narrowly ovate, 5cm/2in long and 2.5cm/1in broad, almost equal at the base (which is unusual for an elm), sharply but regularly toothed around the margin, dark green and smooth on the upper side with some pubescence beneath. The flowers are small and red and are borne in clusters on the shoots in early spring. The fruit is an oval-winged seed.

*Left: The leaves are elliptic to narrowly ovate and regularly toothed.*

*Left: Each seed is contained in a paper-thin oval-shaped wing.*

# ELMS AND NETTLE TREE

*The Ulmaceae family includes, in addition to elms, the North American nettle tree and several zelkovas. All tend to be large trees, perfectly hardy in Europe and quite often planted in extensive landscapes such as parkland and arboreta. In some cases they have been used as a replacement for elms lost to Dutch elm disease.*

## Wych Elm

*Ulmus glabra*

This tough medium-size tree survives particularly well in exposed coastal areas and on mountain slopes. It has strong, dense timber, which is very resistant to decay when immersed in water. For centuries, hollowed-out elm branches were used for water pipes, waterwheel paddles and boat building.

**Identification**: The overall shape is of a short, stocky stem surmounted by a widely spreading, open crown. The bark is grey-brown and in maturity uniformly lined with vertical fissures. The leaves are up to 20cm/8in long and 10cm/4in broad. They are coarsely and doubly toothed, with unequal sides at the base of each leaf where it joins the stalk. The upper surface of the leaf is dull green and rough textured; the lower surface is lighter in colour and heavily furred. The flowers appear in late winter before the leaves. The seed is carried in the centre of a flat, papery, hairy disc, known as a samara. The samaras are 5mm/¼in across and are borne in clusters.

**Distribution**: Europe from Spain to Russia, including western Scandinavia.
**Height**: 30m/100ft
**Shape**: Broadly spreading
**Deciduous**
**Pollinated**: Wind
**Leaf shape**: Oval to obovate

*Left and right: The male flowers have bright red anthers and are very conspicuous in early spring, when copious amounts of pollen are released.*

## Nettle Tree

Common hackberry *Celtis occidentalis*

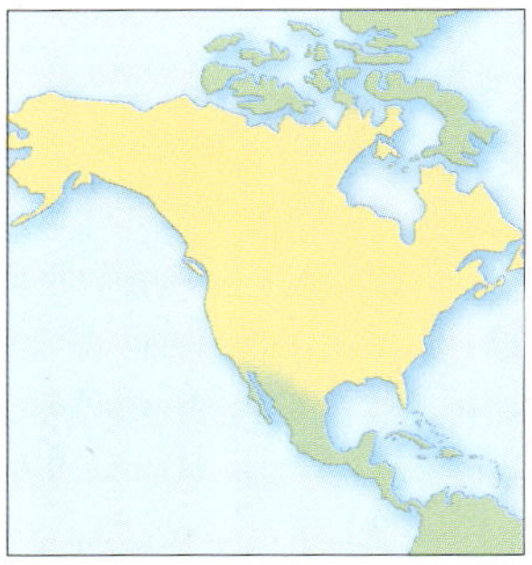

**Distribution**: North America.
**Height**: 25m/82ft
**Shape**: Broadly columnar
**Deciduous**
**Pollinated**: Wind
**Leaf shape**: Ovate

This medium-size tree, which is closely related to the elm, was introduced into Europe in the 17th century. It also grows prolifically across North America. It is also known as the hackberry because it produces a profusion of purple, edible, sweet-tasting berries, which are an important food source for birds.

**Identification**: The bark is light grey, smooth when the tree is young, becoming rough and corky with warty blemishes in maturity. The oval leaves are up to 12cm/4¾in long and 5cm/2in across, pointed, toothed at the tip and rounded at the base, where there are three pronounced veins. They are smooth, glossy rich green on top; lighter green and slightly hairy on the veining underneath. Both the male and female flowers are held separately on the same tree in spring – they are small and green, without petals, and appear in the leaf axils. The reddish fruit, ripening to purple-black, is a rounded berry approximately 1cm/½in across, borne on a thin green stalk, 2.5cm/1in long.

*Right: The nettle tree is named for its leaves, which resemble those of the stinging nettle.*

## Smooth-leaved Elm

*Ulmus minor* subsp. *minor*

As the common name suggests, this is an elm with a smooth, lustrous upper leaf surface, which easily distinguishes it from other European sub-species and varieties. It is still occasionally referred to as the European field elm, *U. carpinifolia*, which indicates the confusion that still surrounds the origins of these European elms. Much of this confusion stems from the way they were widely transported around Europe during the Bronze and Iron Ages. The leaves of this particular species were extensively used as cattle fodder.

**Identification**: The tree is large and broadly conical with an open habit. In later life the lower branches may become pendulous. The bark is light grey, sometimes silvery, and relatively smooth even in maturity. The narrowly oval leaves, up to 10cm/4in long, are unequal at the base, double-toothed, leathery, smooth and shiny. The flowers, which appear in early spring, are small, purple and produced in clusters over much of the tree.

**Distribution**: Europe, North Africa and parts of south-west Asia.
**Height**: 30m/100ft
**Shape**: Broadly conical
**Deciduous**
**Pollinated**: Wind
**Leaf shape**: Oval

*Left: Leaves have 10–13 pairs of veins.*

*Right: The seed is contained in a thin, paper-like wing.*

OTHER SPECIES OF NOTE

**Chinese Elm** *Ulmus parvifolia*
This small rounded tree, deciduous or semi-evergreen, is native to China, Japan, Taiwan and Korea, where it inhabits the lower rocky slopes of mountains. It appears to be resistant to Dutch elm disease but does not have the stature of either the English or American elm, so has not been widely planted in Europe.

**Japanese Elm** *Ulmus japonica*
This graceful, broadly spreading large tree, up to 30m/100ft, originates from north-east Asia, including Japan, and was introduced into Europe in 1895. It has downy twigs that sometimes develop corky wings, pale grey-brown fissured bark, obovate, double-toothed, dark green, rough leaves, 10cm/4in long, and small red flowers, which are borne in clusters in early spring.

**Keaki** *Zelkova serrata*
Native to China, Japan and Korea, this potentially large (40m/130ft), broadly spreading member of the elm family was introduced into Europe by the English nursery Veitch & Sons in 1862. Since then it has been widely planted in parks, gardens and arboreta right across northern and central Europe. It has attractive brown bark, which flakes to reveal fresh orange bark beneath, and deciduous ovate to oblong leaves, up to 12cm/4½in long, which are sharply toothed and taper to a point.

## Caucasian Elm

*Zelkova carpinifolia*

This slow-growing, long-lived forest tree has a very distinctive and pleasing shape. It is similar to hornbeam, having a short trunk and a large, almost mop-head of dense, upright branching. The overall impression is of a tree that was once pollarded but has long since been left to grow unchecked.

**Identification**: The smooth, grey-buff bark flakes in maturity; the trunk has pronounced fluting and a buttressed base. The flowers are small and green and are borne in the leaf axils in spring. The fruit is an insignificant-looking, small, rounded pea-like capsule with pronounced ridging.

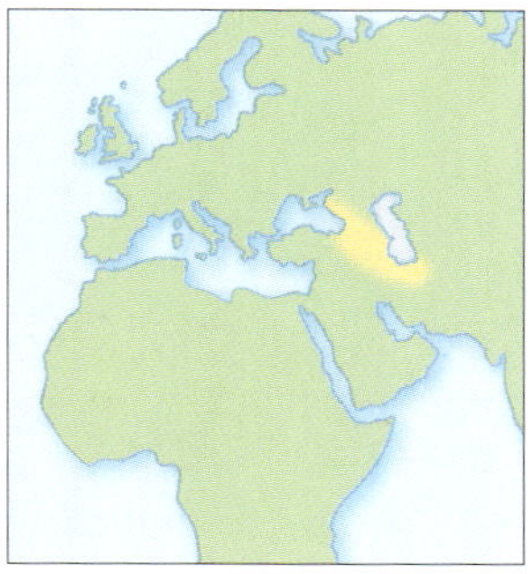

**Distribution**: Caucasus Mountains and northern Iran.
**Height**: 30m/100ft
**Shape**: Rounded
**Deciduous**
**Pollinated**: Wind
**Leaf shape**: Elliptic to oblong

*Left: The leaves are 10cm/4in long and have parallel veining.*

*Right: The flowers are small and easily missed.*

*Right: Bark.*

# MULBERRIES AND WALNUTS

*One of the main characteristics of these two families is that they both contain trees that produce edible fruits. The mulberry family, or Moraceae, includes both mulberries and figs, while the walnut family, or Juglandaceae, includes the common walnut. There are about 800 different species of fig; the majority of them are found in tropical and subtropical regions.*

## Common Walnut

*Juglans regia*

Although the walnut is naturally widespread, it has been distributed further by humans, who have cultivated the tree for its nuts since Roman times. The Romans introduced it to Britain and it spread to the USA in colonial times.

**Distribution**: From Greece in the west to central China and Japan in the east.
**Height**: 30m/100ft
**Shape**: Broadly spreading
**Deciduous**
**Pollinated**: Wind
**Leaf shape**: Pinnate

*Above: Flowers hang from the twigs.*

*Above: Individual leaflets have smooth edges and are up to 13cm/5in long.*

**Identification**: The bark is light grey with black fissures, creating narrow, rough ridges. The aromatic leaves have up to nine leaflets on each leaf stalk. They are bronze-pink when young, becoming deep green with a dull sheen in maturity. When the shoot is cut lengthways, a compartmentalized pith is revealed. Both male and female flowers are green catkins, appearing in late spring. The fruit is a round green husk, 5cm/2in across, not unlike a spineless horse-chestnut husk. Inside is the familiar, brown walnut.

## Japanese Walnut

*Juglans ailantifolia*

This attractive ornamental tree has the largest of all walnut leaves. The sticky fruit husk is poisonous but the nut inside is edible and has been a valuable food source in Japan for centuries. The timber is light and strong. It is grown in parkland across Europe for its ornamental leaves.

**Distribution**: Japan.
**Height**: 25m/82ft
**Shape**: Broadly spreading
**Deciduous**
**Pollinated**: Wind
**Leaf shape**: Pinnate

**Identification**: The grey-brown bark is fissured, segmenting into irregular plates in maturity. The leaves are pinnate and large, each with up to 17 slightly toothed leaflets 15cm/6in long. They are rich green and slightly hairy above; the undersides are very furry. The stout brown shoot is also hairy and very sticky. The male flowers are green, pendulous catkins, up to 30cm/12in long. Both male and female flowers appear on the same tree but are borne separately in late spring. The brown nut is contained in a green, hairy, sticky, rounded husk up to 5cm/2in across.

*Above: The nuts are borne on the tree in clusters of up to 20 in autumn.*

*Right: The female flowers are red upright catkins up to 10cm/4in long.*

## Black mulberry

*Morus nigra*

King James I of England decreed in 1608 that "every Englishman should cultivate a mulberry tree" as a way of establishing a native silk industry. Many took him at his word and dutifully planted mulberries. Unfortunately they planted the black mulberry, *M. nigra*, rather than the Chinese white mulberry, *M. alba*, the only species on which silkworms will flourish. Although not a good start for the silk industry, the decree did result in a land full of delicious black mulberry fruits.

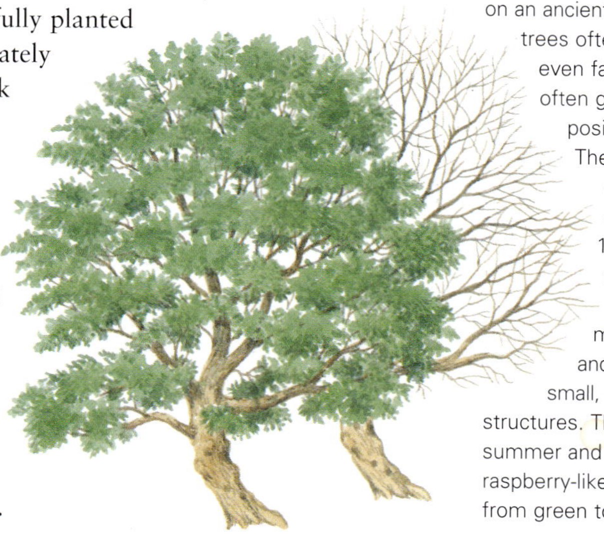

**Identification**: Black mulberry trees often appear older than they are. The orange bark takes on an ancient look early in life and trees often start to lean, or even fall over, when young, often growing in a prostrate position for many years. The leaves are bright green, over 12cm/4¾in long and 10cm/4in wide, rough and hairy with coarsely serrated margins. Both male and female flowers are small, green, soft, cone-like structures. They appear in early summer and are followed by raspberry-like fruits, which ripen from green to red to dark purple.

**Distribution**: Western Asia.
**Height**: 10m/33ft
**Shape**: Broadly spreading
**Deciduous**
**Pollinated**: Insect
**Leaf shape**: Ovate

*Right: A mulberry fruit looks like a small raspberry and appears at the end of a twig.*

OTHER SPECIES OF NOTE

**Common Fig** *Ficus carica*
Originally from south-west Asia, this large shrub or small spreading tree is now cultivated for its fruit throughout the temperate world. It has smooth, grey bark and distinctive, heavily lobed leaves. The flowers, which are fertilized by wasps, are small and green and borne on separate trees. The delicious fruit is heart-shaped and green, becoming purple-brown when ripe.

**Paper Mulberry**
*Broussonetia papyrifera*
This medium-size, broadly spreading tree is a close relative of the mulberries. It was introduced to Europe from Asia in 1750. It has attractive, coarsely toothed, hairy, purple-green leaves, which vary in shape from ovate to rounded and are deeply lobed. The tree's name derives from Japan, where its inner bark was traditionally used to make paper.

**Osage Orange** *Maclura pomifera*
This hardy North American tree, to 15m/50ft, produces large, heavy, rounded, pale yellow and wrinkled orange-like fruits in late summer and early autumn. It has ovate, glossy, bright green leaves, to 10cm/4in long, and orange-brown fissured bark, and was introduced into Europe in 1818.

## White Mulberry

*Morus alba*

This medium-sized Chinese tree has long been cultivated around the world as a food source for silkworms, which eat the leaves. White mulberry is believed to have been introduced into Europe in 1596 or earlier. It does produce an edible fruit but these fruits are normally less sweet and borne in less profusion than those of its more popular cousin the black mulberry, *M. nigra*. White mulberry has spawned a number of cultivars including *M. alba* 'Pyramidalis' which has the form and appearance of a Lombardy poplar.

**Identification**: The bark is orange-brown and the tree's shape becomes wide-spreading and rather ungainly in maturity. The leaves are variable, ovate to rounded, 20cm/8in long, toothed around the margin, bright green, lustrous above and paler beneath, turning bright yellow in autumn. The fruit can vary in colour from white through pink to purplish-red. Although edible it is not as pleasant to eat as the fruit of the black mulberry, *M. nigra*.

**Distribution**: Northern China.
**Height**: 15m/50ft
**Shape**: Broadly spreading
**Deciduous**
**Pollinated**: Insect
**Leaf shape**: Ovate

*Above: The fruit looks like an elongated raspberry.*

*Below: The leaves are used as cattle fodder*

# WING NUTS AND HICKORIES

*The Juglandaceae family contains seven genera and over 60 tree species, which grow throughout the temperate regions of North America, Europe and Asia. They include some of the fastest-growing of all deciduous trees. The leaves of all species are pinnate and the flowers are all catkins. Many of these trees produce edible fruit in the form of nuts.*

## Caucasian Wing Nut

*Pterocarya fraxinifolia*

The natural habitat of the Caucasian wing nut is damp woodland adjacent to rivers and marshland. It is a very fast-growing species, quite regularly achieving 3m/10ft of growth in a single year. Re-growth from coppiced stumps has been known to reach twice that height in a season. This tree has a habit of producing a profusion of sucker shoots around its base.

**Identification**: The bark is light grey, smooth in young trees and becoming fissured in maturity. The leaves are pinnate, up to 60cm/24in long and made of up to 23 slightly toothed, ovate to oblong dark green leaflets, up to 15cm/6in long. Winter buds are naked, with brown hairy bud leaves not unlike those of the wayfaring tree, *Viburnum lantana*. The small flowers, green with a red stigma, are carried in pendulous catkins up to 15cm/6in long in spring.

*Above: The nut-like fruit has a pair of semicircular wings.*

**Distribution**: Caucasus Mountains, the eastern shore of the Black Sea, southern shore of the Caspian Sea and into the northern provinces of Iran.
**Height**: 30m/100ft
**Shape**: Broadly spreading
**Deciduous**
**Pollinated**: Wind
**Leaf shape**: Pinnate

*Left: The nuts are borne in hanging "necklaces" (far left) up to 50cm/20in long. The flowers hang in pendulous catkins.*

## Chinese Wing Nut

*Pterocarya stenoptera*

Native to damp woodland and riverbanks throughout China, the Chinese wing nut was identified by the French missionary Joseph Callery in 1844 and introduced into Europe in 1860. It is easily distinguished from the Caucasian wing nut by the serrated wings on the leaf stalk in the spaces between each pair of leaflets.

**Identification**: This fast-growing tree has grey-brown bark, which becomes deeply fissured as it matures. The bright green, smooth leaves are pinnate, with up to 21 slightly toothed leaflets, from 10–20cm/4–8in long. Both male and female flowers are small and green, in separate pendulous catkins, each up to 5cm/2in long, on the same tree in spring. The seed is a nut, flanked by two green-pink, narrow erect wings. Seeds occur in long pendulous catkins up to 30cm/12in long, throughout the summer.

*Right: The nut-like seed is winged.*

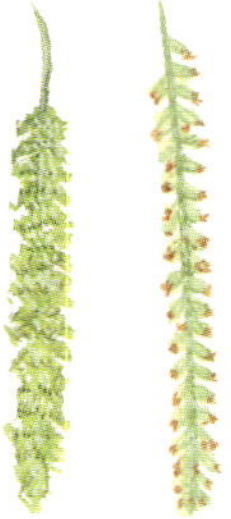

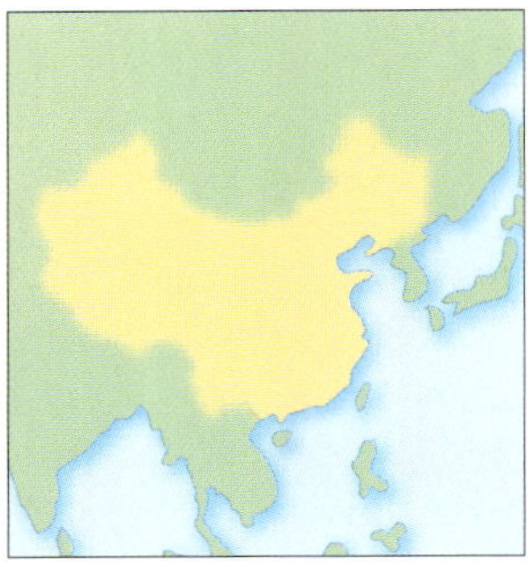

**Distribution**: China.
**Height**: 25m/80ft
**Shape**: Broadly spreading
**Deciduous**
**Pollinated**: Wind
**Leaf shape**: Pinnate

*Left: Flowers are hanging catkins.*

*Right: Nuts occur in necklace-like structures.*

## Bitternut Hickory

Bitternut *Carya cordiformis*

As the name suggests, the nuts are not palatable. They do, however, have a high oil content and were once crushed to produce lamp oil. "Hickory" comes from the Native American word for nut oil, *pawcohiccora*. It was introduced into Europe from North America in 1766. It differs from mockernut by its slender, bright yellow winter buds. It is not commonly planted outside its natural range.

**Identification**: The bark is grey and smooth, becoming thick and heavily ridged in maturity. The pinnate leaves have up to nine heavily serrated leaflets, which may be up to 15cm/6in long. They are deep green above and yellow-green beneath. In autumn they turn a rich golden-yellow. Both the male and female flowers are carried on catkins; the male catkins hang in threes and are up to 7.5cm/3in long; the females are short spikes at the shoot ends. They appear separately on the same tree in late spring. The thin-shelled fruit is a nut with an inedible bitter kernel.

*Right: The nuts are rounded and reddish-brown, in a thin husk that splits along four ridges.*

**Distribution**: East of a line from Minnesota to Texas.
**Height**: 30m/100ft
**Shape**: Broadly columnar
**Deciduous**
**Pollinated**: Wind
**Leaf shape**: Pinnate

### OTHER SPECIES OF NOTE

**Hybrid Wing Nut** *Pterocarya* x *rehderiana*
This hybrid, which was raised at the Arnold Arboretum of Harvard University, Boston, in 1879, is more vigorous than either of its parents, *P. fraxinifolia* and *P. stenoptera*. It has glossy, bright green pinnate leaves with up to 21 leaflets, which turn yellow in autumn. The bark is purple-brown and obliquely fissured. In late summer and autumn pendulous catkins, up to 45cm/18in long, contain winged seeds.

**Mockernut** *Carya tomentosa*
This North American hickory, sometimes called big-bud hickory, is highly valued for its timber, which has a number of uses. Because of the strength of its slender, straight trunk, and its ability to withstand impact, the wood has been used all over the world to make tool handles; it is also used for sports equipment such as hockey sticks. The wood gives off a fragrant smoke when burnt, and is widely used to smoke meats. The pinnate leaves also give off a pleasing aroma when crushed.

**Kingnut** *Carya laciniosa*
Also known as big shellbark, this medium-size, slow-growing, deciduous tree, growing to 20m/65ft, is native to the eastern United States, from where it was introduced into Europe in 1804. It has the largest compound leaves of any hickory, up to 60cm/24in long, and the largest nuts, up to 5cm/2in long, which are sweet and edible. The wood is much used to make tool handles. The botanical name *laciniosa* means 'with flaps' and successfully describes the way the shaggy bark curls away from the trunk.

## Shagbark Hickory

*Carya ovata*

**Distribution**: Eastern North America, from Quebec to Texas.
**Height**: 30m/100ft
**Shape**: Broadly columnar
**Deciduous**
**Pollinated**: Wind
**Leaf shape**: Pinnate

This large, vigorous tree differs from other hickories in having flaking, grey-brown bark, which curls away from the trunk in long thin strips up to 30cm/12in long, but stays attached to the tree at the centre point. This gives the whole trunk a shaggy, untidy but attractive appearance.

**Identification**: The leaves are pinnate, with five to seven leaflets on each leaf. Each leaflet is up to 25cm/10in long, yellowish green with a serrated edge for the top two-thirds. In autumn the leaves turn brilliant yellow. In winter, the bud scales curve away from the bud at the tip. Both the male and female flowers are small, yellowish-green and borne on pendulous catkins clustered in threes in late spring. In North America, the tree produces a profusion of nuts most years. Elsewhere crops are not so prolific. The white, sweet-tasting, kernel is contained in a green husk.

*Right: Fruit occurs at twig ends.*

*Below: The husk has four ridges.*

*Right: Leaves appear finger-like before they fill out.*

# BEECHES

*The Fagaceae family contains ten species of true beech, all of which occur in temperate regions of the world. They can be found in Asia, North America and Europe, including Britain. Beeches are some of the most majestic of all deciduous trees. They typically have smooth, thin, silver-grey bark and can attain heights in excess of 40m/130ft.*

## Common Beech

European beech *Fagus sylvatica*

The name "beech" comes from the Anglo-Saxon *boc* and the Germanic word *Buche*, both of which gave rise to the English word 'book'. In northern Europe early manuscripts were written on thin tablets of beech wood and bound in beech boards. Beech is widely used for hedging because, if trimmed, it retains its dead leaves in winter, providing extra wind protection. In summer, the leaves provide good browse for livestock.

**Identification**: The bark is silver-grey and remains smooth in maturity. The leaves are up to 10cm/4in long and 5cm/2in wide. They have a wavy, but normally untoothed, margin and a rather blunt point at the tip. In spring, juvenile leaves have a covering of hairs and are edible, with a nutty flavour. As they mature, the leaves become tough and bitter. Beech flowers are small: the female flowers are green and the males yellow, and they are borne in separate clusters on the same tree in spring. The fruit is an edible nut. Up to three nuts are contained within a woody husk, which is covered in coarse bristles.

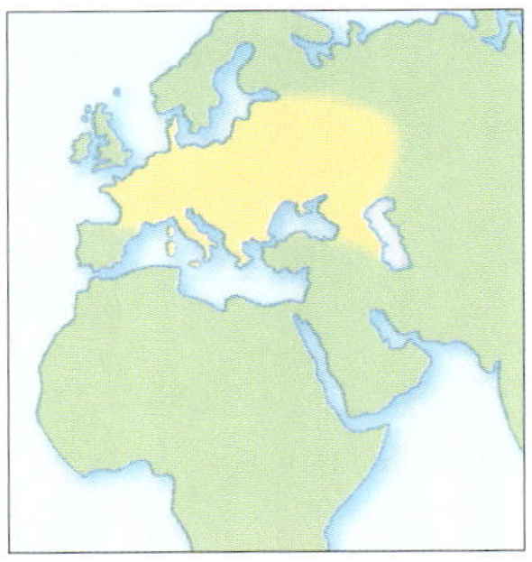

**Distribution**: Europe from the Pyrenees to the Caucasus and north to Russia and Denmark.
**Height**: 40m/130ft
**Shape**: Broadly spreading
**Deciduous**
**Pollinated**: Wind
**Leaf shape**: Ovate to obovate

*Right: The husks open in early autumn.*

*Far right: Mature leaves are smooth and have a rich colour.*

## Copper Beech

*Fagus sylvatica* 'Purpurea'

Neither a true species nor of garden origin, copper or purple beeches are 'sports' or quirks of nature. They were first seen growing naturally near Buchs, Switzerland, and in the Darney forest in the Vosges of eastern France in the 1600s. Seed collected from common beech may produce one in 1,000 seedlings with purple leaves.

**Identification**: The copper beech is similar to common beech. It is reported that copper beech may be slightly slower-growing and not quite as spreading in maturity as the common beech, but this is likely to have more to do with local growing conditions than to be a distinct characteristic of the tree.

**Distribution**: Switzerland and Vosges in eastern France.
**Height**: 40m/130ft
**Shape**: Broadly spreading
**Deciduous**
**Pollinated**: Wind
**Leaf shape**: Ovate to obovate

*Left: The greatest difference between common and copper beech is the leaf colour. The leaves are more oval in shape.*

## Oriental Beech

*Fagus orientalis*

The oriental beech is native to the forests of the Caspian Sea, the Caucasus, Asia Minor, Bulgaria and Iran. It is similar to common beech, and there is undoubtedly some hybridization between the two species on the boundary between their ranges. The oriental beech has larger leaves, with more pairs of veins, than common beech, and, given the right conditions, will develop into a larger tree. It was introduced into cultivation before 1880.

**Identification**: The bark is grey and smooth, occasionally becoming shallowly fissured in maturity. The leaves are elliptic to obovate, 12cm/4¾in long with up to 12 pairs of leaf veins and a wavy margin. They are dark, rich green above and paler beneath, becoming burnished gold in the autumn. Both the male and female flowers are small and yellow-green, carried in separate clusters on the same tree in spring. They are followed by woody, bristly fruit husks, containing up to three edible nuts.

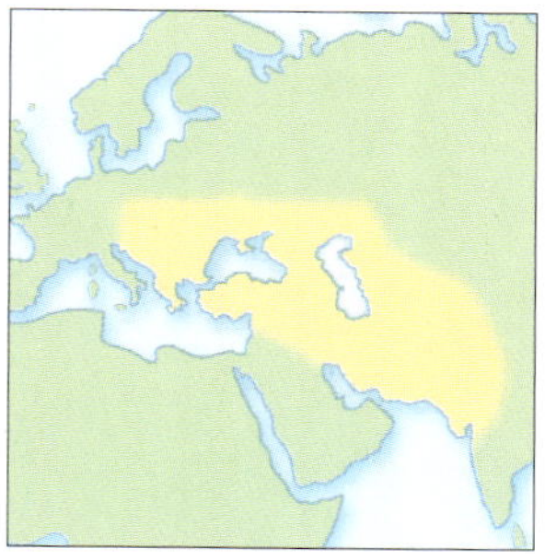

**Distribution**: South-east Europe and south-west Asia.
**Height**: 30m/100ft
**Shape**: Broadly spreading
**Deciduous**
**Pollinated**: Wind
**Leaf shape**: Elliptic to obovate

*Above left: The leaves of the oriental beech turn golden-yellow in autumn.*

*Left: The wavy-edged leaves are often obovate – at their widest near the tip.*

OTHER SPECIES OF NOTE

**Pendulous Beech** *Fagus sylvatica* 'Pendula'
Several cultivars of common beech have weeping foliage, but 'Pendula' has to be the best. It grows into a large tree with enormous pendulous branches, which droop from the main stem rather like elephants' trunks. Where they touch the ground they sometimes take root, sending up another stem that in turn begins to weep. Over time a large tent-like canopy can develop around the original tree. Probably the best example of this is in the grounds of the old Knap Hill nursery near Woking in Surrey, England.

**Dawyck Beech** *Fagus sylvatica* 'Dawyck'
This delightful narrowly columnar tree, shaped like a Lombardy poplar and up to 25m/82ft tall, was discovered on the Dawyck estate in southern Scotland in 1860. It was an overnight success with gardeners and nurserymen, providing the perfect solution for those who wanted a beech tree but lacked space.

**Fern-leaved Beech** *Fagus sylvatica* 'Aspleniifolia'
Sometimes referred to as cut-leaved beech, this is one of the most beautiful of all beech cultivars. It is slightly smaller than the common beech in height and spread, but is not suitable for the domestic garden. However, if there is space, it is well worth growing for its handsome and variable foliage. The leaves may be long and narrow, 10cm x 6mm/4 x ¼in wide, deeply lobed, or anywhere in between. The effect is of hazy, soft green feathery foliage forming an irregular, rounded canopy.

## Japanese Beech

Siebold's beech *Fagus crenata*

This species is sometimes called Siebold's beech after the 19th-century German naturalist Philipp Franz von Siebold, physician to the Governor of the Dutch East India Company's Japanese trading post at Deshima. Siebold first identified the tree in Japan, where it forms considerable forests from sea level to 1,500m/4,900ft. It was introduced to the West in 1892.

**Identification**: Similar to common beech, this tree differs mainly in its more obovate leaf shape and a small leaf-like structure found at the base of each seed husk in early autumn. The bark is silver-grey and smooth, even in older trees. Leaves are up to 10cm/4in long and 5cm/2in wide with a bluntly toothed margin with fine hairs. Each leaf has 7–11 pairs of veins. The leaf stalk is 1cm/½in long. Leaves turn an 'old gold' colour in autumn. The seed husk is 1.5cm/⅔in long and covered in long bristles.

**Distribution**: Japan.
**Height**: 30m/100ft
**Shape**: Broadly spreading
**Deciduous**
**Pollinated**: Wind
**Leaf shape**: Ovate to obovate

*Right: A Japanese beech leaf and mast.*

## Japanese Beech

*Fagus japonica*

Confusingly, the name "Japanese beech" is also applied to *F. crenata*, which is a more widespread native Japanese beech and therefore probably more worthy of the name. The actual trees are rarely confused: *F. japonica* is a smaller tree, often producing multiple stems, which give it a shrubby appearance, and its leaves are more oval and broad than those of *F. crenata*. It was introduced into Europe in 1907, when it was sent to the Royal Botanic Gardens, Kew, from the Arnold Arboretum in Boston. However, a century later it is still relatively uncommon in cultivation.

**Distribution**: Japan.
**Height**: 25m/80ft
**Shape**: Broadly spreading
**Deciduous**
**Pollinated**: Wind
**Leaf shape**: Oval to ovate

**Identification**: The bark is silver-grey and smooth, even in maturity. The leaves are a rich bright green, oval in shape and tapering abruptly to both ends. They are up to 10cm/4in long and 5cm/2in broad and when young are covered with fine hairs, which may persist on the underside of the leaf. There are normally 10–14 pairs of distinctive parallel leaf veins branching from a yellow-green midrib. The fruit is a nut that is contained within a woody, bristly husk, which ripens and opens in autumn. *F. japonica* rarely makes a large tree when grown in Europe and quite often remains as a large shrub, which is probably the reason for its lack of popularity in gardens and parks. It is, however, almost always represented within botanic gardens and arboreta.

*Right: The leaves are bright green, oval and taper towards both ends.*

*Right: The seed is contained in a woody, bristly husk.*

## Chinese Beech

*Fagus lucida*

This small tree was discovered by the Irish physician and plant collector Augustine Henry, in the province of Hubei, China, in 1887. It was almost 20 years later that Ernest Wilson introduced it into Britain and Europe. It has probably the most distinctive leaf surface of any green beech, being bright glossy green on both surfaces. Examples of this species are found in most botanic gardens and arboreta but it is rarely cultivated elsewhere.

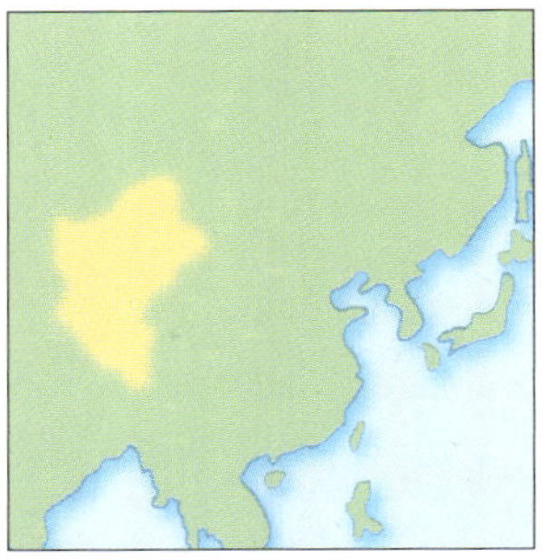

**Distribution**: Western China.
**Height**: 10m/33ft
**Shape**: Broadly spreading
**Deciduous**
**Pollinated**: Wind
**Leaf shape**: Ovate

**Identification**: In maturity this tree has a broad, flattened or rounded crown. The bark is smooth and light grey. The ovate leaves are up to 10cm/4in long and 5cm/2in wide, glossy green on both surfaces and with distinctive leaf veins that end in small teeth at the wavy margin. Both male and female flowers are small and yellow-green, carried in separate clusters on the same tree in spring. The fruit is contained in a woody husk, the surface of which is covered with small, blunt spines.

*Right: The leaves are glossy green on both surfaces and have distinctive leaf veining.*

*Below: The fruit.*

## Engler Beech

Chinese beech *Fagus engleriana*

In the wild, in central China, this medium-sized tree almost invariably divides near the base into multiple stems, but this occurs less in cultivation. It is an attractive tree, which although uncommon in Britain and Europe deserves to be more widely planted. It was introduced into Europe in 1911. There is a fine specimen in southern England, which was planted in 1928 and is now 18m/60ft tall.

**Identification**: The silver-grey bark is smooth, even in maturity, and the tree is quite often multi-stemmed. The leaves are a glaucous sea-green, rounded at the base, pointed at the tip, and up to 10cm/4in long with a wavy margin. There is some hair on the leaf underside and the leaf stalk is longer than on most beeches, being up to 1.25cm/½in long. The fruits, which are nuts contained in a woody, bristly husk, are also borne on a long stalk, up to 5cm/2in long.

*Right: The leaves of the Engler beech turn to a warm russet colour in autumn.*

**Distribution**: Central China.
**Height**: 18m/60ft
**Shape**: Broadly spreading
**Deciduous**
**Pollinated**: Wind
**Leaf shape**: Ovate to oval

OTHER SPECIES OF NOTE

***Fagus sylvatica* 'Zlatia'**
This tree was cultivated in Serbia, and its name is derived from the Serbian word for gold, *zlato*. The original tree was found growing among native beech in the 1880s. It is slow-growing, with soft yellow young leaves, which gradually turn green in summer.

**Round-leaved Beech** *Fagus sylvatica* 'Rotundifolia'
As its name suggests, this upright, neat-looking, medium-sized beech has dainty round leaves, 2.5cm/1in in diameter, which are borne on strongly ascending branches. Bred in England around 1872, it has remained popular in cultivation across Europe ever since.

***Fagus sylvatica* 'Purpurea Tricolor'**
This small, rather attractive tree, developed in France in the 1880s, has purple leaves blotched with cream, with a purple-pink margin. They are smaller and narrower than in common purple beech. The variegation is most striking when the leaves are young. At one time this cultivar was considered identical to the form known as 'Roseomarginata'.

***Fagus sylvatica* 'Rohan Gold'**
This uncommon cultivar is similar to Rohan's beech but with yellow-green leaves, which emerge from bud in spring a bright golden yellow before fading in summer. It is a relatively recent variety, having been raised in Holland in 1970. It is a vigorous small tree, which will reach a height of around 13m/42ft.

## Rohan's Beech

*Fagus sylvatica* 'Rohanii'

This beautiful, purple, cut-leaved beech has leaves reminiscent of those of the fern-leaved beech, *F. sylvatica* 'Aspleniifolia'. It was raised at Sychrov, in Czechoslovakia, in 1888, and was commercially available in 1908. It is a popular addition to parks, gardens and arboreta throughout northern and western Europe. There are a few other purple cut-leaved beeches in cultivation, but this is the best cultivar.

**Identification**: This is a slow-growing, straight tree that will ultimately reach 15m/50ft in height. The bark, as with most beeches, is silver grey and smooth. The leaves are unlikely to be confused with those of any other tree. They are deep red-purple, sometimes tinged with green or brown; the leaf margin is deeply cut into triangular teeth, which may themselves carry serrations. The leaf veins and the leaf stalk are prominently red.

**Distribution**: Originally raised in Czechoslovakia.
**Height**: 15m/50ft
**Shape**: Broadly spreading
**Deciduous**
**Pollinated**: Wind
**Leaf shape**: Ovate cut-leaved

*Below: The leaves are the most distinctive feature.*

# SOUTHERN BEECHES

*This relatively little-known group of trees comes from temperate regions of the Southern Hemisphere, and is known by the genus name* Nothofagus. Nothos *comes from the ancient Greek for "spurious" or "false" and* fagus *means "beech", so the name can be translated as "false beeches". Although the trees are similar to beeches there are some differences – many* Nothofagus *are evergreen and have smaller leaves.*

## Roble Beech

*Nothofagus obliqua*

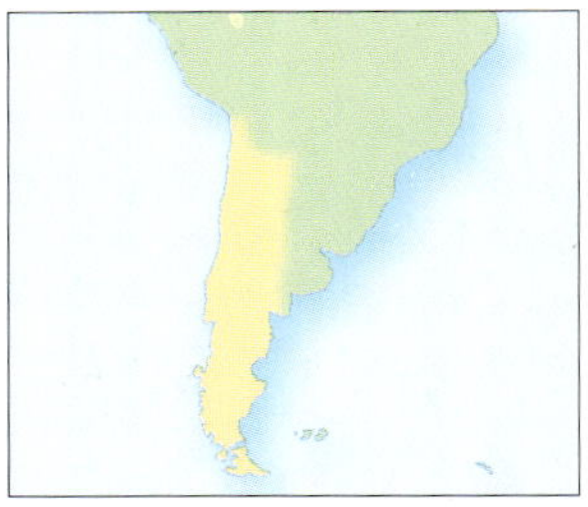

**Distribution**: Argentina and Chile.
**Height**: 35m/115ft
**Shape**: Broadly columnar
**Deciduous**
**Pollinated**: Insect
**Leaf shape**: Ovate

*Roble* is Spanish for oak, and in some respects this large South American tree does resemble a European deciduous oak. Its timber is oak-like, tough and durable, and over the years has been used for ship-building, interior joinery and furniture. It is the most warmth-loving of all southern or false beeches and thrives in the Mediterranean climate.

**Identification**: The grey bark becomes cracked and fissured with age. The tree has dark green ovate to oval leaves; they are blue-green on the underside, roundly toothed, up to 7.5cm/3in long, and have 8–11 pairs of distinct veins. In autumn, the leaves turn golden yellow. The flowers are small and green: the males are borne singly and the females in threes, in late spring.

*Far right: The fruit is a bristly brown husk, similar to that of beech, containing three nuts.*

## Rauli

Raoul *Nothofagus nervosa*

Also known as *N. procera*, this large, deciduous forest tree has upswept branching and heavily veined leaves. The name was given by early Spanish settlers who saw its grey, smooth bark and named it after the Spanish word for beech. It is a fast-growing tree that produces good quality timber and is being planted for forestry purposes in temperate regions of the Northern Hemisphere.

**Distribution**: Central Chile and western Argentina.
**Height**: 25m/80ft
**Shape**: Broadly conical
**Deciduous**
**Pollinated**: Insect
**Leaf shape**: Oblong to ovate

**Identification**: The dark grey bark becomes heavily fissured as the tree matures. The leaves, up to 10cm/4in long and 5cm/2in across, are easily distinguished from those of other *Nothofagus* because they have 14–18 pairs of deep veins, though they could at first glance be mistaken for hornbeam, *Carpinus* spp. The leaves are positioned alternately along the shoots; they are bronze-green above and paler beneath, with some hair on the midribs and veins. The fruit is a four-valved husk about 1cm/½in long, containing three small nuts.

*Left and right: The long, elegant leaves hang heavily. Unlike those of other members of this genus, they have up to 18 pairs of deep veins.*

## Silver Beech

*Nothofagus menziesii*

This slightly tender, graceful tree is native to both islands of New Zealand, where it grows on mountain slopes up to 1,000m/3,280ft above sea level. It is similar to the Australian species, *N. cunninghamii*. Both are far distant from the main populations of the *Nothofagus* genus, which are to be found in South America. Their occurrence here is evidence of the fact that the continents of Australasia and South America were both part of the ancient vast supercontinent. They are well represented in Europe.

*Right: The dark green leaves are distinctively toothed around the margin.*

**Identification**: When young the bark is like that of the *Prunus* species. It starts silvery-white and dulls to grey in maturity. The evergreen leaves are ovate to diamond shaped, doubly round toothed and up to 1cm/½in long. Both the leaf stalk and the shoot are covered in fine yellowish-brown hairs. The flowers are inconspicuous, being small and green. After flowering, four-lobed woody husk-like fruits develop. Each husk normally contains three small nuts.

**Distribution**: New Zealand.
**Height**: 30m/100ft
**Shape**: Broadly columnar
**Evergreen**
**Pollinated**: Insect
**Leaf shape**: Ovate

### OTHER SPECIES OF NOTE

**Dombey's Southern Beech**
*Nothofagus dombeyi*

A native of Chile and Argentina, introduced into Europe in 1916, this vigorous, evergreen tree is one of the most elegant of all South American temperate trees. In maturity it resembles an old cedar in form, and as a juvenile its glossy, dark green leaves in combination with bright red male flowers make it extremely striking.

**Antarctic Beech**
*Nothofagus antarctica*
This elegant, fast-growing, small to medium-sized tree, to 15m/50ft, is native to South America, inhabiting mountainsides from Cape Horn to northern Chile. Also known as *nirre* in Chile, it is deciduous and fast-growing, with ovate, crinkly, glossy, dark green leaves which, when crushed, emit a sweet, honey-like fragrance. In autumn the leaves turn orange-red and yellow. It was introduced into Britain in 1830, though most of the specimens now growing in Europe come from seed brought from Chile at a later date.

**Oval-leaved Southern Beech** *Nothofagus betuloides*
This hardy, South American columnar evergreen tree, to 25m/80ft, has very ornamental, oval, shiny dark green leaves, 2.5cm/1in long, which are carried densely on multiple twisted stems and branches. It was introduced into Europe in 1830 and is fairly common in botanic gardens and arboreta. It produces inconspicuous small green flowers in spring, which are followed by woody husk-like fruits, each containing three small brown nuts.

## Red Beech

*Nothofagus fusca*

This beautiful evergreen tree is native to both islands of New Zealand, from 37 degrees latitude southwards. It reaches large proportions in the wild. In cultivation in the Northern Hemisphere it seldom attains more than 25m/80ft in height. It is quite tender when young, being prone to frost damage, but becomes hardier with age.

**Identification**: The bark of red beech is smooth and dark grey when young, becoming flaky with lighter patches in maturity. The leaves are very distinctive because of their deep, sharply toothed margins. At 4cm/1½in long they are also bigger than those of any other evergreen *Nothofagus* species. They have long leaf veins, normally in three to four pairs. Although evergreen, the old leaves turn coppery-red before they eventually fall, hence the tree's common name. The leaf-stalk is covered in grey-brown hairs. The fruit is a four-lobed husk containing three nuts. The flowers are small, green and inconspicuous. Both male and female flowers are borne on the same tree in late spring.

*Left: The richly coloured, evergreen leaves of red beech are the largest of any southern beech species.*

**Distribution**: North and South Island, New Zealand.
**Height**: 30m/100ft
**Shape**: Broadly spreading
**Evergreen**
**Pollinated**: Insect
**Leaf shape**: Ovate to round

# CHESTNUTS

*The chestnut genus,* Castanea, *contains just 12 deciduous trees, all of which grow wild in temperate regions of the Northern Hemisphere. They are closely related to both the beech,* Fagus, *and oak,* Quercus, *genera. The majority are long-lived, large trees, which are drought-resistant and thrive on dry, shallow soils. They all have strongly serrated leaves and edible fruits in the form of nuts.*

## Japanese Chestnut

*Castanea crenata*

This small tree is native only to Japan but is cultivated across Europe in botanic gardens and arboreta. It was introduced into Europe in 1895, when seedlings were sent to the Royal Botanic Gardens, Kew, near London. In Japan it is a valuable food source and the tree is grown in orchards for its edible nuts, which are slightly smaller than those of the sweet chestnut, *C. sativa*.

**Identification**: The bark is a dull lead grey, smooth when young and becoming vertically fissured in maturity. The young shoots are covered in grey-white down, which persists well into the first winter. The leaves are oblong or lance-shaped, 7.5–17.5cm/3–7in long, up to 5cm/2in wide and heart-shaped at the base. The leaf margin is serrated, with forward-pointing small teeth with bristle-like points. The fruit is a fairly large edible nut, encased in a spiny green husk.

**Distribution**: Japan.
**Height**: 10m/33ft
**Shape**: Broadly spreading
**Deciduous**
**Pollinated**: Insect
**Leaf shape**: Oblong-lanceolate

*Right: The flowers are pollinated by insects.*

*Left: Each husk contains two or three chestnuts.*

## Sweet Chestnut

Spanish chestnut *Castanea sativa*

This fast-growing ornamental tree is native to warm, temperate regions around the Mediterranean and in south-western Asia. It has also been widely cultivated elsewhere, often introduced by the Romans, who valued the tree as a source of food for themselves and their animal stock.

**Identification**: As a young tree the bark is smooth and grey. Older trees develop spiral fissures, which immediately distinguish the tree from oak. The leaves are oblong, up to 20cm/8in long, sharply pointed at the tip and rounded at the base. The leaf margin is edged with coarse teeth, each tooth linking to a strong vein running back to the midrib. Each catkin may be up to 25cm/10in long. The fruit is a spiny greenish-yellow husk, up to 6cm/2½in across, with up to three edible brown nuts.

**Distribution**: Southern Europe, North Africa and south-west Asia.
**Height**: 30m/100ft
**Shape**: Broadly columnar
**Deciduous**
**Pollinated**: Insect
**Leaf shape**: Oblong

*Above left: The male and female flowers are borne on the same upright yellow catkin in summer, making it one of the last trees to come into flower.*

*Left and right: The chestnuts ripen in autumn.*

OTHER SPECIES OF NOTE

**American Chestnut** *Castanea dentata*
This majestic tree, similar to sweet chestnut, *C. sativa*, was once widespread throughout its native North America, but since the 1930s its population has been devastated by chestnut blight. It can be found in botanic gardens in western Europe.

**Golden Chestnut** *Chrysolepis chrysophylla*
Sometimes referred to as the golden chinquapin, this medium-size evergreen tree is native to the USA, from where it was introduced into Europe in 1844. The description 'golden' refers to the underside of the leaf, which is covered with bright golden hairs, a characteristic that immediately distinguishes this tree from other members of the Fagaceae family.

**Tanbark Oak** *Lithocarpus densiflorus*
As the name suggests, this medium-size, pyramidal, evergreen North American tree is closely related to oak and displays some of the characteristics of the oak family, such as the fruit, which is a bristly-cupped acorn. The leaves are, however, more characteristic of chestnut, as are the erect, creamy-yellow, spike-like flowers. It was introduced into Europe in 1874.

**Chinquapin** *Castanea pumila*
This deciduous small tree or large shrub was introduced into Europe in 1699 but is seldom grown outside botanic gardens and arboreta. It is distinguished from other chestnuts by the fact that it throws up suckers from the root system. Both the shoots and the underside of the leaves are covered with a white-grey down. The oblong leaves are coarsely toothed, up to 12cm/4¾in long. The egg-shaped nut is up to 2.5cm/1in long.

## Chinese Chestnut

*Castanea mollissima*

This Chinese tree is common throughout its native land from Beijing westwards, where it is normally found growing in moist woodlands on lower mountain slopes. It was introduced into Europe in 1908.

*Below: The nut and husk.*

**Distribution**: China.
**Height**: 25m/80ft
**Shape**: Broadly columnar
**Deciduous**
**Pollinated**: Insect
**Leaf shape**: Oblong to lanceolate

**Identification**: The bark is dark grey and smooth, becoming deeply fissured in maturity. The leaves are up to 20cm/8in long and 7.5cm/3in broad, rounded at the base and tapering and curving to a sharp point at the tip. They have forward-pointing, coarse teeth around the leaf margin and are glossy dark green above with lighter green undersides. Both male and female flowers are erect, creamy-yellow catkins, to 20cm/8in long, which appear on the tree in mid-summer. These are followed by edible, shiny red-brown fruits, which are contained within a bristly husk.

## Japanese Chinquapin

*Castanopsis cuspidata*

Of a genus of 120 species of evergreen trees, this is the only member that will grow outside the tropics. It is native to China, Korea and Japan, where it is common from Tokyo southwards. It is closely related to both chestnuts and oaks and displays the characteristics of both. It was introduced into Europe in 1830 but seldom attains anything like the dimensions to which it grows in its native lands. The largest specimen in Britain is growing in the grounds of Muncaster Castle, Cumbria, and is 10m/33ft tall.

**Identification**: The dark grey-brown bark is finely fissured from a young age. The oblong to ovate leaves are leathery, 10cm/4in long and 4cm/1½in broad, rounded at the base and drawn out into a slender, blunt tip, dark shiny green above and pale metallic green beneath. The leaf margin is sometimes bluntly toothed or wavy, more so towards the tip. The fruit is acorn-like, with the cup almost completely enclosing the seed, and borne in clusters of six to ten on the same stalk.

**Distribution**: Japan and China.
**Height**: 30m/100ft
**Shape**: Spreading
**Evergreen**
**Pollinated**: Insect
**Leaf shape**: Ovate

*Right: The fruit of the Japanese chinquapin is acorn-like, but the erect catkins resemble those of a chestnut.*

# OAKS

*There are almost 600 different species of oak,* Quercus, *in the world, the majority of which grow in the Northern Hemisphere. The oaks of Europe and Asia are on the whole slower growing and have less dazzling autumn leaf colour than their American cousins, but what they lose in terms of vibrancy and vigour, they more than make up for in diversity and longevity.*

## English Oak

Common oak *Quercus robur*

This majestic tree is one of the most familiar arboreal sights across Europe. It is usually a lowland species, growing best on damp, rich, well-drained soils. It is a long-lived tree, with many recorded veterans over 1,000 years old. In Britain it has hybridized with the sessile oak, *Q. petraea*, which is dominant in upland areas.

**Identification**: The bark is pale grey and smooth when young, quickly developing regular vertical fissures. The leaves, up to 10cm/4in long and 8cm/3in wide, either have no stalk or are on short stalks. The leaf margin is divided into three to six rounded lobes on each side, tapering towards the base where there are normally two smaller lobes. The upper surface is dark green, the underside glaucous. The fruit is an acorn, up to 4cm/1½in long, one-third of which is in a cup that is attached to the shoot by a long, slender stalk up to 10cm/4in long.

**Distribution**: Europe from Ireland to the Caucasus and north to Scandinavia.
**Height**: 35m/115ft
**Shape**: Broadly spreading
**Deciduous**
**Pollinated**: Wind
**Leaf shape**: Elliptic to obovate, lobed

*Right: The male flowers are long catkins that appear in late spring.*

## Turkey Oak

*Quercus cerris*

The Turkey oak is a tall, vigorous, deciduous tree with a straight stem and deeply fissured bark. It is native to central and southern Europe, from France to Turkey, but has been cultivated across the rest of Europe since at least the mid-18th century and has become naturalized in many regions. It is an ornamental species, quite often used for planting in avenues. With its straight, clean stem it might be expected to yield valuable timber; in fact this is not so, as in most cases the timber cracks and splits before it reaches the sawmill.

**Identification**: The bark is silver-grey, thick and deeply fissured in maturity. The leaves are variable in shape, up to 12cm/4¾in long and 7.5cm/3in broad, variably and deeply lobed. They are dark glossy green, which may give the impression that they are evergreen, which they are not. The male flowers are yellow-green pendulous catkins that appear in late spring to early summer, and the seed is an acorn 2.5cm/1in long, carried in a bristly cup.

**Distribution**: Central and southern Europe.
**Height**: 40m/130ft
**Shape**: Broadly spreading
**Deciduous**
**Pollinated**: Wind
**Leaf shape**: Elliptic to oblong and lobed

*Right: The leaves are a less defined version of those of* Q. robur.

## Sessile Oak

Durmast oak *Quercus petraea*

This valuable timber-producing tree is native right across Europe, from Ireland, Spain and Portugal to Turkey. In some countries, such as England, it grows alongside the English oak, *Q. robur*, and intermediate forms between the two species may result. Over the years its timber has been used for everything from barrels for wines and spirits to ship building.

**Identification**: The bark is grey with regular vertical fissures. The leaves are glossy dark green above and paler, and slightly hairy, beneath. Their shape is elliptic, to 13cm/5in long and 7.5cm/3in broad, tapering towards the base and with a margin that has rounded, untoothed lobes. Each leaf is attached to the twig by a yellow stalk up to 1.25cm/½in long. The male flowers appear in late spring and are yellow-green pendulous catkins. The seed is an unstalked acorn, one-third enclosed within a scaly cup.

**Distribution**: Europe.
**Height**: 40m/130ft
**Shape**: Broadly spreading
**Deciduous**
**Pollinated**: Wind
**Leaf shape**: Elliptic and lobed

*Left: The leaves and catkins have yellow stalks.*

*Right: Sessile oak acorns are stalkless and directly attached to the twig by the base of the acorn cup.*

## Holm Oak

Evergreen oak *Quercus ilex*

This domed, densely branched tree is native to southern Europe and the Mediterranean, where it grows to over 1,520m/5,000ft above sea level. It has been cultivated in northern Europe since the 16th century and is important for shelter in coastal areas. It will grow on sand and can withstand prolonged drought. In colder regions up to 50 per cent of its leaves may be shed and replaced in spring.

**Identification**: The bark is charcoal grey, smooth at first but quickly developing shallow fissures which crack into small, square plates. The leathery, tapered leaves, to 7.5cm/3in long and 5cm/2in broad, are occasionally toothed but more often entire; they are dark green above, almost black from a distance, and sage green and covered with fine pale grey hair beneath. The seed is a small, pointed acorn, to 5mm/¼in long, one-third encased in a fawn, scaly cup.

*Below: The male flowers are yellow pendulous catkins, which appear in early summer. The fruit, a nut, appears in the autumn.*

**Distribution**: Southern Europe.
**Height**: 30m/100ft
**Shape**: Broadly spreading
**Evergreen**
**Pollinated**: Wind
**Leaf shape**: Elliptic to ovate

OTHER SPECIES OF NOTE

**Red Oak**
*Quercus rubra*
The red oak is one of the largest and most widespread deciduous trees of eastern North America. It is found growing naturally from Nova Scotia to North Carolina but has been extensively planted throughout Europe. It has deeply cut leaves similar to those of scarlet oak, but not so glossy on the upper leaf surface.

**Pin Oak** *Quercus palustris*
Some specimens of this native of eastern and central North America are known to have exceeded 50m/165ft in height. The pin oak grows naturally in wet, swampy ground and is able to withstand flooding better than any other oak. It has been planted widely as a street tree, both in the USA and in Europe. Its name refers to the profusion of short side shoots that cover the lower part of the crown. The leaves are deeply cut.

**Scarlet Oak** *Quercus coccinea*
Scarlet oak is one of the most ornamental trees of eastern North America, contributing greatly to the autumn leaf-colour spectacle. It was introduced into Europe in 1691 and is common in large gardens, parks and arboreta, although it never quite colours with the intensity seen in North America. It has large, elliptic leaves, 15cm/6in long, which are cut into by angular,

## Cork Oak

*Quercus suber*

This medium-size evergreen tree has thick, corky bark and for centuries it has been used to make cork products such as stoppers for wine bottles. The cork oak is cultivated for this purpose in orchards, particularly in Portugal and Spain, and the outer bark is stripped from the trees on a rotation of 8–10 years. Removing it carefully does no damage to the living tissue beneath the bark, and some trees are known to be over 300 years old.

**Distribution**: Western Mediterranean and southern European Atlantic coast.
**Height**: 20m/65ft
**Shape**: Broadly spreading
**Evergreen**
**Pollinated**: Wind
**Leaf shape**: Ovate to oblong

**Identification**: The obvious distinguishing feature of this tree is its pale grey, prominently creviced, corky bark. The leaves are similar to those of holm oak, *Q. ilex*, although far more variable. They may be anything from oblong to oval in shape, up to 7.5cm/3in long, normally with an entire margin or with occasional serration. They are glossy dark green on top and covered with a grey down underneath. Although the tree is evergreen, the older leaves are shed and replaced in early summer. The acorns are up to 3cm/1¼in long, glossy brown and half encased in their cups.

*Right: The distinctive thick bark.*

*Below left: The male flower is carried on a weeping yellow catkin in spring.*

*Above: The leaves are up to 7.5cm/3in long. The acorn is quite slender and pointed at the tip.*

## Pyrenean Oak

*Quercus pyrenaica*

**Distribution**: South-west Europe and North Africa.
**Height**: 20m/65ft
**Shape**: Broadly columnar
**Deciduous**
**Pollinated**: Wind
**Leaf shape**: Elliptic

The Pyrenean oak, as its name implies, is native to mountain woods in the Pyrenees, where it is found growing up to 1,520m/5,000ft above sea level. It also grows naturally in other parts of Spain, France, Portugal, Italy, and across the Mediterranean in Morocco. It is closely related to the Hungarian oak, *Q. frainetto*. The Pyrenean oak was introduced into cultivation in 1822 and is popular in large tree collections. The cultivar 'Pendula', which has a graceful weeping habit, is often grown in preference to the species.

**Identification**: The tree develops a domed, spreading crown. The bark is pale grey and roughly fissured into knobbly rectangular or square plates. The leaves are variable in size, up to 20cm/8in long and 7.5cm/3in broad, and are deeply indented with wedge-shaped, forward-pointing lobes. The upper surface of the leaf is a glossy deep green and the underside is pale green and covered in dense grey hairs. The male flowers are borne in long, attractive golden catkins, which appear in early summer.

*Left: The leaves are deeply indented with wedge-shaped, forward-pointing lobes. The acorns are broad and rather squat and carried to half their length in a scaly woody cup.*

## Hungarian Oak

*Quercus frainetto*

This large, handsome, deciduous oak is native to south-eastern Europe, including Hungary, and was introduced into cultivation in 1838. It has probably the largest and most distinctive leaves of any oak growing in Europe. They are up to 20cm/8in long and 10cm/4in wide, deeply and regularly cut by many large, forward-pointing lobes. It has long been grown as an ornamental species in parks and arboreta across Europe.

**Identification**: The bark of the Hungarian oak is dark grey, rugged and heavily fissured in maturity. The leaves are the tree's most distinctive feature for identification purposes, and even in winter they still provide clues to its identity because they do not rot easily and persist on the ground around the base of the tree right through to the following spring. The male flowers are green-yellow pendulous catkins, which appear in late spring. The seed is a squat, egg-shaped acorn, with up to half its length encased in a cup.

**Distribution**: South-east Europe.
**Height**: 30m/100ft
**Shape**: Broadly spreading
**Deciduous**
**Pollinated**: Wind
**Leaf shape**: Obovate

*Left: The large leaves have seven lobes.*

*Right: Male catkins.*

OTHER SPECIES OF NOTE

**Algerian Oak** *Quercus canariensis*
Also known as Mirbeck's oak, this handsome tree is native to North Africa and south-western Europe, but it also survives well farther north, where it is regularly planted in gardens. It is columnar when young, broadening in maturity. Although not strictly evergreen, most of its neatly lobed leaves stay on the tree in winter.

**Armenian Oak** *Quercus pontica*
This rare, small tree or large shrub, which grows to 6m/20ft, is native to Turkey, Armenia and the land between the Black and Caspian seas into southern Russia, where it inhabits mountain woodlands. It is fully hardy and quite distinctive, with thick, large, deeply veined leaves, which are glossy and bright green, reminiscent of chestnut leaves, and borne on stout shoots.

**Lucombe Oak** *Quercus* x *hispanica* 'Lucombeana'
This magnificent semi-evergreen tree is a form of the hybrid Spanish oak, *Q.* x *hispanica*, which is itself a variable but natural hybrid between the cork oak, *Q. suber*, and the Turkey oak, *Q. cerris*. The original Lucombe oak was found growing close to a cork oak in the Exeter nursery of William Lucombe in 1762. He propagated the tree by grafting it on to Turkey oaks, and several thousand clones were distributed around Britain and Europe, some of which are still alive today. It is a large, spreading tree, resembling the Turkey oak, with glossy toothed leaves, dark green above and grey beneath. The bark is pale grey and deeply fissured, with a corky texture.

## Downy Oak

*Quercus pubescens*

This small to medium-size, sometimes shrubby-looking tree, which has a large natural range across southern and central Europe, is closely related to the sessile oak, *Q. petraea*, and the two can be quite difficult to tell apart. In general, the downy oak is smaller, with a shrubbier, more ragged outline, and the leaves and twigs tend to carry more pubescence, or down. It grows best in dry soil conditions with plenty of light and warmth. It is relatively uncommon in parks and gardens.

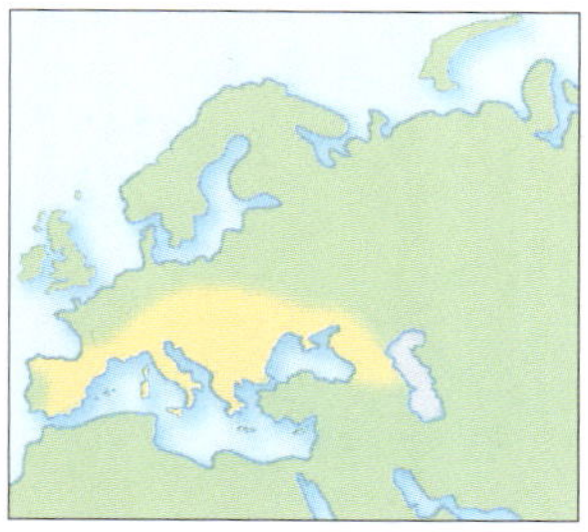

**Distribution**: Spain to the Caucausus.
**Height**: 20m/65ft
**Shape**: Broadly spreading
**Deciduous**
**Pollinated**: Wind
**Leaf shape**: Elliptic

**Identification**: The bark is dark grey and deeply fissured. The leaves are variable, elliptic to obovate, to 10cm/4in long and 5cm/2in broad. They are dark grey-green above and paler green with a covering of soft grey hair beneath, as are the leaf stalks. The leaf lobes are rounded, some ending in a small, sharp point. The male flowers are borne in pendulous, yellow-green catkins. The seed is a stalkless acorn, to 4cm/1½in long and contained, to up to half its length, in a hairy, scaly acorn cup.

*Below: The fruit.*

*Right: The tree is named for the down on its leaves and shoots.*

## Turner's Oak

*Quercus* x *turneri*

This small to medium-size, semi-evergreen, compact, dense-leaved oak is the result of an artificial hybrid cross between the holm oak, *Q. ilex*, and the English oak, *Q. robur*, carried out by Turner's Nursery, Essex, England, in the early 1780s. By the early 19th century it was commonly cultivated in gardens across much of western Europe. More recently it has been planted in urban areas because it can withstand atmospheric pollution and also provides an evergreen screen in the winter.

**Identification**: The bark is dark grey and fissured into rectangular plates. The leaves are oblong to obovate, 13cm/5in long and 5cm/2in broad, shiny dark green above, pale green beneath, borne on hairy shoots and persisting well into winter. The leaves taper towards the base and have variable shallow and triangular lobing around the margin. The male flowers are borne in yellow-green weeping catkins in early summer. The seed is an acorn, one half of which is enclosed in a scaly cup. The acorns do not always develop into mature fruits.

**Distribution**: Of UK garden origin.
**Height**: 20m/65ft
**Shape**: Broadly spreading
**Semi-evergreen**
**Pollinated**: Wind
**Leaf shape**: Oblong to obovate

*Left: The leaves take many of their characteristics from the holm oak parent. They are shallow to un-lobed and in mild areas of Europe the leaves may persist on the tree all winter.*

## Sawtooth Oak

*Quercus acutissima*

As the name suggests, this distinctive oak has leaves that are regularly edged with saw-like teeth. It is native to a broad region that runs from the eastern Himalayas through China to Japan, and it is a close relation of the Turkey oak, *Q. cerris*. The sawtooth oak was introduced into European cultivation in 1862 by an Englishman, Richard Oldham. It is said that in Japan, silkworms were at one time fed on its leaves.

**Identification**: The bark of the sawtooth oak is dull grey-brown, smooth at first, becoming vertically and deeply fissured in maturity. The leaves are chestnut-like, oblong, to 20cm/8in long and 5cm/2in broad, with many distinctive parallel leaf-veins running from the midrib and ending in slender-tipped teeth. They are glossy dark green above and pale green beneath. The male flowers are borne on yellowish pendulous catkins in late spring. The seed is a rounded acorn, to 2.5cm/1in long.

**Distribution**: South-east Asia.
**Height**: 15m/50ft
**Shape**: Broadly spreading
**Deciduous**
**Pollinated**: Wind
**Leaf shape**: Oblong

*Right: Two-thirds of the acorn is set within a bristly cup.*

*Above and above left: The leaves persist on the tree into winter.*

## Golden Oak of Cyprus

*Quercus alnifolia*

This interesting small oak takes its name from the golden undersides of its leaves, which are covered with a dense golden- or mustard-yellow felt quite similar to the indumentum found on some Himalayan rhododendrons. In the wild, in Cyprus, the golden colouring can be extremely vibrant; it is less so in cultivation in western Europe. The tree was introduced into Europe in 1885 and, although perfectly hardy, it is not as common in cultivation as might be expected.

**Distribution**: Cyprus.
**Height**: 8m/26ft
**Shape**: Broadly spreading
**Evergreen**
**Pollinated**: Wind
**Leaf shape**: Rounded

**Identification**: The bark is dark grey and smooth with pale orange flecks. The leaves are thick and leathery, rounded, up to 5cm/2in in diameter, edged with sparse small teeth, glossy deep green above and golden-brown beneath. Young leaves will also have some pubescence on the top side. The male flowers are borne in drooping, yellow-green catkins in late spring. The acorns (not always present) are elongated, broadening from the base upwards and shaped like a policeman's truncheon. When ripe these are also a rich golden colour.

*Left: The leaves have a distinctive golden covering to the underside. Acorns are shaped like a policeman's truncheon.*

OTHER SPECIES OF NOTE

**Caucasian Oak** *Quercus macranthera*
This handsome, medium-size to large, rounded, deciduous tree is native to the Caucasus Mountains and northern Iran, south of the Caspian Sea. It has strongly ascending branches and verdant foliage. The leaves are large, and can be up to 15cm/6in long and 10cm/4in wide. They taper towards the base, with up to 11 rounded, forward-pointing lobes on each side.

**Daimo Oak**
*Quercus dentata*
This 20m/65ft spreading and heavily branched deciduous oak is native to Japan and mainland north-east Asia, from where it was introduced into Europe and cultivation in 1830. Its main claim to fame is that it produces some of the largest leaves of any oak: they may be anything up to 30cm/12in long and 15cm/6in broad. Each leaf has between five and nine lobes on each side.

**Gall Oak** *Quercus infectoria*
This small semi-evergreen tree, 6m/20ft tall, is native to the region around the Aegean Sea and Cyprus. Its twigs often bear galls, which are collected as a source of tannin for the dyeing industry. It was cultivated in western Europe from 1850 but has never been widely grown there. It has leathery, oblong to elliptic leaves, 7.5cm/3in long and 5cm/2in wide, which are serrated around the margin with sharp spines.

## Portuguese Oak

*Quercus faginea*

At one time it was thought this broad-headed tree was native to North Africa as well as the Iberian peninsula. It was cultivated in northern Europe, including Britain, in 1824, but has never become popular, despite its ability to grow well on all soil types including shallow chalk. The leaf form is very variable and may be anything from oval to elliptic with either regular pronounced teeth around the margin, or sometimes no teeth at all. In mild winters the leaves will persist on the tree right through to the following spring.

**Distribution**: Spain and Portugal.
**Height**: 20m/65ft
**Shape**: Broadly spreading
**Semi-evergreen**
**Pollinated**: Wind
**Leaf shape**: Oval to obovate

**Identification**: The bark is grey-brown, smooth at first, becoming vertically cracked in maturity. The leaves are rather leathery, up to 7.5cm/3in long and 5cm/2in broad, grey-green above and covered with a dense grey felt beneath. The male flowers are borne in early summer on pendulous yellow-green catkins. The seed is a squat, rounded acorn, which is enclosed to one third of its length in an urn-shaped cup.

*Right: Up to 12 pairs of parallel leaf veins end in sharp, regular teeth.*

*Right: The fruit.*

# BIRCHES

*The birches,* Betula, *are a group of catkin-bearing, alternate-leaved, deciduous trees, native to northern temperate regions of the world. There are more than 60 species, spread right across the region, from Japan to Spain and across North America. They are particularly well known for their attractive bark, which, depending on species, varies from pure white to red.*

## White Birch

Downy birch *Betula pubescens*

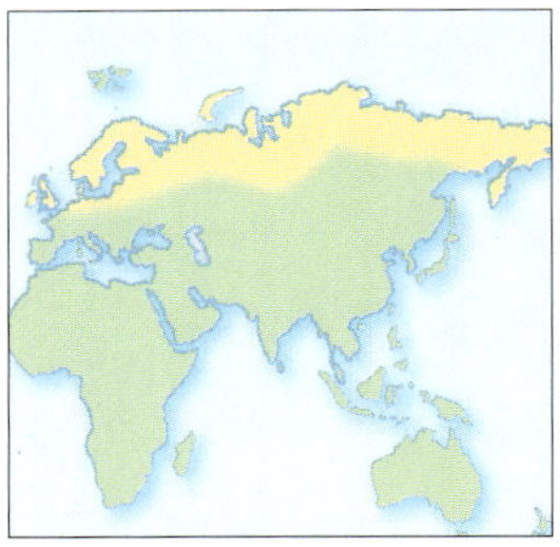

**Distribution**: Northern Europe to northern Asia.
**Height**: 25m/80ft
**Shape**: Broadly conical
**Deciduous**
**Pollinated**: Wind
**Leaf shape**: Ovate to round

This common birch thrives in wet conditions and poor soils right up into the Arctic region. It is sometimes known as downy birch because of its hairy shoots. Although related to silver birch, *B. pendula*, it is distinct, and the two species rarely hybridize. White birch is not as popular as silver birch, lacking its graceful pendulous form.

**Identification**: The bark on young trees may be dull grey, but it matures white. Branches and shoots are glossy copper-brown. The male flowers are yellow catkins and the females green upright catkins, both borne on the same tree in early spring. When ripe, the female catkins break up on the tree, shedding copious small, papery winged seeds.

*Right: The dark green leaves are 5cm/2in long and wide, with toothed edges, ending in a pointed tip.*

## Silver Birch

*Betula pendula*

**Distribution**: Europe and northern Asia, from the Atlantic to the Pacific.
**Height**: 30m/100ft
**Shape**: Narrowly weeping
**Deciduous**
**Pollinated**: Wind
**Leaf shape**: Ovate to triangular

The silver birch is one of the toughest of all trees, able to withstand intense cold and long periods of drought. It seeds prolifically and will quickly establish on cleared ground. Birch was one of the first trees to colonize northern Europe after the last ice age 12,000 years ago. It is just as much at home in semi-tundra regions of northern Scandinavia as it is in temperate southern France.

**Identification**: The bark is a distinctive white to silver colour. Mature trees develop corky black fissures between white irregular plates. The leaves are almost triangular in shape, 6cm/2½in long and 4cm/1½in across, tapering to a point. They are glossy grass green above and paler beneath, normally turning butter yellow before falling in autumn. The leaf stalk is usually red-brown, 2.5cm/1in long and attached to a slim, hairless, warty, weeping shoot. Both male and female flowers are catkins, appearing separately on the same tree in early spring. The male flower is yellow and drooping, up to 6cm/2½in long; the female flower is green, erect and smaller.

*Above: The silver birch is often planted for the ornamental value of its distinctive bark.*

*Right: The catkins are thick and the leaves coarsely serrated.*

## Himalayan Birch

*Betula utilis*

**Distribution**: Himalayas from Kashmir to Sikkim and western and central China.
**Height**: 25m/80ft
**Shape**: Broadly conical
**Deciduous**
**Pollinated**: Wind
**Leaf shape**: Ovate

The exact distribution of the Himalayan birch is difficult to define, because several other Asiatic birches occur in the same region, and cross-pollination and hybridization do occur. At one stage the Chinese red-barked birch, *B. albo-sinensis*, was considered to be a form of this birch and not a separate species. The colour of Himalayan birch bark varies from white, in the west of its natural range, to pink-brown or orange-red in the east. The tree grows in parks and arboreta all over Europe.

**Identification**: The variably coloured bark is paper-thin and peels off in long, ribbon-like, horizontal strips. The serrated leaves, up to 10cm/4in long and 6cm/2½in wide, taper to a point. They have strong veining with up to 12 pairs of parallel veins, each vein ending in a sharp tooth. The leaves are shiny dark green above and paler beneath, with some hair on the midrib, and are attached to hairy shoots by a short leaf stalk, 1.25cm/½in long. The flowers are catkins: the male long, yellow and drooping, up to 12.5cm/5in; the female smaller, green and upright. Both flowers are borne on the same tree in early spring.

*Right: A long, yellow male catkin.*

*Left: The leaf is teardrop-shaped.*

### OTHER SPECIES OF NOTE

**River Birch** *Betula nigra*
Extensively grown in Europe, where it grows on river banks and low-lying, swampy ground, this tree has ovate, toothed leaves up to 10cm/4in long and peeling, pink-brown bark, that becomes fissured in maturity. It regenerates from the stump and is often coppiced to encourage multiple stems, displaying the peeling effect.

**Canoe Birch** *Betula papyrifera*
Also known as the paper birch, this creamy white-barked large birch grows extensively across Europe and is extremely hardy. Native Americans used the timber to make canoes. It has ovate, toothed leaves that turn orange-yellow in autumn.

**Erman's Birch** *Betula ermanii*
This magnificent, long-lived birch, to 25m/80ft tall, has a natural range from Japan to Siberia. It was introduced into cultivation in 1890. It has creamy white to pink bark, which peels in long papery strips and then hangs from the main branches, giving the tree a ragged appearance.

**Monarch Birch** *Betula maximowicziana*
Also known as Japanese red birch and sometimes Maxim's birch, this magnificent large Japanese birch has long been cultivated across Europe. Its heart-shaped leaves are larger than those of any other birch, 15cm/6in long and 13cm/5in broad, and its reddish-brown bark peels to reveal greyish-yellow, or grey-pink, new bark beneath. The leaves turn yellow in autumn.

## Chinese Red-barked Birch

*Betula albo-sinensis*

**Distribution**: Western China.
**Height**: 25m/80ft
**Shape**: Broadly conical
**Deciduous**
**Pollinated**: Wind
**Leaf shape**: Ovate

This beautiful, medium-sized birch is best known for its attractive coppery to orange-red bark, which peels to reveal cream-pink bark beneath, on both the trunk and branches. The bark and catkins make this tree a spectacular sight. It grows high in the mountains in mixed woodland on relatively impoverished, shallow soils. It was introduced into Europe in 1901 by the English plant collector Ernest Wilson and has been a favourite for planting in gardens and arboreta ever since.

**Identification**: The leaves are ovate, 7.5cm/3in long and 4cm/1½in across. The margin is cut by forward-pointing, sharp serrations, and the leaf ends in a long, fine, slightly curved tip. When the leaves emerge from small, slightly sticky buds, they are covered with a soft down. This quickly disappears to reveal glossy grass-green leaves, which turn golden yellow in autumn before falling. Both male and female flowers are catkins, and are very similar to those of the Himalayan birch. Fertilized female flowers develop into brown papery catkins containing hundreds of winged seeds.

*Right: The male and female catkins appear on the same tree.*

## Japanese Cherry Birch

*Betula grossa*

This medium-size Japanese birch, native to the mountains of Honshu, Shikoku and Kyushu, was introduced into Europe in 1896. It is comparatively uncommon within cultivation but normally found in botanic gardens and arboreta. It is closely related to the American cherry birch, *Betula lenta,* and has similar characteristics, such as bark colouring and aromatic foliage and twigs.

*Below: The leaves turn yellow in the autumn.*

*Left: The tree can grow to be as wide as it is tall.*

**Identification**: The bark on young trees is shiny red-brown, becoming grey-brown and flaking in maturity into horizontal strips. The leaves are ovate, heart-shaped at the base, rather like hornbeam, with prominent veins and a coarsely-toothed margin, up to 10cm/4in long and 5cm/2in broad, dark green above, paler beneath with some fine pubescence. Both male and female flowers are borne in catkins, male are yellow and pendulous, female green and upright. Female catkins ripen and break up on the tree releasing hundreds of paper-winged seed.

**Distribution**: Japan.
**Height**: 20m/65ft
**Shape**: Broadly conical
**Deciduous**
**Pollinated**: Wind
**Leaf shape**: Ovate

*Below: The bark is an attractive shiny red-brown when the tree is young, and becomes grey-brown and flaking in maturity.*

## Transcaucasian Birch

*Betula medwediewii*

The transcaucasian birch is a very hardy species native to the Caucasus Mountains, between the Black Sea and the Caspian Sea, where it grows up to sub-alpine elevations. It is a small, spreading tree or large shrub, which was introduced into cultivation in Germany and Britain around 1897. In parks and gardens it is mainly grown for the attractive bright yellow autumn colour of its large leaves, and for its handsome, prominent catkins in spring. The species is recognizable in winter by its large, glossy green sticky buds, which are borne on stout upright branches.

**Identification**: The bark, on what is often a multi-stemmed tree, is shiny pale grey-brown, and on older trees has a tendency to flake and peel. The leaves are ovate to rounded, up to 10cm/4in long and 7.5cm/3in wide and sometimes slightly heart-shaped at the base. They are heavily indented, almost to the point of corrugation, with 8–11 pairs of sunken veins. The top surface of the leaf is dark green and the lower surface is paler green with some hairiness along the leaf veins. The male and female flowers are borne on separate catkins, both on the same tree, in early spring.

**Distribution**: Caucasus.
**Height**: 4.5m/15ft
**Shape**: Broadly spreading
**Deciduous**
**Pollinated**: Wind
**Leaf shape**: Ovate to round

*Right: The large corrugated leaves turn bright yellow in autumn.*

## Young's Weeping Birch

*Betula pendula* 'Youngii'

There are few weeping trees that surpass Young's weeping birch for sheer beauty and elegance. Consequently it is commonly planted throughout Europe just about everywhere, from shopping centre car parks to large arboreta. It was raised in a nursery at Milford, Surrey, England, in the latter part of the 19th century. The trees are seen at their very best when planted as a group, close enough to allow their domed mushroom heads to intermingle, so that the fine tracery of branches produces a hanging "curtain" of foliage.

*Above: The leaves are coarsely serrated.*

**Distribution**: Of garden origin.
**Height**: 10m/30ft
**Shape**: Broadly spreading and weeping
**Deciduous**
**Pollinated**: Wind
**Leaf shape**: Ovate to triangular

**Identification**: Young's weeping birch is relatively easy to identify because its grey-white bark instantly suggests birch (though on very young trees the bark is glossy brown), and there is no other birch with such graceful weeping branches. Mature trees invariably develop crooked trunks that have a tendency to lean to one side. The overall effect could be described as untidy but it is very charming. The leaves are the same as those of the silver birch, *B. pendula*, or slightly smaller, and are borne on fine, thin, weeping branches.

OTHER SPECIES OF NOTE

**White-barked Himalayan Birch**
*Betula utilis* var. *jacquemontii*
The Himalayan birch has a long natural range, from Afghanistan to China. Across this range its characteristics vary considerably; depending on the region in which forms have been raised, their bark colour may vary from pure white through pink to copper-brown. *Jacquemontii* has dazzling white bark and has spawned several popular cultivars, including 'Grayswood Ghost', 'Jermyns' and 'Doorenbos'.

**Creamy Bark Birch** *Betula costata*
This vigorous, attractive birch, of conical habit, is native to north-east Asia and deserves to be more widely planted than it is. It has been suggested that it may be a form of Erman's birch, *B. ermanii*: if not, it is certainly closely related. However, it differs in having narrower leaves, which are wedge-shaped at the base and more pointed at the tip. It has beautiful pink to creamy-fawn peeling bark and in autumn the leaves turn clear yellow.

**Asian Black Birch** *Betula davurica*
This medium-size hardy tree from northern China and Korea takes its common name from the American black birch, *B. nigra*, which it resembles in having the same peeling dark silver-grey bark, which gives the tree a rough, rugged, almost shaggy appearance. In the wild it may reach 30m/100ft, but is more likely to achieve 20m/65ft in cultivation. It comes into leaf earlier than most birches grown in northern Europe, and the foliage can be damaged by frost.

## Swedish Birch

*Betula pendula* 'Dalecarlica'

This distinctive cultivar of silver birch, sometimes wrongly called *B. pendula* 'Laciniata', has slender leaves that are deeply cut, almost back to the midrib. It was found in Ornas, north of Stockholm, Sweden, in 1767, and in 1810 grafts were cultivated in the Stockholm Botanical Garden, from where it was made commercially available. Today, it is a reasonably common sight in botanic gardens and arboreta across northern and western Europe.

**Distribution**: Of garden origin.
**Height**: 25m/80ft
**Shape**: Narrowly conical
**Deciduous**
**Pollinated**: Wind
**Leaf shape**: Ovate and deeply cut

**Identification**: The bark is similar to that of the silver birch, *B. pendula*. The overall form is of an elegant, light and airy, tall, narrow tree with weeping twigs and shoots borne on ascending branches. The leaves are slenderly oval and then cut almost to the midrib. The lobes are themselves lance-shaped, finely toothed and have a long slender point. They are glossy dark green above, paler beneath and turn butter yellow in autumn. Both male and female flowers are borne on catkins in early spring.

*Below: The deeply cut leaves give the tree a delicate appearance.*

# ALDERS

*Alders are a group of 36 deciduous species within the Betulaceae family. They are primarily native to northern temperate regions of the world, where they grow in damp conditions, quite often alongside rivers and other watercourses. In boggy ground they can form inpenetrable thickets, and when growing in the open they have a broad open shape.*

## Common Alder

Black alder *Alnus glutinosa*

This tree has always been associated with water. It thrives in damp, waterlogged conditions, growing close to rivers and marshy ground, where it creates its own oxygen supply. Alder timber is waterproof and has been used to make everything from boats to water pipes, including wooden clogs. It also forms the foundations of many of the buildings in Venice, Italy.

**Identification**: The bark is dark grey-brown and fissured from an early age. The leaves are obovate to circular, finely toothed with up to ten pairs of pronounced leaf veins and a strong central midrib. Up to 10cm/4in long and 8cm/3in wide, they are dark green and shiny above and pale grey-green beneath, with tufts of hair in the leaf axils. Both male and female flowers are catkins: the male is greenish yellow, drooping and up to 10cm/4in long; the female is a much smaller, red, upright catkin which, after fertilization, ripens into a distinctive small brown cone. Spent cones persist right through to the following spring.

**Distribution**: All of Europe, extending into western Asia and south to North Africa.
**Height**: 25m/80ft
**Shape**: Broadly conical
**Deciduous**
**Pollinated**: Wind
**Leaf shape**: Obovate

*Left: The catkins appear in early spring, before the leaves open. Alder cones begin to grow in summer, by which time the rounded leaves are thick on the branches.*

## Grey Alder

*Alnus incana*

**Distribution**: Europe (not Britain) to the Caucasus.
**Height**: 20m/65ft
**Shape**: Broadly conical
**Deciduous**
**Pollinated**: Wind
**Leaf shape**: Ovate

This medium-size, fast-growing tree reaches elevations of up to 1,000m/3,300ft in the Caucasus Mountains, and gets its name from the dense covering of grey hairs on the undersides of the leaves. It has been in cultivation in Europe since 1780 and, due to its exceptional hardiness, is a popular tree for planting in cold, wet conditions and in areas being reclaimed following industrial activities, such as mining and landfill. It has spawned several ornamental cultivars, including the handsome, cut-leaved 'Laciniata'.

**Identification**: The bark is dark grey and smooth, even in maturity. The leaves are ovate, matt dark green above and paler with grey hairs beneath. They are 10cm/4in long and 5cm/2in broad, with a double-toothed margin and a pointed tip. The flowers are borne in catkins: the male catkin is 10cm/4in long, orange-yellow and pendulous, the female catkin is erect and red. Both are carried separately on the same tree in late winter. The fruit persists on the tree long after ripening.

*Right: The fruit is a woody cone-like structure containing tiny winged seeds.*

## Italian Alder

*Alnus cordata*

This handsome tree originated in southern Italy, parts of Albania and Corsica, where it grows at up to 1,000m/3,300ft on dry mountain slopes. It is less dependent on damp conditions than most alders, thrives on chalk, and is by far the largest of its genus, sometimes attaining heights over 30m/100ft in the wild. It was named *cordata* because of its heart-shaped leaves, and was introduced into cultivation in 1820.

**Identification**: The bark is dull grey, smooth at first becoming shallowly fissured with age. The 10cm/4in leaves are glossy bright green above, paler with some hairs beneath. The male catkins are yellow and pendulous, 7.5cm/3in long; the females are smaller, red and erect. Both appear on the same tree in early spring and are followed by egg-shaped, woody, cone-like fruits.

**Distribution**: Southern Italy and Corsica.
**Height**: 28m/90ft
**Shape**: Broadly conical
**Deciduous**
**Pollinated**: Wind
**Leaf shape**: Broadly ovate

*Right: Female and male catkins, and the heart-shaped leaf.*

OTHER SPECIES OF NOTE

**Red Alder** *Alnus rubra*
This pioneer species is native to the west coast of North America, where it grows on just about any available space not already colonized by other species. It was introduced to Britain in the mid-19th century. Although not a large tree, it grows particularly fast until it reaches its optimum height of 15m/50ft. Red alder has light brown-grey bark, which is rough and warty, and an ovate leaf up to 10cm/4in long with rusty-red hairs on the underside.

**Oriental Alder** *Alnus crientalis*
This medium-size tree, to 15m/50ft tall, is native to the eastern Mediterranean region, including Syria and Cyprus, from where it was introduced into cultivation in western Europe in 1924. It has oval to ovate, glossy green, irregularly toothed leaves, bright green sticky buds in winter, and clusters of yellow, pendulous male catkins in early spring. It is related to the Italian alder, *A. cordata*, but is distinguished from this species by its small wingless seeds.

**Caucasian Alder** *Alnus subcordata*
This attractive, fast-growing, medium-size tree, to 20m/65ft, is native to the Caucasus region, between the Black and Caspian Seas and south into Iran, where it grows naturally in damp woodland on flood plains. It has large, oval to ovate, heart-shaped, matt green leaves, which may be up to 15cm/6in long. It is one of the earliest alders to flower: its long, 15cm/6in, male catkins may appear as early as the beginning of winter. It was introduced into Britain in 1838 and into France in 1861.

## Green Alder

*Alnus viridis*

This small spreading tree, or large shrub, which produces clumps of long, erect stems, is native to the central and southern European Alps, where it can be found growing naturally up to elevations around 1,000m/3,300ft. It was introduced into cultivation in 1820 but has not been widely planted as an ornamental since. However, when planted on poor soils its root system has the ability to fix nitrogen in the soil and thereby improve soil fertility, both for itself and for other plants growing around it.

**Distribution**: Central and south-east Europe.
**Height**: 6m/20ft
**Shape**: Broadly spreading
**Deciduous**
**Pollinated**: Wind
**Leaf shape**: Ovate

**Identification**: The bark is grey-green and smooth and the tree regularly sends up olive green, cane-like suckers. The rounded leaves, up to 10cm/4in long and 7.5cm/3in broad, end in a pointed tip. They are dark green and smooth above, paler beneath with some hair along the midrib and veins. Male and female flowers are borne in separate catkins on the same tree in spring.

*Above: The woody fruit.*

*Right: The leaves are unevenly toothed.*

## *Alnus firma*

**Distribution**: Japan.
**Height**: 9m/30ft
**Shape**: Broadly conical
**Deciduous**
**Pollinated**: Wind
**Leaf shape**: Ovate-lanceolate

This alder is native to all the main islands of Japan, where it is found primarily in mountainous regions. Elsewhere, particularly in the agricultural land outside Tokyo, it has been widely cultivated along field boundaries. At one time these trees were used to support poles erected to dry freshly cut rice. The species was introduced into Europe in 1893, and although not common in cultivation it is represented in most botanic gardens and arboreta.

**Identification**: This is a small tree, clothed rather densely but gracefully in long slender branches. The leaves are similar to those of the common hornbeam, *Carpinus betulus* – rounded or wedge-shaped at the base and with a long slender tip. They are 11cm/4½in long and 5cm/2in wide, with distinctive parallel veining, and are finely toothed around the margin. The lower surface may be covered with fine hairs, as is the shoot. An interesting characteristic is that the leaves remain green until they fall in early winter.

*Right: The seeds are contained in a cone-like structure.*

*Far right: The leaves are hornbeam-like with fine regular serrations around the margin and distinctive parallel veining.*

*Below: The catkins are lime-yellow in colour.*

## Manchurian Alder

*Alnus hirsuta*

**Distribution**: North-east Asia and Japan.
**Height**: 20m/65ft
**Shape**: Broadly conical
**Deciduous**
**Pollinated**: Wind
**Leaf shape**: Ovate

The Manchurian alder is native to Eastern China and Japan, from where it was introduced into Europe in 1879, when propagation material was collected and sent to James Veitch. The species is closely related to the grey alder, *A. incana*, and effectively extends that tree's natural range east into Asia. *A. hirsuta* is distinguished from *A. incana* by its larger leaves, which are occasionally more round than ovate and have more pronounced indentations around the margin. It is a vigorous and handsome tree that, like many other alders, thrives in wet conditions.

**Identification**: The bark is dark grey-brown and smooth even in maturity. Large trees may have a trunk girth up to 1.8m/6ft. Young shoots are covered in fine hairs and the winter buds are purple and shaped like boxing gloves. The leaves are broadly ovate, rounded at the base and at the apex, with shallow lobes and double serrations around the margin. They are 12.5cm/5in long and nearly as wide, dull dark green and heavily corrugated, with between 9 and 12 pairs of leaf veins. The fruit is a large woody cone 2.5cm/1in long.

*Below: The leaves are ovate with shallow lobes and double serrations, and the catkins.*

## Japanese Alder

*Alnus japonica*

This is a striking large alder with a distinctive pyramidal shape and unusually narrow leaves. In maturity it makes a fine, shapely specimen. It has never been widely cultivated and is relatively rare in gardens and arboreta across Europe.

**Identification**: *A. japonica* has dull grey, smooth bark, and the young shoots may be covered in fine hairs. The leaves are narrowly elliptic to lanceolate, tapering at both ends, with a slender tip. They are up to 12.5cm/5in long and 5cm/2in wide, and are glossy dark green with minimal veining branching from a yellow midrib. The male flowers are borne in terminal clusters of yellow-brown catkins. The seed is borne in an oval, woody cone-like structure.

**Distribution**: China, Japan, Korea and Taiwan.
**Height**: 25m/80ft
**Shape**: Pyramidal
**Deciduous**
**Pollinated**: Wind
**Leaf shape**: Lanceolate to elliptic

*Left: The leaves are glossy dark green, narrowly elliptic, tapering at both ends. The flowers are catkins.*

## *Alnus maximowiczii*

This small tree, or large shrub, with thick shoots and twigs, is native to high mountain ranges. It is closely related to the Sitka alder, *A. sinuata*, a native of north-west North America, to which 250 million years ago Asia would have been joined in the supercontinent Pangaea. *A. maximowiczii* was first cultivated in Europe just before World War I.

*Far right: Clusters of short-fat yellow catkins appear in spring.*

*Right and above left: The leaves are broadly ovate to heart-shaped and reminiscent of some lime (Tilia) species.*

**Distribution**: South-east Asia and Japan.
**Height**: 9m/30ft
**Shape**: Broadly spreading
**Deciduous**
**Pollinated**: Wind
**Leaf shape**: Broadly ovate

**Identification**: This is a shrubby tree with thick shoots and grey bark. The leaves are bright green on both surfaces, with a fine point and fine, slender teeth around the margin, giving a slightly ragged, fringe-like appearance. They are up to 10cm/4in long and 7.5cm/3in wide, with tufts of hair in the leaf axils on the undersides. Clusters of stout yellow male catkins are produced in spring.

### OTHER SPECIES OF NOTE

***Alnus glutinosa* 'Imperialis'**
This cultivar of the common alder, *A. glutinosa*, is one of the most distinctive and graceful alders, on a par with other cut-leaved trees for ornamentation. Its delicate leaves are cut into long slender lobes, which run right into the midrib and give a light, feathery effect. It was raised in 1859 and has been widely planted across Europe.

***Alnus glutinosa* 'Pyramidalis'**
Sometimes referred to by its former name, 'Fastigiata', this cultivar has a rather conical shape with branches at an acute angle. It has been in cultivation since the late 1800s and is commonly planted across Europe, particularly in damp soil close to water.

***Alnus x spaethii***
This is a 1908 hybrid between the Caucasian alder, *A. subcordata*, and the Japanese alder, *A. japonica*. It is fast-growing, up to 20m/65ft tall, and has leaves that are purple when they first emerge in spring. It is grown in European arboreta for its outstanding display of male catkins, which appear in late winter.

**Himalayan Alder** *Alnus nitida*
This tall alder, to 25m/80ft, has bark that is dark brown or almost black, and thin, ovate toothed leaves up to 15cm/6in long. It is unique among alders in having male catkins that appear in autumn rather than spring. It was introduced into cultivation from the north-west Himalayas in 1882.

# HORNBEAMS AND HAZELS

*Hornbeams are attractive, hardy trees, which have long been cultivated in Europe. Common hornbeam is used for hedging and, as a hedge, has similar characteristics to beech, including retaining its dead leaves into winter and forming a solid windbreak. The genus* Corylus *includes hazel, which is widely distributed throughout Europe and west Asia.*

## Heartleaf Hornbeam

*Carpinus cordata*

This medium-sized attractive tree was introduced into Europe from Japan in 1879. In its natural habitat it grows in mixed woodland in mountainous regions. Since then it has become widely distributed in botanic gardens, but is still fairly uncommon in parks and gardens. The name *cordata* means "heart-shape" and refers to the leaves, which are heart-shaped at the base. It is a slow-growing tree and hardy in northern Europe and Great Britain.

**Identification**: It has a broadly columnar form, smooth charcoal-grey bark at first, becoming fluted at the base and shallowly fissured in maturity. The leaves are deeply veined, relatively broad when compared to other hornbeams, oblong to ovate, heart-shaped at the base, dark green and up to 10cm/4in long and 5cm/2in wide. Male and female flowers are borne in yellow-green catkins in spring. The fruit is a nut that is covered with several small, overlapping green bracts.

**Distribution**: Japan, North-east Asia, north and west China.
**Height**: 15m (50ft)
**Shape**: Broadly columnar
**Deciduous**
**Pollinated**: Wind
**Leaf shape**: Oblong-ovate

*Left: The catkins.*

*Above right: The leaves are heart-shaped at the base and deeply veined.*

## Hornbeam

European hornbeam *Carpinus betulus*

**Distribution**: Central Europe, including southern Britain, to south-west Asia.
**Height**: 30m/100ft
**Shape**: Broadly spreading
**Deciduous**
**Pollinated**: Wind
**Leaf shape**: Ovate

*Left: The fruit is a ribbed nut, which is held in a three-lobed bract.*

Hornbeam is sometimes confused with beech because of its silver-grey bark and similar leaf. However, hornbeam bark is far more angular than beech bark. Hornbeam leaves also have obvious serrations around the margin, which are not present on beech. Hornbeam timber is dense and hard and has a clean, white, crisp appearance. It was traditionally used to make ox yokes and butchers' chopping blocks.

**Identification**: The leaves are oval to ovate, up to 10cm/4in long and 5cm/2in across, double-toothed around the margin and tapering to a long point. There are normally between 10 and 13 pairs of leaf veins. The upper leaf surface is dark green, the underside a paler green. In autumn the leaves turn rich yellow before falling. Male and female catkins are borne separately in spring on the same tree. The fruit is a distinctive three-lobed bract with a small, ribbed, brown nut at the base of the centre bract. The bract is green in summer, ripening to fawn in autumn and persisting on the tree until the following spring.

*Right: The male catkins are up to 5cm/2in long. Female catkins (not shown) are much smaller.*

## Turkish Hazel

*Corylus colurna*

This magnificent tree has a very stately, symmetrical form. It normally develops a straight, short stem holding a pyramid of foliage, and is ideal for planting as an avenue tree. It was introduced into central and western Europe, including Britain, in the middle of the 16th century and has been popular in cultivation ever since. There are particularly good specimens in Hanover, Germany, and Vienna, Austria. It produces a pink-brown timber, much prized for furniture.

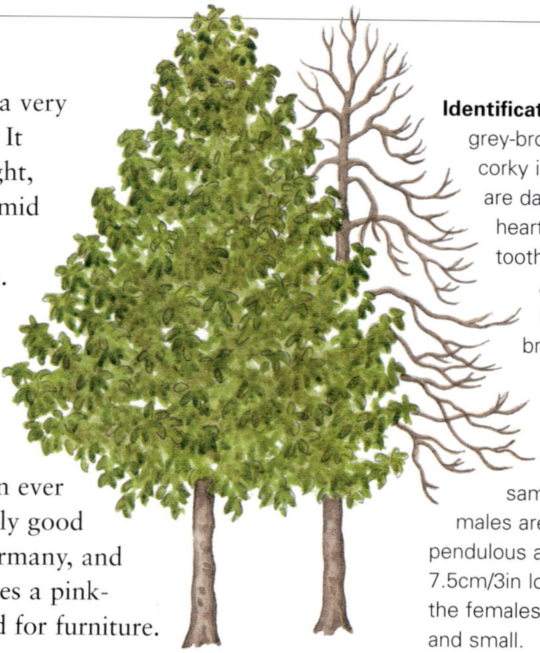

**Identification**: The bark is light grey-brown and distinctly corky in maturity. The leaves are dark green, broadly oval, heart-shaped at the base, toothed around the margin and up to 15cm/6in long and 10cm/4in broad. The male and female flowers are carried in separate catkins on the same tree. The males are yellow, pendulous and up to 7.5cm/3in long, while the females are red and small.

*Above: The leaf.*

*Above left: The fruit is an edible nut, the husk of which is covered with bristles.*

**Distribution**: South-east Europe and western Asia.
**Height**: 25m/80ft
**Shape**: Broadly conical
**Deciduous**
**Pollinated**: Wind
**Leaf shape**: Broadly oval

### OTHER SPECIES OF NOTE

**Japanese Hornbeam** *Carpinus japonica*
This is a handsome, widely spreading Japanese tree, to 15m/50ft tall, which was first cultivated in Europe in 1895. It has narrow, ovate leaves with up to 24 pairs of prominent, parallel, leaf veins, which give the whole leaf a corrugated effect. In spring the male catkins are bright yellow and pendulous. The nut is covered with green bracts, which resemble the fruit of the ironwood, *Ostrya virginiana.*

**Oriental (Caucasian) Hornbeam** *Carpinus orientalis*
This small tree or large shrub, to 15m/50ft tall, is native to south-east Europe, the Caucasus Mountains and south-west Asia. It grows on hot, dry sites, and was abundant on the battlefields of the Crimean War. It produces dense, almost impenetrable branches and foliage. The leaves are ovate, up to 5cm/2in long and sharply toothed, with 12–15 pairs of parallel veins.

**Ironwood** *Ostrya virginiana*
Otherwise known as the American hop hornbeam, this medium-sized tree, to 20m/65ft, is native to eastern and southern USA, from where it was introduced into Europe in 1692. It is an attractive tree with ovate, heavily veined leaves, which turn a rich warm yellow in autumn. The name ironwood refers to the timber, which is very heavy and strong and is often used to make tool handles.

## Common Hazel

European filbert *Corylus avellana*

There is much discussion about whether hazel is a tree or a shrub. In theory, a tree has 1m/3ft of clear stem before it branches or forks, and has the potential to grow to more than 6m/20ft in height. Hazel can certainly achieve the latter, but has a tendency to fork low down. Its status as a tree is not helped by the fact that for centuries, right across Europe, it has been regularly coppiced to ground level. Its ability to regrow after such harsh pruning has made it a popular plant for agricultural hedging.

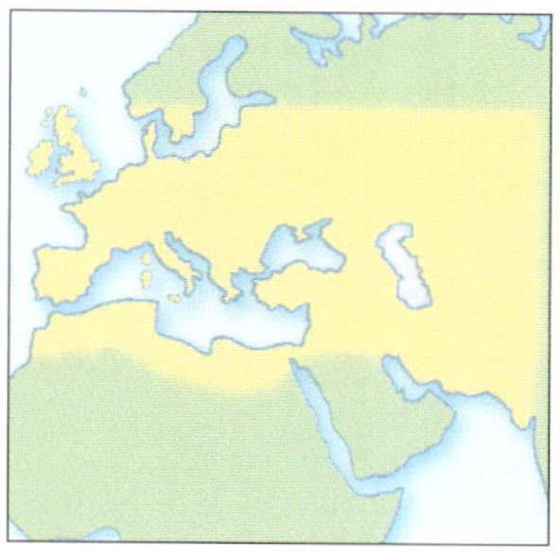

**Distribution**: Europe, western Asia and North Africa.
**Height**: 6m/20ft
**Shape**: Broadly spreading
**Deciduous**
**Pollinated**: Wind
**Leaf shape**: Orbicular

*Left: Hazel leaves are almost round in shape. This plant produces the popular hazelnut.*

**Identification**: The bark is smooth and silver-grey to pale brown, even in maturity. The trunk seldom exceeds 20cm/8in in diameter. The leaves are up to 10cm/4in across, with double teeth around the margin, and are thick and rough to the touch. There are coarse hairs on the leaf, bud and shoot. The male flower is a yellow catkin, up to 10cm/4in long, which ripens in early spring to release copious amounts of pollen to the wind. The female flower is a tiny red floret, which is borne on the end of what looks like a leaf bud but will develop into the fruit – a round to ovoid, matt, light brown, edible nut, which is half encased in a green calyx.

*Right: Hazel catkins are a familiar sight across Britain and Europe in early spring.*

# LIMES

*There are about 45 species within the* Tilia *genus. All are deciduous and are found in northern temperate regions. Limes, also known as lindens, are handsome trees, and many grow into large, ornamental specimens. Several species have been used for urban tree planting, as they respond well to pollarding and hard pruning in street situations. They look good planted in avenues and formal vistas.*

## Small-leaved Lime

*Tilia cordata*

This tall, attractive column-like tree has heart-shaped leaves and fragrant yellow flowers. When found growing wild in woodland across its natural range it is a good indicator that the woodland is very old or even ancient. There are some coppiced trees in Britain that are believed to be over 2,000 years old. The inner bark (known as bast) was at one time used to make rope.

**Identification**: The bark is grey and smooth when young, becoming vertically fissured in maturity. The leaves are cordate, almost rounded, glossy bright green above and glaucous beneath, with some hairiness in the vein axils. In autumn they turn a rich butter-yellow. In summer, highly scented yellow flowers are borne in drooping clusters of up to ten, and are accompanied by a long, 10cm/4in, pale green bract.

*Left: The heart-shaped leaf tapers abruptly to a blunt tip.*

*Right and left: The flowers are followed by round, hard, grey-green felted fruits.*

**Distribution**: Europe from Portugal to the Caucasus.
**Height**: 30m/100ft
**Shape**: Broadly columnar
**Deciduous**
**Pollinated**: Insect
**Leaf shape**: Cordate

## Weeping Silver Lime

Pendent silver lime *Tilia tomentosa* 'Petiolaris'

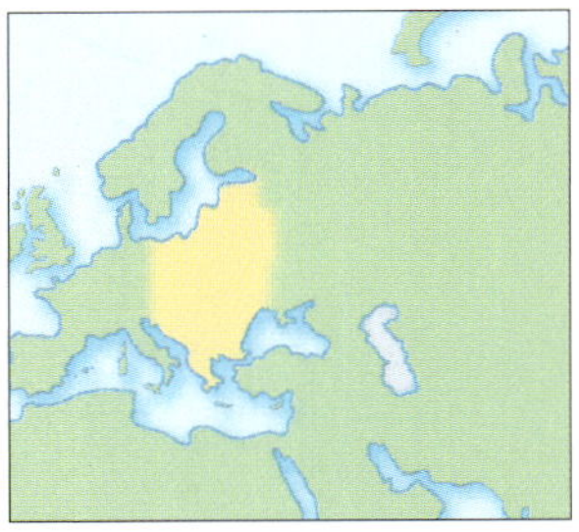

**Distribution**: Of garden origin from an eastern European species.
**Height**: 30m/100ft
**Shape**: Broadly columnar
**Deciduous**
**Pollinated**: Insect
**Leaf shape**: Broadly ovate to rounded

This is without doubt one of the most beautiful of all large trees growing in Britain and Europe. It is a cultivar of the silver lime, *T. tomentosa*, and was first described in Switzerland in 1864, although the original plant is believed to have been raised 20 years earlier. The structure of the tree is superb, with long arching branches creating what has been described as a "vaulted cathedral ceiling". The flowers are beloved by bees, and their humming is a characteristic feature of this tree in summer.

**Identification**: The bark is grey and vertically fissured in maturity. Quite often a graft union will be evident anywhere between 2–5m/6½–16½ft from the ground. Weeping shoots hang from arching branches and are covered with almost rounded leaves, which are dark green above and pure silvery-white beneath. Close inspection reveals that this whiteness is made up of thousands of tiny white hairs.

*Right: The creamy-white flowers are borne in fragrant clusters in summer.*

## European Lime

Common lime *Tilia* x *europaea*

**Distribution**: Most of Europe.
**Height**: 40m/130ft
**Shape**: Broadly columnar
**Deciduous**
**Pollinated**: Insect
**Leaf shape**: Broadly ovate

The European lime is a hybrid between the small-leaved lime, *T. cordata*, and the large-leaved lime, *T. platyphyllos*, and occurs naturally wherever the ranges of the two parents overlap. It does not have the elegance of either parent, producing unsightly suckering around the base. It is also prone to aphid attack in summer, which results in a coating of sticky aphid excrement, known as honeydew, appearing on anything that lingers beneath its boughs. Surprisingly, despite all this, it is regularly planted in towns and avenues in preference to its parents.

**Identification**: The bark is grey to grey-brown, smooth when young and developing shallow vertical fissures in maturity. The leaves are broadly ovate and up to 10cm/4in both in length and across. They are heart-shaped at the base with a sharply toothed margin and end in a tapering point at the tip. The colour of the leaves is a rather flat green above and slightly paler beneath, and they have hairy tufts in the main vein axils. The flowers are yellow and fragrant, borne in drooping clusters of up to ten under a pale green bract in summer.

*Right: The clusters of flowers hang down beneath the leaves in summer.*

*Left: Like that of other limes, the fruit is small and pea-like.*

## Large-leaved Lime

*Tilia platyphyllos*

This splendid, large, domed tree has a clean, straight trunk and graceful arching branches. Unlike the European lime, *T.* x *europaea*, it does not produce suckers or suffer from aphid attack, so does not shed sticky honeydew in summer.

*Left: The leaves have pointed tips.*

**Distribution**: Europe into South-west Asia.
**Height**: 30m/100ft
**Shape**: Broadly columnar
**Deciduous**
**Pollinated**: Insect
**Leaf shape**: Broadly ovate

**Identification**: The bark is light grey with shallow fissures in maturity. The large leaves are rounded, up to 15cm/6in across. They are deep green above with some hair, and light green on the underside, with dense hairs along the midrib and in the vein axils. The leaf stalk, which can be up to 5cm/2in long, is also covered in soft white down. The flowers are pale yellow, very fragrant and produced in weeping clusters of up to six in early summer. Each flower cluster is accompanied by a downy floral bract, up to 12.5cm/5in long and 1.5cm/⅝in wide. The fruit is a pale green, downy "pea", borne on a stalk in late summer and early autumn.

### OTHER SPECIES OF NOTE

**American Basswood** *Tilia americana*
Commonly known as American lime, this attractive tree has a natural range from Maine to North Carolina and west to Missouri. It was introduced to Britain in 1752. It has two distinguishing features: rough, almost corky bark in maturity and the largest leaf of any lime, up to 20cm/8in long and almost as broad.

**Crimean Lime** *Tilia* x *euchlora*
This elegant tree is believed to be a hybrid cross between the small-leaved lime, *T. cordata*, and *T. dasystyla*, which occurred naturally in the wild somewhere in the Crimea region around 1860. Crimean lime is a "clean" lime, which means it does not suffer attack by aphids and so escapes their sticky "honeydew" excretions, which fall on to cars parked beneath the tree. It does, however, have yellow fragrant flowers that are extremely attractive to bees. The bees gorge on the nectar until they become intoxicated and then stumble around beneath the tree.

**Oliver's Lime** *Tilia oliveri*
This elegant, medium-sized, broadly spreading lime, otherwise known as Chinese white lime, was discovered in central China in 1888 by the English plant hunter Ernest Wilson, who introduced it into Europe in 1900. It has silver-grey smooth bark and heart-shaped leaves, which are dark green above and covered with fine white hairs beneath. It is a popular tree, widely cultivated across western Europe in parks, gardens and arboreta.

## Begonia-leaved Lime

*Tilia begoniifolia*

**Distribution**: Southern Russia and Northern Iran.
**Height**: 25m/80ft
**Shape**: Broadly columnar
**Deciduous**
**Pollinated**: Insect
**Leaf shape**: Broadly ovate

This hardy, vigorous, medium-sized tree is native to the lands bordering the eastern side of the Caspian Sea – where it may attain heights around 25m/80ft. It is too early to say how tall it will grow in cultivation, as the tree was introduced into western Europe only in 1972. Since its introduction, specimens at the Hillier Arboretum in Hampshire, England, have already exceeded 10m/30ft in height.

**Identification**: The bark is grey and smooth when young, becoming shallowly and sparsely fissured as the tree matures. The leaves are broadly ovate, large (up to 15cm/6in long), heart-shaped at the base and regularly toothed around the margin. They are deep grass green above and paler beneath, with some hair in the leaf vein axils. In autumn, the leaves turn a strong yellow before falling. The winter shoots are orange-red and an attractive feature when caught by low winter sun.

*Above: The leaves are heart-shaped at the base and regularly toothed around the margin.*

## Caucasian Lime

*Tilia caucasica*

**Distribution**: Northern Iran to the Caucasus.
**Height**: 30m/100ft
**Shape**: Broadly columnar
**Deciduous**
**Pollinated**: Insect
**Leaf shape**: Rounded to broadly ovate

Although cultivated in Europe since at least 1880, this vigorous tree from the Caucasus and northern Iran has never been widely planted in gardens or parks across Europe. This is a shame, because it is a regular-shaped, tidy-looking tree, which does well in urban areas where space is not a limiting factor. It was identified and described by the German botanist Franz Joseph Ruprecht, and is probably more widely cultivated in Germany than elsewhere in Europe.

**Identification**: The bark is grey and smooth; the shoots and twigs are olive green and when young are covered with fine hairs. The leaves are almost round, up to 15cm/6in across, with a sharply bristle-toothed margin running to a short point. They are glossy deep green above and pale green with tufts of cream hairs in the vein axils beneath. Clusters of fragrant, pale yellow flowers, each with a single long, narrow, pale green bract, are produced in summer, followed by rounded fruits.

*Left: In autumn the leaves turn golden-brown.*

*Right: The flowers and fruits hang beneath a pale green bract.*

## Japanese Lime

*Tilia japonica*

There is a school of thought that suggests the Japanese lime may represent merely an eastern extension of the natural range of the small-leaved lime, *T. cordata*, to which it is closely related. In fact, there is a difference in leaf shape: the Japanese lime has larger leaves, with a distinctive pointed tip that is not so evident on the small-leaved lime. It is an attractive ornamental tree, which was introduced into Europe in 1875.

**Identification**: This medium-sized lime is conical in form with a densely leaved crown. Its bark is silver-grey and smooth. The leaves are broadly ovate, heart-shaped at the base and up to 7.5cm/3in long. They are regularly serrated around the margin, running to a long, pointed tip. They are bright matt green above, pale green beneath and appear relatively early in the spring. Flowering tends to be much more prolific than on the small-leaved lime, with large flower clusters appearing all over the crown in summer.

**Distribution**: Japan and eastern China.
**Height**: 20m/65ft
**Shape**: Broadly conical
**Deciduous**
**Pollinated**: Insect
**Leaf shape**: Broadly ovate

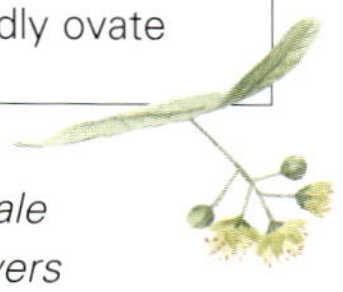

*Right: Clusters of pale yellow fragrant flowers appear in summer.*

*Left: The heart-shaped leaves run to a long pointed tip.*

OTHER SPECIES OF NOTE

***Tilia insularis***
This lime is native to the South Korean island of Cheju Do (Daghelet) and in the wild it may grow up to 35m/115ft tall. However, in cultivation it seldom exceeds a height of 20m/65ft. It is an attractive tree, which produces masses of fragrant pendulous flowers in summer. It has heart-shaped, coarsely toothed leaves, to 10cm/4in long.

***Tilia kiusiana***
On first inspection, this small tree or large shrub, which comes from southern Japan, looks more like a birch than a lime. It has fine, slender branches and small, narrowly ovate leaves, to 5cm/2in long. It produces multiple stems and is commonly grown in gardens and urban areas in some parts of central Europe. It is not cultivated so widely in Britain, where it was introduced from Japan in 1930.

***Tilia mandshurica***
This small to medium-sized lime is native to north-east Asia, from where it was introduced into Europe in 1860. It is related to the American lime, *T. americana*, and has similarly large, 20cm/8in long and 15cm/6in broad, coarsely toothed, heart-shaped leaves. It differs in having hair on both surfaces of the leaf and by the fact that it comes into leaf much earlier in the spring, as a result of which it is sometimes damaged by late frosts.

## Mongolian Lime

*Tilia mongolica*

The Mongolian lime was first seen by Europeans in 1864, when the French Jesuit missionary Abbé Armand David discovered it in central China, but it was not cultivated in Europe until 1880, when specimens were planted at the Jardin des Plantes in Paris. It is a slow-growing small lime, ideal for the smaller garden, and perfectly hardy. Unusually for a lime, its new leaves are reddish-bronze on emerging in early spring.

**Identification**: The bark is grey and smooth becoming slightly and vertically fissured in maturity. The leaves are broadly ovate, glossy dark green above and glaucous beneath with tufts of hairs in the leaf vein axils. They are up to 7.5cm/3in long and have between three and five irregular lobes. At first glance they look like maple (*Acer*) leaves, but are not borne in opposite pairs. In autumn they turn golden yellow before falling.

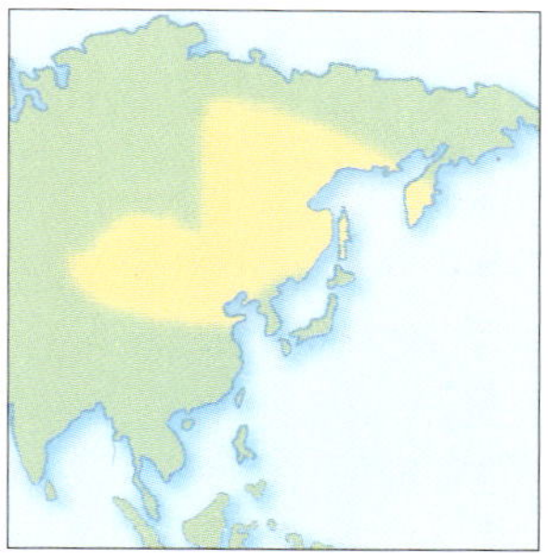

**Distribution**: Mongolia, northern China and eastern Russia.
**Height**: 15m/50ft
**Shape**: Broadly columnar
**Deciduous**
**Pollinated**: Insect
**Leaf shape**: Broadly ovate

*Below: The leaves are lobed and coarsely toothed.*

# POPLARS

*The poplars, of the family Salicaceae, are a genus of over 35 species of deciduous trees found throughout the northern temperate regions of the world. They produce small male and female flowers, which are borne in catkins on separate trees and pollinated by wind. Poplars are fast-growing trees, many of which can withstand atmospheric pollution and salt spray from the ocean.*

## Chinese Necklace Poplar

*Populus lasiocarpa*

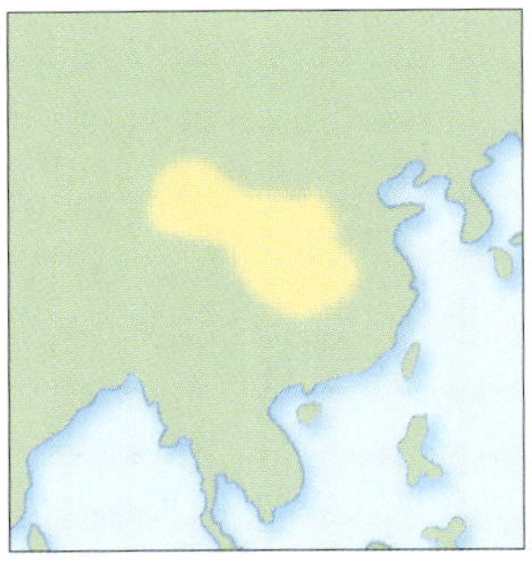

**Distribution**: Central China.
**Height**: 20m/65ft
**Shape**: Broadly spreading
**Deciduous**
**Pollinated**: Wind
**Leaf shape**: Broadly ovate

*Right: The large leaves have toothed margins.*

This striking, medium-sized tree has among the largest and thickest leaves of all poplars. The word 'necklace' in its name refers to the long, hanging, green seed capsules that appear in mid-summer. In late summer these ripen and burst, shedding phenomenal amounts of seed, over a wide area. It grows all over Europe.

**Identification**: The overall shape of the tree tends to be open and gangly; it can look particularly dishevelled in winter. The bark is grey-brown and fissured in maturity. The leaves are large and broadly ovate to heart-shaped; on some trees they are up to 35cm/14in long and 20cm/8in across. They have a leathery feel and are attached to the chunky shoots by red leaf stalks up to 10cm/4in long. The leaf-stalk colouring appears to 'bleed' into the midrib and main veins of the leaves. In autumn the leaves turn light brown and crisp, making a distinctive sound on windy days. The male and female flowers are stiff, yellow-green catkins, borne on separate trees in spring.

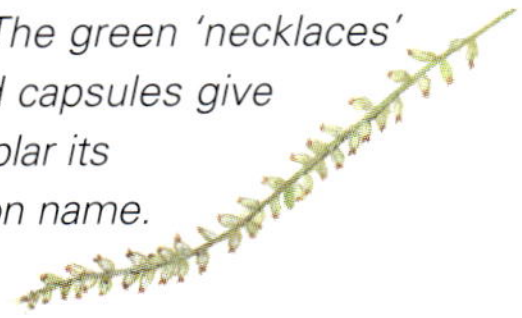

*Right: The green 'necklaces' of seed capsules give this poplar its common name.*

## Black poplar

*Populus nigra*

This is a large, broadly spreading tree, with long heavy branches. It inhabits river valleys and damp regions right across western Asia and much of Europe, but in western Europe, including Britain, it gives way to the Atlantic sub-species *P. nigra* subsp. *betulifolia*. Black poplar has long been cultivated outside its natural range.

**Identification**: The bark is pale grey and when young it is covered with darker diamond-shaped fissures. In maturity it develops a characteristic craggy appearance, with deep furrows and large burrs. The leaves are glossy deep green above, pale sage green beneath, and densely covered in silver-grey down. They are 7.5cm/3in long and broad. The male and female flowers are borne in drooping catkins up to 10cm/4in long, which appear on separate trees in early spring. The female flowers are green, the males are grey with red anthers. The fruit is a small green capsule; it ripens to release hundreds of tiny seeds, which are covered in white hairs that resemble cotton wool, and are carried on the wind.

**Distribution**: Western Asia and Europe.
**Height**: 30m/100ft
**Shape**: Broadly spreading
**Deciduous**
**Pollinated**: Wind
**Leaf shape**: Triangular to ovate

*Right: The leaves are ovate to triangular, sometimes rounded.*

OTHER SPECIES OF NOTE

**Grey Poplar** *Populus* x *canescens*
This large tree is a fertile hybrid between the white poplar, *P. alba*, and the aspen, *P. tremula*. It has become widespread throughout Europe and western Asia. Its striking leaves are deep sage-green above and grey-green, heavily felted, beneath. The bark is pale grey to cream, developing distinctive diamond-shaped dark plates in maturity.

**White Poplar** *Populus alba*
Native to Europe, western Asia and parts of North Africa, this tree is regularly planted alongside motorways, where it can withstand localized atmospheric pollution. It has light grey bark and distinctive white down on both the shoots and the underside of the leaves.

**Balsam Poplar** *Populus balsamifera*
This distinctive North American poplar was introduced into Europe before 1689 and has been widely cultivated. It has large, sticky leaf buds that emit a pleasant balsam scent as the leaves emerge in spring. It is a large (30m/100ft), broadly columnar, upright tree with ovate to triangular leaves, to 10cm/4in.

**Berlin Poplar** *Populus* x *berolinensis*
This is a popular hybrid between the Lombardy poplar, *P. nigra* 'Italica', and the Asian balsam poplar, *P. laurifolia*, which was raised at the Botanic Garden of Berlin around 1865. It is elegant, with a slender, columnar shape. It is used for street planting in central Europe and for windbreaks on the North American prairies.

## Lombardy Poplar

*Populus nigra* 'Italica'

The Lombardy poplar is probably one of the most easily recognizable of all trees because of its slender, columnar outline and upright branching. It is not a true species but a distinctive variety of the black poplar, *P. nigra*, which is believed to have originated beside the River Po in northern Italy in the early 1700s. Since then the Lombardy poplar has been propagated by cuttings and planted as an ornamental tree around the world.

*Below: Leaves may be 10cm/4in wide. Catkins have an orange tinge.*

**Distribution**: Originated as a 'sport' (variety) of black poplar in northern Italy.
**Height**: 30m/100ft
**Shape**: Narrowly columnar
**Deciduous**
**Pollinated**: Wind
**Leaf shape**: Ovate

**Identification**: The tree is immediately recognizable by its upright form. Strongly ascending branches grow from near ground level. The trunk, with its dark grey bark, is normally fluted and buttressed at the base. Each leaf is ovate to diamond-shaped and glossy bright green. Lombardy poplars are predominantly male trees, hence the need to propagate from cuttings. The male catkins are up to 7.5cm/3in long and borne in mid-spring. This poplar is not long-lived and is susceptible to bacterial canker and fungal diseases.

## Aspen

European aspen *Populus tremula*

This tree is properly known as European aspen, so as not to confuse it with its American cousin, *P. tremuloides*. The botanical name *tremula* is derived from the fact that the leaves, which are borne on slender flattened leaf stalks, tremble and quiver in even the slightest breeze. 'To tremble like an aspen leaf' is a phrase that goes back to the time of the 16th-century English poet, Edmund Spenser.

**Identification**: This medium-sized suckering tree has grey, smooth bark, becoming ridged at the base in maturity. The leaves are 7.5cm/3in long and equally wide. The leaf margin is edged with rounded teeth and there are three distinct, light-coloured veins at the base of each leaf. The leaves emerge a pink-bronze colour from the bud in spring, gradually turning dull green by early summer. Both male and female flowers are catkins, up to 5cm/2in long, borne in early spring before the leaves emerge. Tiny seeds carried in white, cotton wool-like hairs are released from green catkin-like capsules in late spring.

**Distribution**: Europe from the Atlantic to the Pacific, south to North Africa.
**Height**: 20m/65ft
**Shape**: Broadly spreading
**Deciduous**
**Pollinated**: Wind
**Leaf shape**: Broadly ovate

*Right: Male and female catkins appear on separate trees.*

*Left: Leaves are rounded.*

## Western Black Poplar

Atlantic black poplar *Populus nigra* subsp. *betulifolia*

This fascinating and confusing sub-species of black poplar, *P. nigra*, is known as the Manchester poplar in that English city, where the entire population originated from two or three individual trees. The true sub-species is considered rare, or possibly even endangered, in the wild. This is due to the cultivation of faster and straighter-growing hybrids, which have been planted in preference, as well as years of cross-pollination and hybridization with other cultivated poplars, such as *P. x canadensis*. Most of the mature pure specimens left are male

**Identification**: The bark is pale grey, smooth at first, becoming deeply fissured in maturity. The leaves are triangular to ovate, 7.5cm/3in long and broad, glossy dark green above and pale green beneath. This tree differs from the typical form of *P. nigra* in that its shoots, leaf stalks, young leaves and flower stalks are all covered with a soft pubescent down. Both male and female flowers are borne in long, pendulous catkins, which appear in early spring on separate trees.

**Distribution**: Britain, western France, Belgium and Holland.
**Height**: 30m/100ft
**Shape**: Broadly spreading
**Deciduous**
**Pollinated**: Wind
**Leaf shape**: Triangular to ovate

*Right: Male and female catkins appear on separate trees in early spring.*

*Left: The leaves are almost triangular in shape.*

## Asian Balsam Poplar

Laurel-leaf poplar *Populus laurifolia*

This slow-growing, medium-sized tree was introduced into western Europe from Siberia around 1830. Although it has an elegant form it has never been widely planted in parks and gardens and is normally confined to arboreta and botanic gardens. It has a rather lax growth, with semi-pendulous branching and distinctive, angular, grey young shoots. It is a balsam poplar and has sticky winter buds, which emit a strong balsam-like fragrance in early spring.

**Identification**: The bark is silver-grey and smooth. The leaves are lanceolate, or narrowly ovate, tapering at both ends, rounded at the base and pointed at the tip with a margin that has small, fine, regular serrations. Each leaf is up to 12.5cm/5in long and 5cm/2in wide, dark green above and grey-green beneath with some down. The underside displays a fine network of leaf veins. Both the male and female flowers are borne in catkins. The male catkins are erect at first, becoming pendulous and up to 5cm/2in long when ripe.

*Left: Asian balsam poplar has an open, spreading crown.*

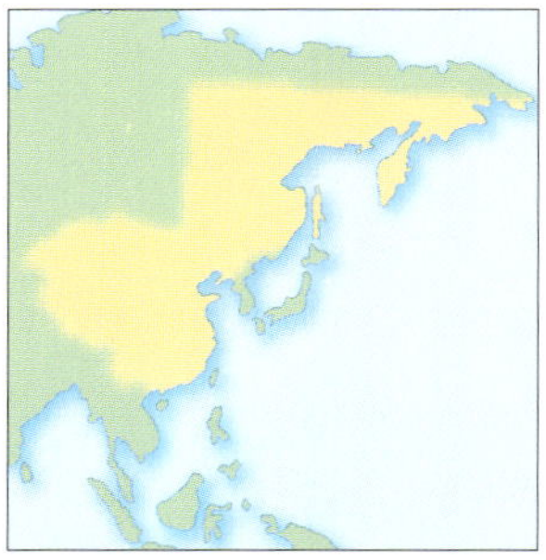

**Distribution**: Eastern Siberia, Mongolia and parts of China.
**Height**: 21m/70ft
**Shape**: Broadly spreading
**Deciduous**
**Pollinated**: Wind
**Leaf shape**: Narrowly ovate to lanceolate

*Right: The leaves are narrowly ovate, rounded at the base, pointed at the tip with small serrations around the margin.*

OTHER SPECIES OF NOTE

**Balm of Gilead** *Populus* x *candicans* 'Aurora'
This is a popular cultivar of a hybrid that is believed to have developed naturally between two North American species, the balsam poplar, *P. balsamifera*, and the eastern cottonwood, *P. deltoides*, in the 18th century. It was introduced into Europe in 1773. It is strongly balsam-scented when young. The cultivar 'Aurora' has been commercially available since the 1920s and has distinctive, broadly ovate leaves, which are strikingly marked in summer with pink, white and cream blotches.

***Populus szechuanica***
This is a fast-growing, handsome, large tree, growing to 30m/100ft, which is native to western China, from where it was introduced into Europe in 1908. It has dramatic, large, ovate leaves, 30cm/12in long and 20cm/8in broad, and a distinctive red midrib and leaf stalk, which stand out well against the pale green-white lower surface of the leaf.

***Populus wilsonii***
This attractive, medium-sized poplar takes its species name from the English plant collector Ernest Wilson, who introduced the species into Europe from Central China in 1907. It is similar to the Chinese necklace poplar, *P. lasiocarpa*, bearing long, pendulous strings of ripe fruit in summer, but with smaller leaves. On this tree the ovate leaves are 20cm/8in long and 15cm/6in wide, bright blue-green and borne on thick, rigid shoots.

## Korean Poplar

*Populus koreana*

This handsome poplar is native to Korea, parts of Russia and possibly Japan, where it naturalized in western regions of Honshu. It was introduced to Europe in 1918 by Ernest Wilson. It is a balsam poplar, and emits the distinctive balsam-like fragrance in early spring. It is one of the first trees to come into leaf in spring.

**Distribution**: Korea and southern Russia.
**Height**: 30m/100ft
**Shape**: Broadly columnar
**Deciduous**
**Pollinated**: Wind
**Leaf shape**: Ovate to oval

**Identification**: The bark is grey, smooth at first becoming heavily fissured in maturity. The leaves are up to 15cm/6in long and 10cm/4in broad, ovate to oval, borne on thick, strong shoots. They are fresh grass-green on the upper surface and almost white beneath, with some down. The central midrib is red. The male and female flowers are borne in long, pendulous catkins on separate trees in spring.

*Right: The female catkins may be up to 25cm/10in long.*

## Hybrid Black Poplar

*Populus* x *canadensis*

This hybrid between the European black poplar, *P. nigra*, and the North American cottonwood, *P. deltoides*, occurred naturally in western Europe, soon after the American species was introduced in the early 18th century. It was recognized as being superior, both in form and vigour, to each parent, and by the 1750s was widely planted throughout Europe. Since then, many fast-growing and ornamental cultivars of the original hybrid have been developed including 'Serotina', 'Robusta' and the glorious yellow-foliaged 'Serotina Aurea'.

**Identification**: The bark is grey, smooth at first, then developing deep, vertical fissures. The leaves are glossy green above, paler beneath, broadly triangular, to 10cm/4in long and broad with a short point at the tip. Both male and female flowers are borne in pendulous catkins on separate trees in early spring. The female flowers are green. The fruit is a small green capsule, which ripens to release numerous seeds coated in fluffy white hairs.

**Distribution**: Of garden origin in western Europe.
**Height**: 35m/115ft
**Shape**: Broadly columnar
**Deciduous**
**Pollinated**: Wind
**Leaf shape**: Broadly triangular

*Right: The male flowers are grey-green and up to 10cm/4in long.*

# WILLOWS

*There are more than 300 species of willow in the world, varying from large, spreading, ornamental specimens to diminutive, creeping, tundra-based alpines. The majority are native to northern temperate regions of the world. Willows are mainly deciduous, although one or two subtropical species have leaves that persist into winter. Male and female flowers are normally borne on separate trees.*

## Crack Willow

*Salix fragilis*

**Distribution**: Asia and Europe.
**Height**: 25m/80ft
**Shape**: Broadly spreading
**Deciduous**
**Pollinated**: Wind
**Leaf shape**: Lanceolate

This fast-growing riverside tree is one of the largest willows, and regularly attains heights in excess of 20m/65ft. It takes its name from the distinctive crack heard when a twig is snapped off the tree. The wood is brittle, and large branches are quite often broken off by high winds. Where such a branch touches the ground roots may grow out from the bark, keeping the branch alive and eventually making another tree genetically the same as the parent.

**Identification**: The bark is dull dark grey, becoming heavily fissured in maturity. Stems may be up to 2m/6½ft in diameter. The leaves are glossy dark green above and blue-green beneath, to 15cm/6in long and 2.5cm/1in broad, tapering to a fine point at the tip. Both the male and female flowers are very small and are borne in pendulous catkins, 5cm/2in long, on separate trees in spring.

*Right: The leaves appear in mid-spring.*

*Right: The male catkins are yellow, the females green.*

## White Willow

*Salix alba*

White willow thrives in damp soils and grows naturally alongside rivers and in watermeadows. The dense coating of hairs on the leaves gives the tree an attractive silvery appearance, particularly when the leaves are stirred by the wind.

**Identification**: The bark is brown-grey, becoming deeply fissured in maturity. The leaves are lanceolate, tapering at both ends, and are up to 10cm/4in long and 1.25cm/½in wide. Young leaves are covered in silver hairs on both sides; mature leaves retain this pubescence on the underside but the top side becomes smooth and bright green. Both male and female flowers are catkins, borne on separate trees in spring as the leaves emerge. The fruit is a green capsule, ripening to release numerous white-haired seeds, which are dispersed by the wind.

*Right: The long, slender leaves hang from the tree and move easily in the wind.*

**Distribution**: Europe and western Asia.
**Height**: 25m/80ft
**Shape**: Broadly columnar
**Deciduous**
**Pollinated**: Insect, and occasionally wind
**Leaf shape**: Lanceolate

*Right: Male catkins are 5cm/2in long and have yellow anthers; female catkins are green and smaller.*

## Goat Willow

Pussy willow *Salix caprea*

This small tree, sometimes a large shrub, of bushy habit, is common throughout Britain, Europe and western Asia. Through prolific seed dispersal it will, if left unchecked, colonize vacant or cultivated land for vast distances. Male trees produce the golden catkins known as "palm"; females produce the soft, silvery catkins called "pussy willow".

**Identification**: The bark is dull grey and smooth, with some shallow fissures in maturity. Young shoots are covered with fine grey down. The leaves vary from oval to obovate or lanceolate and are up to 10cm/4in long (usually 5–7.5cm/2–3in). They are dull green and wrinkled above and grey-green with dense hair beneath. Male and female flowers are produced in early spring on separate trees. The fruit is a grey-green capsule, covered with down, from which copious amounts of tiny seed are released to the wind.

**Distribution**: Europe and western Asia.
**Height**: 10m/33ft
**Shape**: Broadly spreading
**Deciduous**
**Pollinated**: Wind
**Leaf shape**: Variable

*Left: The leaf shape varies from oval to lanceolate and the male and female flowers are produced on separate trees in early spring.*

OTHER SPECIES OF NOTE

**Grey Willow** *Salix cinerea*
Sometimes known as grey sallow, this is a large shrub or small spreading tree, to 4.5m/15ft. It has obovate leaves, the undersides of which are covered in grey down. The grey willow is native to Europe, western Asia and parts of North Africa and is a pioneer species, meaning that it will colonize vacant land before other tree species. It grows well in wet, marshy conditions. There is a subspecies called *oleifolia*, which has slightly smaller leaves with copper-coloured down, and is more prevalent in western Europe.

**Violet Willow**
*Salix daphnoides*
Native to central Europe, this attractive small tree, to 10m/33ft tall, has long been cultivated because of its plum-coloured shoots, which when young are covered with a white bloom, and for its conspicuous male catkins in early spring. The cultivar 'Aglaia' produces masses of bright white male catkins in the late winter.

**Golden Weeping Willow** *Salix* x *sepulcralis* 'Chrysocoma'
This is the familiar ornamental weeping willow, grown alongside countless rivers and streams throughout western Europe. It is a cultivar of a hybrid between the white willow, *S. alba*, and the Chinese weeping willow, *S. babylonica*, and was raised in 1888. No other weeping willow has such bright yellow twigs and young leaves. It is prone to attack by canker disease.

## Bay Willow

*Salix pentrandra*

This is a beautiful small to medium-sized tree, with bay-like leaves that emit an attractive fragrance when crushed. In Scandinavia, bay willow is used as a culinary substitute for true bay, *Laurus nobilis*. Although widely distributed in both Europe and Asia, it prefers cooler conditions and does not grow naturally around the Mediterranean. It is grown as an ornamental tree in parks, gardens and arboreta, and is considered one of the most handsome willows in cultivation.

**Identification**: The bark is grey to grey-brown, smooth at first, becoming shallowly and vertically fissured in maturity. The leaves and shoots are a deep glossy green, up to 10cm/4in long and 5cm/2in broad. Both male and female flowers are small and borne in catkins on separate trees in late spring to early summer. The male flowers are yellow, the females are green.

**Distribution**: Europe and Asia.
**Height**: 15m/50ft
**Shape**: Broadly spreading
**Deciduous**
**Pollinated**: Wind
**Leaf shape**: Elliptic to narrowly ovate

*Below: The leaves are finely toothed around the margin and taper to a short point.*

## Almond-leaved Willow

*Salix triandra*

**Distribution**: Europe to eastern Asia.
**Height**: 12m/40ft
**Shape**: Broadly spreading
**Deciduous**
**Pollinated**: Wind
**Leaf shape**: Lanceolate

This small spreading tree is unusual among willows in having grey bark, which flakes to reveal lighter brown bark beneath. It has long been cultivated right across Europe and Asia for its vigorous, strong new growth, which is used to make heavy-duty basketware. In Britain it is grown on the Somerset Levels for this purpose. The stems of *S. triandra* may grow more than 3m/10ft in one season. Over the years many cultivars, including 'Black Hollander', 'French' and 'Pomeranian', have been developed specifically for basketry.

**Identification**: Almond-leaved willow has distinctive grey bark, flaking to reveal brown patches beneath. It has lanceolate, glossy green leaves, to 10cm/4in long, with a leaf margin that is finely toothed and tapering to a long point. The lower surface of the leaf is a glaucous blue-green. Small flowers are produced in catkins in mid- to late spring. The male catkins are up to 6cm/2½in long, pendulous and yellow; the female catkins are smaller and green. Each gender is produced on separate trees.

*Right and far right: The leaves are lanceolate-shaped, glossy green above and matt blue-green beneath.*

*Left: Yellow male catkins appear in early spring.*

## Kilmarnock Willow

*Salix caprea* 'Kilmarnock'

**Distribution**: Of Scottish garden origin.
**Height**: 3m/10ft
**Shaped**: Domed and pendulous
**Deciduous**
**Pollinated**: Wind
**Leaf shape**: Obovate

The original cultivar, called *S. caprea* 'Pendula', is said to have come from a single male tree, found by James Smith, "an old and enthusiastic Scottish botanist", growing on the banks of the River Ayr in Scotland, in the mid-1800s. By 1853 it was being sold commercially by the nurseryman Thomas Lang of Kilmarnock, hence its current name. Today, it is one of the most popular trees for growing in small gardens throughout Europe.

**Identification**: Unmistakable and unlikely to be confused with anything else, the Kilmarnock willow is a weeping form of the goat willow, *S. caprea*. It is a neat, umbrella-shaped tree with numerous weeping branches, covered in spring with beautiful "pussy willow" catkins that start silver-grey and gradually turn golden yellow. The catkins are borne on the tree before the obovate, wrinkled, grey-green leaves emerge from bud.

*Right: The leaves are a matt grey-green colour and quite often wrinkled.*

*Left: The female flowers are borne in less showy green catkins.*

*Left: In spring the silvery male catkins develop showy yellow anthers.*

## Corkscrew Willow

*Salix babylonica* 'Tortuosa'

This very popular willow is widely planted as a curious ornamental in gardens across Europe. It is a quirk of nature that was cultivated in northern China, where it is known as the dragon's claw willow. It was introduced into France and the rest of Europe in the 1920s. Everything about this tree is contorted – branches, twigs and leaves. Cut twigs and small branches are very often used by flower arrangers.

**Identification**: Instantly recognizable, this tree is unlikely to be confused with anything else (other than, possibly, in winter, the corkscrew hazel, *Corylus avellana* 'Contorta'). It is a vigorous small tree of roughly columnar habit, which can normally reach a height of 10–15m/33–50ft in less than 30 years. It is relatively short-lived, with few specimens surviving beyond 50 years of age. The contortions begin with the major branches and are carried through the twigs to the leaves, which are linear, glossy grass-green above, slightly paler beneath and up to 10cm/4in long.

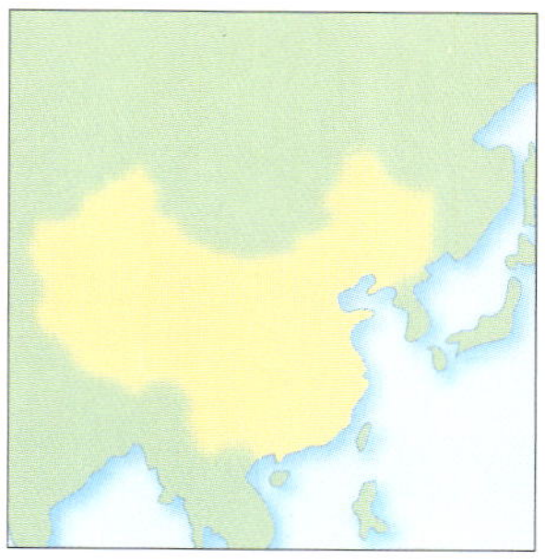

**Distribution**: China.
**Height**: 15m/50ft
**Shape**: Columnar and contorted
**Deciduous**
**Pollinated**: Wind
**Leaf shape**: Linear to lanceolate

*Left: Both the leaves and the twigs have a distinctive twisted appearance.*

OTHER SPECIES OF NOTE

**Silver Willow** *Salix alba* var. *sericea*
This is a smaller, more round-headed, less vigorous tree than the white willow, *S. alba*, but it is grown for its superb bright silver foliage, which is particularly striking when the whole tree is viewed from afar. This form occasionally occurs in the wild, but most trees available for sale from garden centres and nurseries have been selectively cloned in cultivation.

**Coral-bark Willow** *Salix alba* 'Britzensis'
This cultivar of the white willow, *S. alba*, is grown for its brilliant orange-scarlet winter shoots, which are reminiscent of the new shoots of the dogwood shrub, *Cornus alba* 'Sibirica'. The coral-bark willow has narrow, glaucous, lanceolate leaves and an upright columnar form. For best effect it should be coppiced to almost ground level every two to three years to stimulate bright new growth. It was raised at Britz, near Berlin, in 1878.

**Caspian Willow**
*Salix acutifolia*
This elegant small tree, to 10m/33ft, is native to Russia, and was first cultivated in western Europe around 1890. It is closely related to the violet willow, *S. daphnoides*, and has similar plum-red shoots, which are covered in white bloom when young. The leaves are narrowly lanceolate with a long point at the tip, rather lax and up to 15cm/6in long. The male flowers are borne in erect bright yellow catkins in spring.

## Cricket Bat Willow

*Salix alba* var. *caerulea*

Sometimes referred to as the blue willow, this subspecies is a variant of the white willow, *S. alba*, and was discovered as a single specimen growing in Norfolk, England, around 1700. As the common name suggests, the timber of this tree is highly prized for the production of cricket bats. All the trees cultivated worldwide for this purpose have been propagated vegetatively (by cuttings), so effectively the original 1700 tree is still being grown today.

**Distribution**: England.
**Height**: 30m/100ft
**Shape**: Broadly conical
**Deciduous**
**Pollinated**: Wind
**Leaf shape**: Lanceolate

**Identification**: Cricket bat willow has a conical or pyramidal form with ascending branches. The grey-brown bark is smooth at first and becomes vertically fissured in maturity. The lanceolate leaves are a distinctive blue-grey-green and lose their downy covering much earlier than those of the white willow, *S. alba*. They are up to 10cm/4in long. The winter buds are long, pointed, held close to the shoot and a bright chestnut brown.

*Right: The leaves are long and thin and a distinctive blue-grey-green colour.*

# PLANES AND WITCH HAZELS

*Although the trees under this heading belong to two different genera, they have two things in common. They are all deciduous, dropping their leaves in autumn and producing replacements the following spring. And, before they lose their leaves, they all take on spectacular autumn leaf colour. Because of their beauty, many are planted as ornamental trees.*

## London Plane

*Platanus* x *hispanica*

**Distribution**: Of garden origin.
**Height**: 40m/130ft
**Shape**: Broadly columnar
**Deciduous**
**Pollinated**: Insect
**Leaf shape**: Palmate lobed

This tree is a hybrid between the oriental plane, *P. orientalis*, and the American buttonwood, *P. occidentalis*. It is widely planted in cities across the world as it is able to withstand atmospheric pollution and severe pruning. It is more vigorous than both its parents, its leaves have shallower lobes and it has a lighter bark, which peels to reveal cream patches.

*Above: The fruits are mace-like, bristly spheres, hanging on long stalks.*

**Identification**: The leaves are glossy bright green above and paler green beneath, with conspicuous yellow leaf veins, and are covered with rust-coloured down when young. They are 20cm/8in long and up to 25cm/10in broad, with three to five large toothed lobes. Both the yellow male and red female flowers are small and are borne in clusters on the same tree in late spring to early summer. The fruit persists well into winter after the leaves have fallen.

## Oriental Plane

*Platanus orientalis*

**Distribution**: Albania, Greece, Crete, Cyprus, Lebanon, Syria and Israel.
**Height**: 30m/100ft
**Shape**: Broadly spreading
**Deciduous**
**Pollinated**: Insect
**Leaf shape**: Palmate lobed

The oriental plane is a majestic tree. Hippocrates, the ancient Greek 'father of medicine', is said to have taught his medical students under the great oriental plane tree that still stands on the island of Cos. Another large oriental plane, beside the Bosphorus near Buyukdere, is known as 'the plane of Godfrey de Bouillon', because tradition states that the French nobleman and his knights camped under it during the first Crusade in 1096.

**Identification**: One of the largest of all deciduous temperate trees, this plane can reach heights in excess of 30m/100ft, with a great spreading canopy and a trunk girth of up to 6m/20ft. It has attractive buff-grey bark, which flakes to reveal cream-pink patches. The leaves are palmate, 20cm/8in long and 25cm/10in across, deeply cut into five narrow lobes, shiny green above, pale green below, with brown tufts of hair along the veins. They are attached alternately to the shoots by yellow-green leaf stalks up to 7.5cm/3in long. In autumn the leaves turn clear yellow, then old gold. The fruits are globular and mace-like, 2.5cm/1in across, attached in clusters of two to six on pendulous stalks.

*Right: Fruit remains on the tree through the winter.*

## Chinese Witch Hazel

*Hamamelis mollis*

This small, widely spreading ornamental tree, or large shrub, was introduced into Europe in 1879. It is prized for its bright yellow, fragrant winter flowers, and is widely cultivated in parks, gardens and arboreta. In autumn the leaves turn clear yellow before falling. Chinese witch hazel is best grown in acid soil.

**Identification**: The bark is dark grey and smooth, occasionally flaking to reveal lighter bark beneath. The leaves are up to 13cm/5in long (more usually 10cm/4in) matt mid-green above with some hair beneath, unequally heart-shaped at the base and shallowly toothed around the margin, which runs to an abrupt point. The fragrant yellow flowers are borne in crowded clusters, directly on the twigs, in winter.

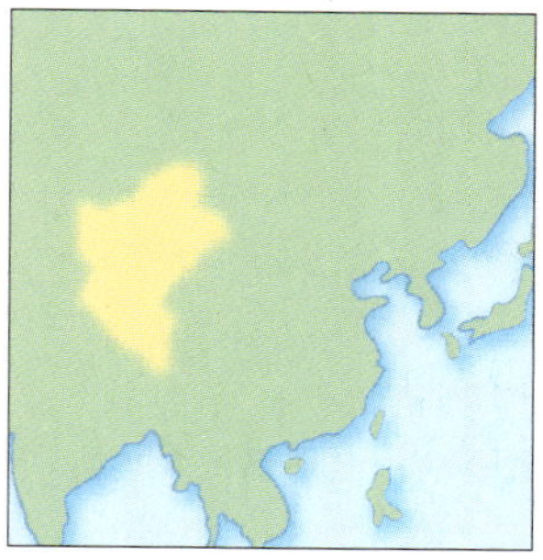

**Distribution**: Western China.
**Height**: 4m/13ft
**Shape**: Broadly spreading
**Deciduous**
**Pollinated**: Wind
**Leaf shape**: Broadly ovate

*Right: The leaves are broadly ovate to rounded.*

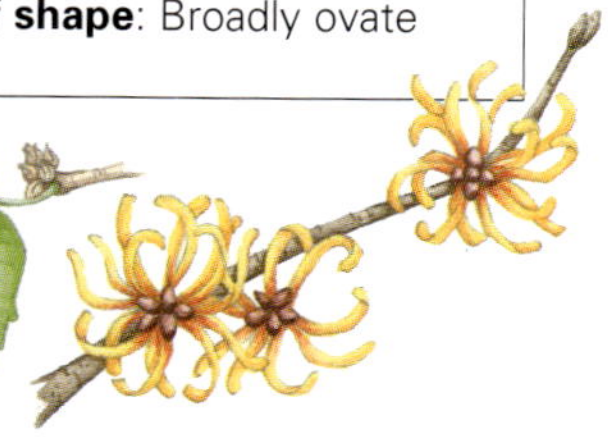

*Far right: The flowers are made up of rich sulphur-yellow, spidery petals.*

**Distribution**: Japan.
**Height**: 4m/13ft
**Shape**: Broadly spreading
**Deciduous**
**Pollinated**: Wind
**Leaf shape**: Oval to ovate

*Right: Each dark green leaf has five to eight pairs of parallel leaf veins, which run forward from the midrib.*

## Japanese Witch Hazel

*Hamamelis japonica*

This beautiful winter-flowering tree, or large shrub, was introduced into Europe from Japan in 1862. It is widely cultivated across Europe. It is small, spreading and sparsely branched and has smaller, more variable leaves than Chinese witch hazel, *H. mollis*. They are sometimes almost diamond-shaped and have a glossier upper surface. Japanese witch hazel also tends to flower more prolifically than its Chinese cousin.

**Identification**: The bark is dark grey and smooth. The leaves are variable, to 7.5cm/3in long and 5cm/2in broad. They have a wavy margin and are slightly heart-shaped at the base. The lower surface of the leaf is covered with fine grey down, which disappears by autumn, when the leaves turn bright yellow before falling. The flowers are bright yellow and fragrant, with four spidery, crimped petals.

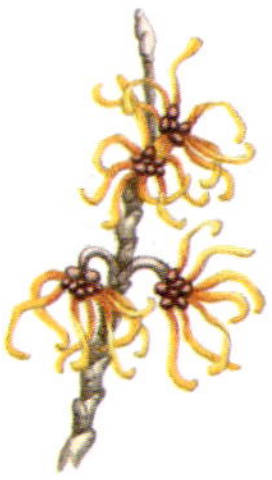

*Right: The flowers are borne directly on the bare branches in winter.*

OTHER SPECIES OF NOTE

**Buttonwood** *Platanus occidentalis*
Sometimes called the American sycamore, this large tree occurs right across eastern North America. Specimens have been grown in Europe since the 17th century, and it is one of the parents of the London plane, *P.* x *hispanica*. The leaf is heart-shaped at the base, with more, but shallower, lobing than either the London plane or the oriental plane. The bark is grey-brown and has the same flaking characteristic as the other planes.

**Virginian Witch Hazel**
*Hamamelis virginiana*
This small North American deciduous tree, to 6m/20ft tall, has spreading zig-zag branches and a short, thick trunk. It is native to eastern North America, from Nova Scotia to Tennessee, from where it was introduced into Europe in 1736. It is not grown so widely as the Asian witch hazel, mainly because the beauty of its sulphur-yellow flowers is masked by it still being in full leaf when the flowers appear in autumn.

**Hybrid Witch Hazel** *Hamamelis* x *intermedia*
This is a hybrid between the Chinese and Japanese witch hazels, and is of variable nature, taking on some of the characteristics of both parents. From this hybrid many of the common cultivars commercially available today have been raised. These include 'Arnold Promise', 'Diane', which has red flowers, and 'Jelena', which has copper-coloured flowers.

# SWEET GUMS AND TREES WITH STUNNING AUTUMN FOLIAGE

*The trees in this section all belong to different families, but are grouped together here because of the stunning leaf colours they produce in autumn. Because of their beauty many are planted in Europe as ornamental trees in parks.*

## Katsura Tree

*Cercidiphyllum japonicum*

The single representative of the family Cercidiphyllaceae, this graceful tree, with its attractive foliage and the unusual "sweet caramel" aroma emitted by the leaves as they begin to decompose, is a favourite for ornamental planting. Once established, growth may be anything up to 60cm/24in per year.

*Right: Male and female flowers are on separate trees.*

**Identification:** The bark is grey-brown, freckled with lenticels, and becoming fissured and flaking in maturity. The thin leaves are heart-shaped, slightly toothed around the margin and up to 7.5cm/3in long and wide, turning orange-apricot in late autumn. The male flowers are bright red, appearing on side shoots before the leaves emerge in early spring. Female flowers develop in late spring in clusters of four to six.

*Left: Leaves fade to blue-green in summer and then vibrant shades of butter yellow to purple-pink in autumn.*

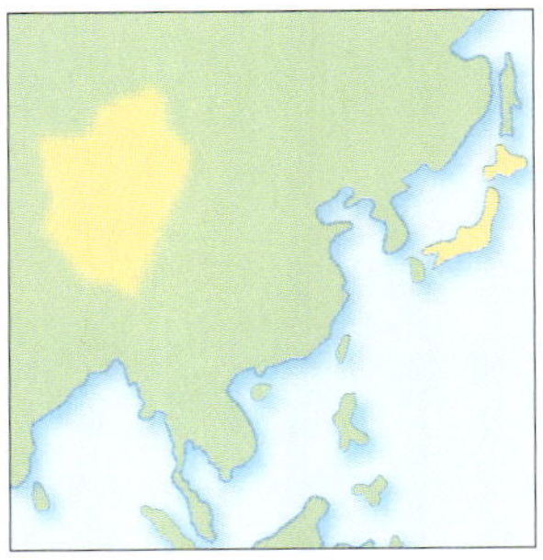

**Distribution**: Western China, and Hokkaido and Honshu in Japan.
**Height**: 30m/100ft
**Shape**: Broadly spreading
**Deciduous**
**Pollinated**: Insect
**Leaf shape**: Cordate

## Chinese Tupelo

*Nyssa sinensis*

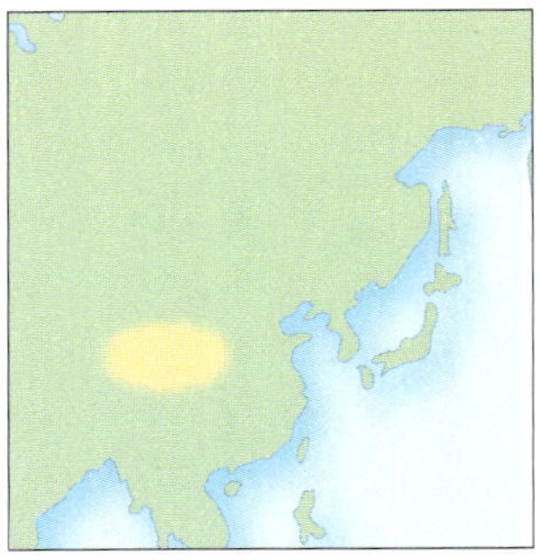

**Distribution**: Central China.
**Height**: 15m/50ft
**Shape**: Broadly conical
**Deciduous**
**Pollinated**: Insect
**Leaf shape**: Oblong to lanceolate

Closely related to both the tupelo, *N. sylvatica*, and the water tupelo, *N. aquatica*, this attractive small tree grows in woodland and along stream banks in central China. It is comparatively rare in cultivation, although under the right conditions it produces superb autumn leaf colours, ranging from red to yellow. Seeds of the tree were first sent from China to Britain in 1902 by the English plant collector Ernest Wilson.

**Identification**: The bark is grey-brown and smooth, becoming cracked and flaking with age. The narrow, pointed leaves, up to 20cm/8in long and 6cm/2½in wide, emerge from bud a purplish-red but quickly fade to shiny deep green above and paler and matt beneath. Both the male and female flowers are small and green. They are borne in separate clusters in the leaf axils on the same tree in summer. The fruit is a purple-blue berry about 2.5cm/1in long.

*Right: In autumn the leaves turn red and yellow.*

*Left: The narrow pointed leaves are a shiny deep green colour on the top side.*

## Chinese Sweet Gum

*Liquidambar formosana*

This large tree inhabits mountainous woodland areas, growing at altitudes of up to 2,000m/6,560ft, in Taiwan (formerly Formosa) and southern and central China. It was introduced into Europe in 1884, but in cultivation rarely attains heights in excess of 20m/65ft. It is grown mainly for its autumn colour, which in good years ranges from vibrant scarlet to burgundy. In spring, when the leaves first emerge from bud, they are bronze-purple before turning dark green.

**Identification**: The bark on young trees is grey-white, becoming darker and vertically fissured in maturity. The palmate leaves, to 12.5cm/5in long and 15cm/6in broad, are carried on long, purple-red stalks. They normally have three lobes but occasionally five, each being regularly toothed and running to a fine point. The male and female flowers are small and green, borne in separate rounded clusters on the same tree in spring.

**Distribution**: China and Taiwan.
**Height**: 40m/130ft
**Shape**: Broadly conical
**Deciduous**
**Pollinated**: Insect
**Leaf shape**: Palmate lobed

*Left: The deep green leaves take on a range of vibrant red tints in autumn.*

OTHER SPECIES OF NOTE

**Tupelo** *Nyssa sylvatica*
This slow-growing, medium-sized North American tree, also known as the black gum, was introduced into Europe in 1750 and has been widely cultivated, mainly for its spectacular autumn colours, which range from yellow and orange to red and burgundy. It has obovate to elliptic leaves up to 15cm/6in long.

**Sweet Gum** *Liquidambar styraciflua*
This giant tree, with maple-like leaves, is known to have reached 45m/150ft in height in its native eastern North America. In cultivation in Europe it rarely exceeds 30m/100ft. Sweet gum takes its name from the sweet, viscous sap it exudes when the bark is damaged. This sap dries into a gum, which was at one time used as chewing gum in the USA. The tree was introduced into Europe in the 17th century.

**Oriental Sweet Gum** *Liquidambar orientalis*
This slow-growing, medium-sized tree, to 20m/65ft, is native to south-west Turkey, from where it was introduced into Europe around 1750. It has leaves that are similar to those of the field maple, *Acer campestre*, with leaves deeply cut into three or five lobes. In autumn the leaves turn a glorious orange-marmalade colour.

## Persian Ironwood

*Parrotia persica*

A member of the Hamamelidaceae family, this is one of the finest trees for autumn colour, turning copper to burgundy in mid-autumn. In the wild it tends to be a broad, upright tree, but in cultivation it becomes a sprawling mass, seldom attaining a height greater than 15m/50ft. Its botanical name is derived from the German climber F. W. Parrot, who made the first ascent of Mount Ararat in 1829.

*Above: Small ruby-red flowers appear in winter.*

**Identification**: The tree is quite often seen in large gardens and arboreta as a dense low-spreading mound, which is quite difficult to penetrate because of criss-cross branching. The bark is dark brown, flaking to reveal light brown patches. The leaves are obovate, sometimes elliptic, 12cm/4¾in long and 6cm/2½in wide, becoming progressively shallowly toothed and wavy towards the tip. They are glossy bright green above and dull green with slight hairiness beneath. The flowers are tiny, clothed in soft velvet-brown casing but emerging a startling ruby-red, which stands out dramatically on the bare branches in mid-winter. The fruit is a nut-like brown capsule, 1cm/½in across.

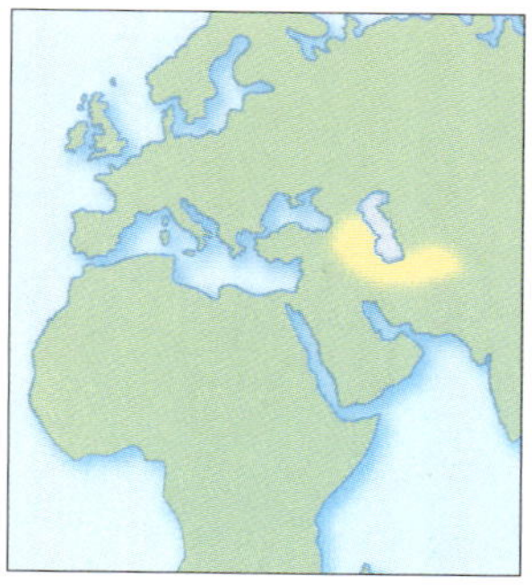

**Distribution**: Mount Ararat, eastern Caucasus to northern Iran.
**Height**: 20m/65ft
**Shape**: Broadly spreading
**Deciduous**
**Pollinated**: Insect
**Leaf shape**: Obovate

*Below: The leaves are dark and glossy on top and lighter beneath.*

# STRAWBERRIES, SNOWBELLS AND STEWARTIAS

*Strawberries, persimmons and snowbells form part of the camellia subclass, known as Dilleniidae. They are a mixture of deciduous and evergreen, acid-loving and lime-tolerant species. They all have an ornamental appeal, which has secured their places in gardens and arboreta across the temperate world.*

## Strawberry Tree

*Arbutus unedo*

This small, attractive, evergreen tree is a member of the heather family but, unlike most heathers, it grows perfectly well on limy soil. It is sometimes called the 'Killarney strawberry tree' because it is so common in that part of Ireland, where it is often used for hedging. The strawberry-like fruits, which are borne on the tree at the same time as the flowers, are edible but not enjoyable.

**Identification**: The young bark is a rich red-brown, fading to grey-brown and becoming rough and fissured in maturity. The tough, leathery leaves are elliptic to obovate, up to 10cm/4in long, glossy dark green above and paler beneath, with uniform serrations around the margin. The small creamy-white flowers are urn-shaped and are borne in drooping clusters in autumn. The strawberry-like fruit is a warty berry, approximately 2.5cm/1in across and ripening red in autumn from flowers borne in the previous year.

**Distribution**: Southern Ireland, western France, Iberian peninsula and countries bordering the Mediterranean.
**Height**: 10m/33ft
**Shape**: Broadly spreading
**Evergreen**
**Pollinated**: Insect
**Leaf shape**: Elliptic

*Left: Drooping clusters of creamy-white flowers are borne in autumn.*

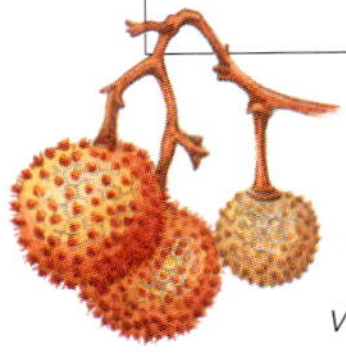

*Left: The fruit is a red, strawberry-like, warty berry.*

## Grecian Strawberry Tree

Cyprus strawberry tree *Arbutus andrachne*

**Distribution**: South-east Europe.
**Height**: 10m/33ft
**Shape**: Broadly spreading
**Evergreen**
**Pollinated**: Insect
**Leaf shape**: Elliptic to obovate

This beautiful small tree has been in cultivation across warm temperate regions of Europe since 1724. It is rather tender when young but hardier in maturity. It has particularly attractive bark, which is a rich red-brown to cinnamon colour and peels to reveal bright orange or cream new bark beneath. Where it grows in the wild alongside the strawberry tree, *A. unedo*, the two species naturally cross-pollinate to produce the hybrid strawberry tree, *A.* x *andrachnoides*.

**Identification**: This is a dome-shaped tree, very often with multiple stems growing from just above ground level. The oval leaves are thick and leathery, up to 10cm/4in long and 5cm/2in broad, glossy dark green above and pale green beneath, with no serrations around the leaf margin. The flowers are small and green, turning to white or creamy-white as they open. They are heather-like and are borne in upright clusters at the end of the shoots in late spring.

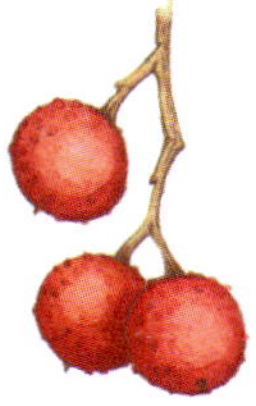

*Right: The fruits are small, rounded, orange-red berries and the leaves are glossy dark green.*

## Japanese Snowbell Tree

*Styrax japonica*

This beautiful, small, spreading tree deserves to be much better known and more widely planted than it is. It was introduced to the West in 1862, when Richard Oldham collected a specimen from Japan, which was then planted in the Royal Botanic Gardens, Kew, London. Since then the tree has been grown in various botanic collections and arboreta, but is comparatively rare as a garden specimen. It is an ideal tree for a small garden: it seldom reaches more than 7m/23ft tall and is perfectly hardy even in cooler regions, although late frosts may injure the flower buds.

**Identification**: The bark is orange-brown, smooth at first and becoming fissured in maturity. The leaves are oval to elliptical, tapering at both ends, to 10cm/4in long and 5cm/2in across. The leaf margin is set with small, shallow teeth. The leaves are a rich shiny green above and a paler green beneath, and are normally arranged in groups of three on the shoot. The slightly fragrant, open, bell-shaped flowers are creamy white with yellow anthers, and hang, either on their own or in small clusters, all along the branches on long slender stalks in early summer. The overall effect is delightful. The fruit is an egg-shaped green-grey berry containing a single seed.

**Distribution**: Japan, Korea and China.
**Height**: 10m/33ft
**Shape**: Broadly spreading
**Deciduous**
**Pollinated**: Insect
**Leaf shape**: Oval to elliptic

*Below: The leaves are reminiscent of bay. Once pollinated, the hanging flowers form green berries.*

OTHER SPECIES OF NOTE

**Mountain Snowdrop Tree** *Halesia monticola*
This magnificent spreading tree grows wild in upland areas, reaching altitudes in excess of 1,230m/4,000ft in North Carolina, Tennessee and Georgia, USA. It is widely cultivated for its masses of hanging, bell-shaped white flowers, which appear in late spring before the leaves.

**Hybrid Strawberry Tree** *Arbutus* x *andrachnoides*
This is a naturally occurring hybrid between the strawberry tree, *A. unedo*, and the Grecian strawberry tree, *A. andrachne*, which is found most commonly in Greece. It is widely planted elsewhere across Europe as an ornamental species in gardens and parks. It inherits its attractive flaking red-brown bark from one parent and its hardiness from the other.

**Hemsley's Storax** *Styrax hemsleyana*
This beautiful small Chinese tree, or large shrub, to 8m/26ft tall, was introduced into Europe in 1900. It has ovate to obovate leaves, to 13cm/5in long and 10cm/4in broad, and small white flowers with bright golden anthers, which are borne in upright terminal racemes in early summer.

## Japanese Stewartia

*Stewartia pseudocamellia*

This small to medium-sized, tender Japanese tree is related to the camellia, which is reflected not only in its species name, but also in its beautiful camellia-like flowers. It was introduced into cultivation in Europe in the 1870s. It takes its generic name from John Stuart, a patron of the Royal Botanic Gardens in Kew, London.

*Below: The flowers have five frilly petals and conspicuous bright orange stamens.*

*Below: The leaves are ovate.*

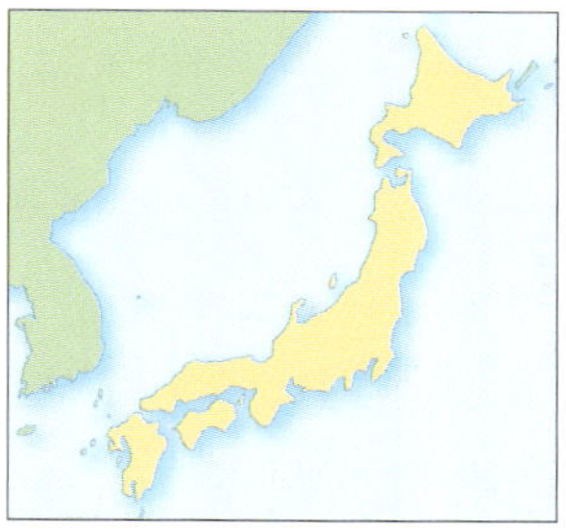

**Distribution**: Japan.
**Height**: 20m/65ft
**Shape**: Broadly columnar
**Deciduous**
**Pollinated**: Insect
**Leaf shape**: Elliptic to ovate

**Identification**: The bark is smooth, red-brown and thinly flaking to reveal patches of cream, fawn or pink-grey fresh bark. The leaves, alternately borne on the twig, are dark green above, paler beneath with some hair, and up to 10cm/4in long. In autumn they turn orange-red to burgundy-purple before falling. The white flowers appear in mid-summer; each one falls all in one piece when it is finished.

# LABURNUMS, BROOMS AND ACACIAS

*The pea family, Leguminosae, contains over 15,000 species of trees, shrubs and herbaceous plants in 700 genera. They are found growing wild throughout the world, in both temperate and tropical conditions. Most have compound leaves, pea-like flowers and seed pods, and root systems that have the ability to use bacteria to absorb nitrogen from the soil.*

## Mount Etna Broom

*Genista aetnensis*

Although it is native only to Sardinia and Sicily, where it grows prolifically in volcanic ash on the slopes of Mount Etna, this hardy, spreading broom has become naturalized along much of the coastline of southern and western Europe. In mid- to late summer every shoot is usually covered with fragrant, bright yellow pea-flowers.

**Identification**: This elegant tree has grey-brown bark, which may be very heavily fissured around the base. The bright green leaves, sparsely borne on green shoots, are narrowly linear and small, to 1cm/½in long. By summer, when the tree comes into flower, most of the leaves have fallen.

*Right: The flowers are 1.5cm/½in long, golden yellow and borne in profusion.*

**Distribution**: Sardinia and Sicily.
**Height**: 15m/50ft
**Shape**: Broadly spreading
**Deciduous**
**Pollinated**: Insect
**Leaf shape**: Linear

*Right: Shoots are grass-like.*

OTHER SPECIES OF NOTE

**Voss's Laburnum** *Laburnum* x *watereri* 'Vossii'
This beautiful small tree, with long racemes of sulphur yellow pea-flowers, is a cultivated form of *L.* x *watereri*, which is the result of a natural cross between common laburnum, *L. anagyroides*, and the Scotch laburnum, *L. alpinum*. The cross is believed to have occurred in the Austrian Tyrol in the 1850s. Voss's laburnum, which is of garden origin, is the most planted of all laburnums.

**Scotch Laburnum** *Laburnum alpinum*
Native to the mountains of central and southern Europe, this hardy, small, widely spreading tree produces long, weeping racemes of bright yellow, fragrant flowers in late spring to early summer. After flowering, brown, flat, pea-like pods are formed that contain hard-coated (poisonous) black seeds.

**Adam's Laburnum** + *Laburnocytisus adamii*
This strange and remarkable small tree is an inter-generic graft hybrid (chimaera) between common laburnum, *L. anagyroides*, and purple broom, *Cytisus purpureus*. It was produced by a Paris nursery in 1852. Grown widely in parks and gardens across Europe for its curiosity value, it produces flowers of both parents on the same tree.

## Judas Tree

*Cercis siliquastrum*

This small, sparsely branched tree is said to be the one from which Judas Iscariot hanged himself, but the name may derive from "Judea's tree", after the region encompassing Israel and the Palestinian state, where the tree is commonplace. It has long been cultivated across Europe for its bright pink flowers. It requires a sunny position to flower well.

**Identification**: Mature trees may develop a pronounced lean, or even fall to the ground, but carry on growing. The leaves quite often fold in on themselves along the midrib. The flowers are borne in clusters on old wood, just as the leaves emerge.

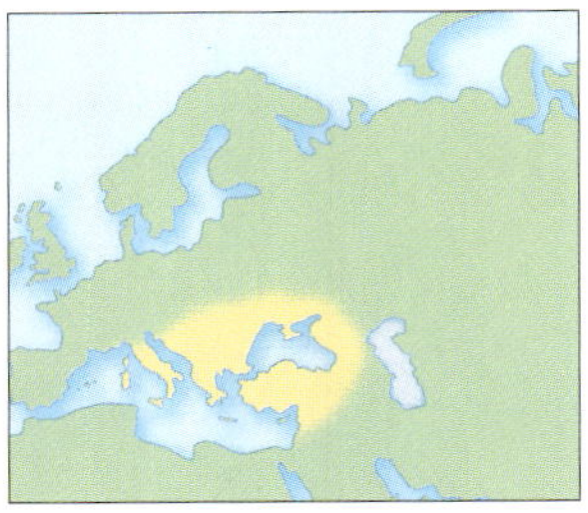

**Distribution**: South-east Europe and western Asia.
**Height**: 10m/33ft
**Shape**: Broadly spreading
**Deciduous**
**Pollinated**: Insect
**Leaf shape**: Rounded to cordate

*Above: Lilac-pink pea-like flowers appear in spring.*

*Right: The leaves are heart-shaped.*

## Mimosa

Silver wattle *Acacia dealbata*

This tender, temperate tree is prized by florists and flower arrangers the world over for its delicate, feathery foliage and fragrant yellow flowers. It has been known to reach 30m/100ft in height in the wild but seldom attains this in cultivation. It is extremely popular for planting as an ornamental street tree in Mediterranean countries but does not grow as well in cooler climates further north.

**Identification**: The bark is green-grey to almost glaucous with pale vertical striations. It becomes darker in maturity. The lax leaves, up to 12cm/4¾in long, are doubly pinnate and have countless small, linear, blue-green hairy leaflets, giving the whole tree a soft, feathery effect. The small, rounded flowers are sulphur yellow and fragrant. They are clustered on rounded panicles up to 10cm/4in across. In the Southern Hemisphere flowers appear in summer; in Europe they bloom from late winter to early spring. The fruit is a flat, blue-white seed pod, ripening to brown. It contains several round brown seeds.

*Below: The flat seed pod measures up to 7.5cm/3in long.*

*Left: The scented flowers open in summer in the Southern Hemisphere. North of the Equator they appear earlier.*

**Distribution**: South-east Australia and Tasmania.
**Height**: 25m/80ft
**Shape**: Broadly conical
**Evergreen**
**Pollinated**: Insect
**Leaf shape**: Bipinnate

## Common Laburnum

*Laburnum anagyroides*

This beautiful tree occurs naturally in mountainous regions of central Europe at elevations up to 2,000m/6,560ft. A small, spreading, short-lived tree, it grows particularly well on lime-rich soils and is best known for its profusion of pendulous golden yellow flowers in late spring. All parts of the tree contain an alkaloid that is poisonous if eaten; the green, unripe seed pods are particularly toxic.

*Above: The bright yellow flowers appear in spring.*

**Distribution**: Central and southern Europe from France to Hungary and Bulgaria.
**Height**: 9m/30ft
**Shape**: Broadly spreading
**Deciduous**
**Pollinated**: Insect
**Leaf shape**: Trifoliate

*Below: The seedpods hang down from the tree after flowering. The leaves are trifoliate.*

**Identification**: The bark is dark brown-grey and smooth, becoming shallowly fissured in maturity. The new shoots are olive green and the winter buds are covered with silver hairs. The elliptic leaflets, borne in threes, are up to 10cm/4in long, rich green above, grey-green beneath and covered with silver hairs when young. The flowers are golden yellow, pea-like, 2.5cm/1in long, in dense, hanging sprays up to 30cm/12in long. They are followed by green, hairy, pea-like seed pods, which ripen brown and contain several small, round, black seeds.

# CHERRIES

*The cherry genus,* Prunus, *contains over 400 different species of tree, the majority of which are deciduous and native to northern temperate regions of the world. They include some of the most beautiful spring-flowering trees, many of which have been cultivated in parks, gardens and arboreta for centuries. The genus is distinguished by having fruit that is always a drupe surrounding a single seed.*

## Wild Cherry

Gean, Mazzard *Prunus avium*

**Distribution**: Europe.
**Height**: 25m/80ft
**Shape**: Broadly columnar
**Deciduous**
**Pollinated**: Insect
**Leaf shape**: Elliptic

This is the largest of Europe's native cherries. It has been widely grown across Europe and in western Asia and is the parent of many cultivated sweet cherries that are grown for their edible fruit. The botanical name *avium*, meaning "of the birds", is a reference to the fact that the fruit is loved by birds. The wild cherry is used as a rootstock on to which many other ornamental members of the Rosaceae family are grafted.

**Identification**: The bark is rich red brown and glossy; in maturity distinctive horizontal light brown bands of lenticels, interspersed with peeling red-brown bark, ring the stem. Suckering around the base is common. The leaves are elliptic, occasionally oblong, up to 15cm/6in long and 5cm/2in broad, deep green on both sides and sharply toothed around the margin. The flowers are single, white with five petals, 2.5cm/1in across and borne in clusters in spring just as the leaves emerge from bud.

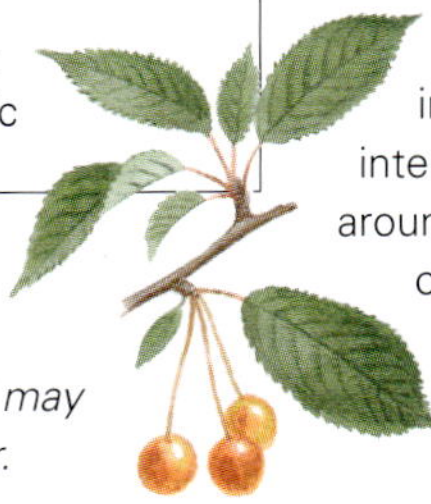

*Right: The fruit of the wild cherry is small and shiny and may taste sweet or bitter.*

*Right: The white flowers are sweetly scented.*

## Bird Cherry

*Prunus padus*

This hardy, medium-sized tree has a natural range that extends from the British Isles to Japan and grows on both acid and alkaline soils, on the edge of woodland or in small clearings. It has small, black, bitter-tasting fruits, which are relished by birds, hence its common name. The tree has long been grown in parks and gardens across Europe and has given rise to a number of ornamental cultivars, including the beautiful 'Colorata', which has dark purple stems and pink flowers.

**Identification**: The bark is dark grey and smooth and has a strong acrid smell when cut or scratched. The elliptic leaves are dark matt green, to 10cm/4in long, finely serrated around the leaf margin and ending in a short point. In autumn they turn red or orange before falling. The flowers, which appear in mid- to late spring, are small, white, fragrant and borne in 10–15cm/4–6in long, upright racemes, which become lax and pendulous as they mature. Small, round to egg-shaped berries, to 1cm/½in across, are produced in summer.

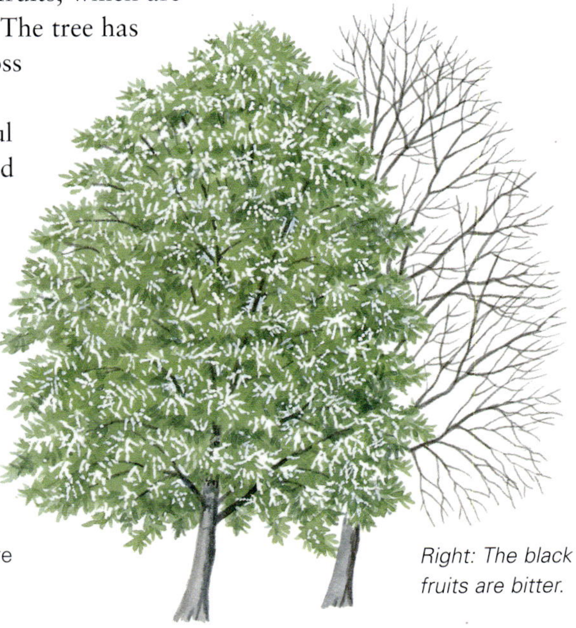

**Distribution**: Europe and northern Asia.
**Height**: 15m/50ft
**Shape**: Broadly spreading
**Deciduous**
**Pollinated**: Insect
**Leaf shape**: Elliptic

*Right: The flowers appear in spring.*

*Right: The black fruits are bitter.*

## Cherry Laurel

*Prunus laurocerasus*

This common plant is widely cultivated in gardens throughout Europe. It is particularly favoured for hedging and can be clipped into a dense, windproof barrier. Despite its appetizing-sounding name, both the fruit and leaves of cherry laurel contain cyanide and are poisonous. Where it is allowed to grow to tree size little grows beneath its dense, evergreen canopy. It was introduced into western Europe at the end of the 16th century and was in cultivation in Britain early in the 17th century.

**Identification**: The bark of cherry laurel is dark grey-brown and smooth, even in maturity. The leaves are thick and leathery, glossy dark green above, and pale or even yellow-green beneath. They are oblong to elliptic, up to 20cm/8in long and 6cm/2½in broad, with occasional shallow teeth around the margin, which ends in a short point. The flowers are small, dull white, fragrant and borne in upright racemes to 13cm/5in long in mid- to late spring.

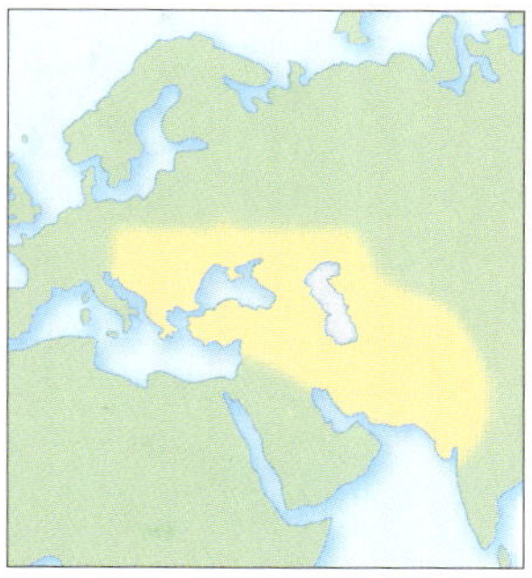

**Distribution**: South-east Europe and south-west Asia.
**Height**: 10m/33ft
**Shape**: Broadly spreading
**Evergreen**
**Pollinated**: Insect
**Leaf shape**: Oblong

*Far left: The fruit is a shiny rounded berry 1.25cm/½in across, red becoming jet black.*

### OTHER SPECIES OF NOTE

**Great White Cherry** *Prunus* 'Tai Haku'
Hundreds of ornamental garden cherries have been cultivated in Japan. Most are either forms or hybrids of two native Japanese cherries: the Oshima cherry and the mountain cherry. The great white cherry is one of the finest hybrids. It has large, pure white, single flowers, which open at the same time as its bronze-pink leaves emerge from the bud.

**Pissard's Purple Plum** *Prunus cerasifera* 'Pissardii'
This is a well-known variety of the myrobalan or cherry plum, *P. cerasifera*. It is widely grown in parks and gardens for its shiny, deep purple leaf colour, which is very distinctive and provides great contrast with the green leaves of other trees. It has abundant, delicate, small flowers, which are pink in bud and open white, well before the leaves emerge in spring.

**Tibetan Cherry** *Prunus serrula*
This small, spreading cherry tree is widely grown in parks and gardens because of its striking, highly polished, deep mahogany-red bark, which becomes a real feature in winter. It originates from western China and may attain heights of around 15m/50ft, but more often 10m/33ft. The white flowers are relatively inconspicuous and are produced after the lanceolate leaves appear in mid-spring.

## Japanese Cultivated Cherry

*Prunus* 'Kanzan'

Sometimes labelled 'Sekiyama', *Prunus* 'Kanzan' is the most popular and commonly cultivated ornamental cherry in Europe. It was introduced into Europe from Japan immediately before World War I, and since then has become a familiar springtime feature in cities, towns and parks.

**Identification**: The overall form of this small to medium-sized tree is vase-shaped, with ascending branches when young becoming more spreading as the tree matures. The bark is brown-pink with distinctive horizontal banding caused by the presence of light brown lenticels. The oval leaves, to 13cm/5in long, have a shallow serrated margin and end in a drawn-out point at the tip. In spring, when they appear, they are at first a beautiful bronze colour, which changes to bright green within a few days.

**Distribution**: Of Japanese garden origin.
**Height**: 10m/33ft
**Shape**: Broadly spreading
**Deciduous**
**Pollinated**: Insect
**Leaf shape**: Obovate

*Below: Bright pink, frilly double flowers are densely borne throughout the canopy just as the leaves emerge in spring.*

## Apricot

*Prunus armeniaca*

The apricot was being cultivated for its edible fruit in China over 2,000 years ago. It gradually spread west into Europe and was being cultivated in Britain by at least the 16th century. It is now naturalized in many warmer parts of Europe, but it requires protection in northern Europe. The species name, *armeniaca*, indicates its possible Armenian origins. There are a number of free-fruiting cultivars that have been developed over the centuries.

**Identification**: The bark is smooth and a warm chestnut brown with a low sheen. The leaves are broadly ovate to roundish, 7.5cm/3in long and 5cm/2in wide, and abruptly pointed. They are deep lustrous green above and a similar colour beneath, but with tufts of hair in the leaf axils. There are fine, even serrations around the leaf margin and the leaf is borne on a 2.5cm/1in long stalk which may feel "warty" to the touch. The single flowers are white to very pale pink, 2.5cm/1in across and produced on the previous year's wood. The fruit is rounded, golden-yellow when ripe, 3–5cm/1¼–2in across, with sweet edible flesh surrounding a single hard, slightly flattened stone.

**Distribution**: Northern China and sporadically throughout central Asia.
**Height**: 10m/33ft
**Shape**: Broadly spreading
**Deciduous**
**Pollinated**: Insect
**Leaf shape**: Ovate

*Left: The flowers are small and delicate.*

*Right: The leaves are lustrous green.*

*Left: The seed is within a golden yellow, sweet fruit.*

## Sour Cherry

*Prunus cerasus*

There is much discussion as to whether this is a cultivated tree of garden origin from which many orchard varieties of fruiting cherry have been raised, or a native tree of South-east Asia. Either way, it has become naturalized over much of Europe, including some parts of Britain. The tree is certainly related to the wild cherry, *P. avium* (which is native to Europe), and is one of the parents of the Morello cherry, which has been in cultivation in European orchards for centuries.

**Identification**: The sour cherry is a small rounded tree, quite often suckering at the base. It has purple-brown bark with horizontal, peeling orange-brown stripes. The leaves are elliptic to oval, abruptly pointed, up to 7.5cm/3in long and 5cm/2in wide. They are dark green above, slightly paler beneath and sharply, but finely, double-toothed around the margin. The flowers are double, pure white, up to 2.5cm/1in across and appear in clusters in mid-spring. The cherry fruit is red, ripening to black, and up to 2cm/¾in across. Although edible, it has a sour, acid taste.

**Distribution**: Possibly South-east Asia.
**Height**: 8m/26ft
**Shape**: Broadly spreading
**Deciduous**
**Pollinated**: Insect
**Leaf shape**: Elliptic

*Below: The double flowers are pure white and are borne in clusters. The fruit is a sour red cherry.*

## Plum

*Prunus domestica*

**Distribution**: Unknown but probably a hybrid of garden origin.
**Height**: 10m/33ft
**Shape**: Broadly spreading
**Deciduous**
**Pollinated**: Insect
**Leaf shape**: Elliptic to obovate

*Above: The plum fruit is a succulent drupe with a single large seed.*

The origins of the garden plum are lost in the mists of time. It is probably a hybrid, possibly between the sloe, *P. spinosa*, and the cherry plum, *P. cerasifera*. Both species are native to the Caucasus and known hybrids have occurred naturally here. In the future, DNA examination may be able to unravel the mystery.

**Identification**: There are countless cultivars of plum, all developed to enhance the quality of the fruit, which is of course not just edible, but delicious. If it were not for the fruit, it is unlikely that this tree would be planted widely for ornamental purposes. It has brown-grey bark, fissured in maturity, and bluntly serrated leaves, up to 7.5in/3in long, which are a dull grass-green. The flowers are white, slightly fragrant and about 2.5cm/1in across, and are borne in spring before the leaves emerge.

*Right: There are various plum cultivars, which produce egg-shaped fruit in a variety of different colours. Plum leaves and flowers are always the same, however.*

OTHER SPECIES OF NOTE

**Common Almond** *Prunus dulcis*
This much planted tree, originally from Asia and North Africa, has been widely cultivated for its nuts for at least 2,000 years and has become naturalized right across southern Europe. It is a small tree with lanceolate, long-pointed leaves and beautiful pink flowers in spring.

**Ground Cherry** *Prunus fruiticosa*
This small, mop-headed miniature tree, or large spreading shrub, is native to central and northern Europe and Siberia. It was introduced into western Europe in the 16th century and is often grown as a top-grafted plant on wild cherry, *P. avium*. It has glossy, deep green obovate leaves, small white flowers and purple red fruits about the size of a large pea.

**Willow Cherry** *Prunus incana*
This is a small erect-branched tree or large shrub, with slender willow-like leaves. It is native to the Caucasus, from where it was introduced into western Europe in 1815. It has deep rosy-red flowers and small red fruits.

**Fuji Cherry** *Prunus incisa*
This slow-growing, small flowering cherry, a native of south-west Japan, has been popular for ornamental planting in Europe since its introduction in the early 20th century. Its small, doubly and sharply toothed leaves emerge a beautiful bronze and colour up well in autumn. The flowers, which open before the leaves, are white or pale-pink and emerge from pink buds in mid-spring.

## *Prunus conradinae*

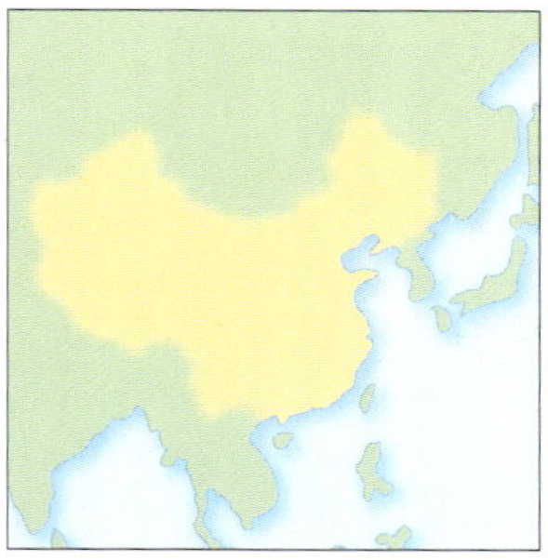

**Distribution**: China.
**Height**: 10m/33ft
**Shape**: Broadly spreading
**Deciduous**
**Pollinated**: Insect
**Leaf shape**: Oval

This beautiful elegant cherry is one of the earliest trees to come into flower. It regularly produces masses of fragrant, pale pink or white flowers in late winter, and in warmer climates they may appear even earlier. The flowers are long-lasting and provide a welcome hint that spring is not far away. It is native to western Hubei, China, from where it was introduced into Europe by Ernest Wilson in 1907. Since then it has been widely cultivated in parks and gardens throughout the continent.

**Identification**: The bark is red-brown, with a sheen that is particularly evident in low winter sunshine. The serrated leaves are rounded at the base and slender-pointed, serrated around the margin, mid-green, with some hair on the underside. They are up to 11cm/4½in long and 5cm/2in wide. The 2.5cm/1in flowers, white or pale pink and fragrant, are produced in clusters. The fruit is a small, ovoid, red cherry.

*Left: The leaves.*

*Far left: Flowers appear on bare branches.*

## Portugal Laurel

Laurel *Prunus lusitanica*

This is one of the most handsome and useful evergreen trees. Ideally, it should be grown as an isolated specimen in order to show off its rich green, glossy foliage and elegant form to its best. However, the Portugal laurel can be grown as a bushy, dense shrub that regrows readily after pruning, so it also makes an effective and luxuriant hedge or screen, and this is how it is mostly seen. In the wild it can grow significantly taller than the cultivated height. It has been cultivated outside its natural range since the mid-17th century. There is also a sub-species known as subsp. *azorica*, which is native only to the Azores and has larger and thicker leaves than the species, with reddish-purple shoots and red emerging leaves.

**Identification**: The bark is dark slate-grey to grey-brown and smooth, even in maturity. The leaves are ovate to elliptic, around 10cm/4in long and 5cm/2in across. They are a rich, glossy, deep green above and paler beneath, tapering at the point and with some shallow, round serrations along the margin. The flowers are small, creamy-white, scented (not necessarily pleasantly) and borne in long, slender racemes, 25cm/10in long, in early to mid-summer. The fruit is egg-shaped, red at first, ripening to purple-black.

**Distribution**: Portugal, Spain and south-west France.
**Height**: 10m/33ft
**Shape**: Broadly spreading
**Evergreen**
**Pollinated**: Insect
**Leaf shape**: Ovate to elliptic

*Right: After flowering, long open clusters of purple-black berries are produced.*

## Autumn Cherry

*Prunus* x *subhirtella* 'Autumnalis'

As the name suggests, this beautiful small cherry has flowers that initially open in autumn and then appear intermittently through winter into early spring. It was cultivated in Japan in the late 19th century and introduced into Europe around 1900. It is a popular small tree that is commonly found in parks, gardens, town squares and precincts. It is a cultivar of the spring cherry, *P. subhirtella*, which it is believed by some may be a hybrid of two other Japanese cherries.

**Identification**: The bark is smooth and grey-brown, becoming horizontally banded in maturity. The leaves are elliptic to ovate, 7.5cm/3in long and 5cm/2in across, sharply toothed around the margin and running to a long, tapered point. They emerge bronzy-pink from bud in early to mid-spring and turn golden yellow before falling in autumn. The flowers are this tree's most distinctive feature, appearing in autumn rather than spring. They are white flushed with pale pink, opening from pink buds and grouped in small clusters along the bare branches, occasionally among the autumn foliage.

**Distribution**: Of Japanese garden origin.
**Height**: 6m/20ft
**Shape**: Broadly spreading
**Deciduous**
**Pollinated**: Insect
**Leaf shape**: Elliptic to ovate

*Far left: The leaves are sharply toothed.*

*Left: The fruit is a small dark red cherry, seldom produced in great numbers upon the tree.*

## Hill Cherry

*Prunus serrulata*

This was one of the first Asian cherries to be introduced into Europe, arriving in Britain from China in 1822. Its original natural distribution is obscure, and due to the excessive natural variability of the population, it may be that this species is itself a variation of the Chinese hill cherry, *P. serrulata* var. *hupehensis*. Today it is rarely found outside botanic gardens and old collections, since its popularity has been surpassed by the Japanese 'Sato Zakura' free-flowering cherries.

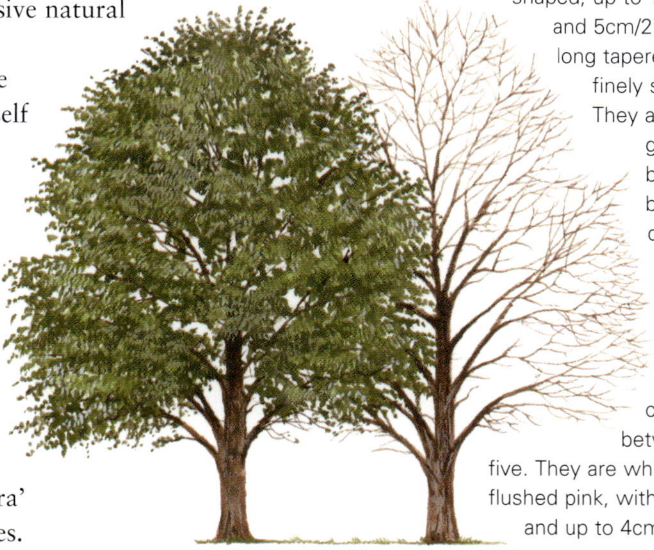

**Identification**: The bark is grey-brown with horizontal banding. The leaves are oval to lance-shaped, up to 13cm/5in long and 5cm/2in wide, with a long tapered point and a finely serrated margin. They are bright grass-green above and blue-green beneath. The double-petalled flowers open in mid- to late spring, held in short-stalked clusters of between two and five. They are white or white flushed pink, without fragrance and up to 4cm/1½in across.

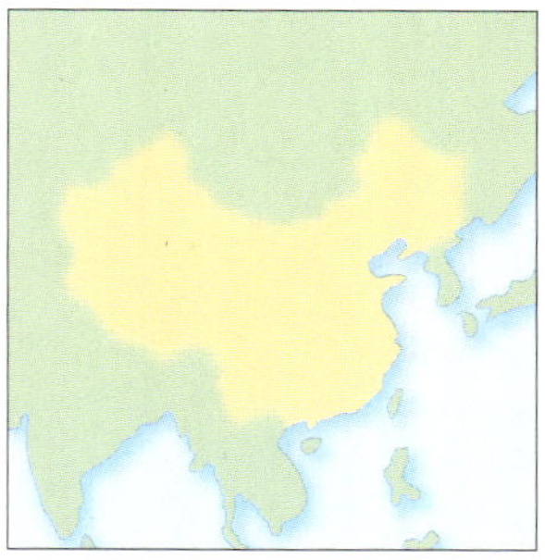

**Distribution**: China.
**Height**: 5m/16½ft
**Shape**: Broadly spreading
**Deciduous**
**Pollinated**: Insect
**Leaf shape**: Ovate to ovate-lanceolate

*Left: The leaves have a long tapering point. The fruit is inconspicuous and seldom produced in any quantity.*

OTHER SPECIES OF NOTE

**Sargent's Cherry** *Prunus sargentii*
Sargent's cherry is one of the loveliest of all cherries, producing a profusion of rich pink, single flowers, coupled with bronze emerging leaves in spring, and brilliant orange-red leaves in autumn. It is named after Professor Charles Sargent, one-time director of the Arnold Arboretum, Boston, USA. It was introduced into Europe in 1890 and has been commonly cultivated ever since.

**St Lucie Cherry** *Prunus mahaleb*
Native to central and southern Europe, this attractive, fast-growing cherry, to 12m/40ft tall, has been cultivated since 1714. It produces masses of small, pure white and very fragrant flowers in mid- to late spring. The wood was at one time used to make cherry-wood pipes for smoking tobacco and for walking sticks.

**Peach** *Prunus persica*
Probably native to China, the peach has been cultivated for its sweet, edible fruit throughout Asia and southern Europe for centuries. It has glossy deep green, narrow elliptic to lanceolate leaves, which are up to 15cm/6in long and 5cm/2in across. Flowers may vary from white, through pink to red (normally pale pink). The fruit is round, normally to 7.5cm/3in across, orange, yellow and red and surrounding a stone (pit).

## Japanese Apricot

*Prunus mume*

This beautiful small tree with olive-green twigs and sweetly fragrant flowers has been cultivated in China for at least 1,500 years. It was introduced into Japan very early on and it rapidly found popularity there, becoming widely used for bonsai. Up to 300 cultivars have been raised in Japan. It was introduced into Europe in 1841 and is one of the earliest cherries to flower, before the leaves appear. In Japan, the fruits are used to make a sweet liqueur.

**Distribution**: China and Korea but not Japan.
**Height**: 10m/33ft
**Shape**: Broadly spreading
**Deciduous**
**Pollinated**: Insect
**Leaf shape**: Rounded to broadly ovate

**Identification**: The bark is red-brown, becoming banded in maturity, and the twigs are a striking green to black-green. The leaves have a long tapering point a are up to 10cm/4in long. They are finely serrated around the margin, mid-green with some sporadic hairs along the midrib beneath. The flowers, borne either singly or in pairs, are pale pink, 2.5cm/1in across and almond-scented. The fruits are round, 2.5cm/1in across, yellow or yellow-green and bitter to the taste.

*Right: The flowers are pale pink and almond scented.*

*Far right: A leaf.*

# HAWTHORNS

*The Hawthorns are among the hardiest and most adaptable of all trees. Many species will withstand very low temperatures, exposure, pollution and salt spray in coastal regions. They are also extremely attractive trees, particularly when in flower, and as such are widely planted as ornamental specimens throughout northern and western Europe.*

## Midland Hawthorn

*Crataegus laevigata*

**Distribution**: Europe.
**Height**: 10m/33ft
**Shape**: Broadly spreading
**Deciduous**
**Pollinated**: Insect
**Leaf shape**: Ovate to obovate, lobed

This hardy small tree takes its name from an area in central England, where it is found growing in much profusion and is an accepted indicator of sites of ancient woodland. It is native to north-west and central Europe and differs from common hawthorn, *C. monogyna*, in having less deeply lobed leaves. The popular garden cultivar 'Paul's Scarlet' was raised from it in the mid-1800s. It is a common tree of hedgerows and woodland edges.

**Identification**: The bark is grey-brown, smooth at first becoming cracked into small plates in maturity. The leaves are glossy dark green above, paler beneath, shallowly lobed and toothed, up to 5cm/2in across and long. The flowers are scented, white with five petals and pink-orange anthers, and occur in small clusters in late spring and early summer. They are followed by oval, glossy red fruits (haws), up to 2.5cm/1in long, each containing two hard seeds.

*Left: The leaves are most attractive in summer.*

*Right: The fruit is a glossy red in colour.*

## Hungarian Hawthorn

*Crataegus nigra*

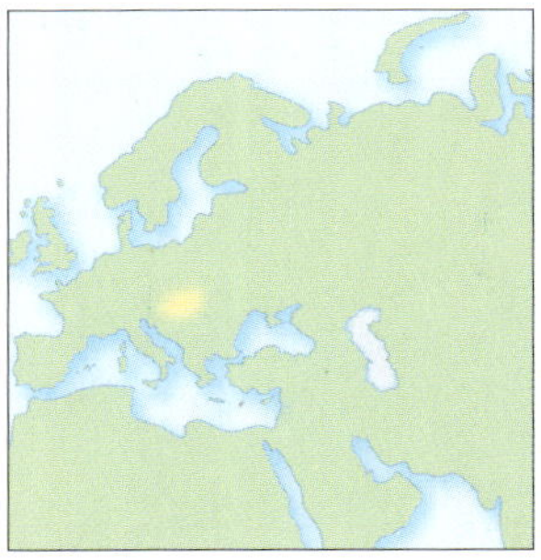

**Distribution**: Hungary.
**Height**: 6m/20ft
**Shape**: Broadly spreading
**Deciduous**
**Pollinated**: Insect
**Leaf shape**: Triangular

*Right: The fruits are borne on the tree in autumn.*

*Right: The leaves are known in folklore for their medicinal use.*

Although not regularly found in parks and gardens, this Hungarian native has been cultivated in other parts of Europe since 1819. Its species name, *nigra*, is derived from the fact that it has black fruits (haws) rather than red, found on most hawthorns. Another distinctive feature is that the young shoots are covered with a felt-like grey down. In maturity, the Hungarian hawthorn forms a round-headed tree made up of stiff, angular branches.

**Identification**: The bark is grey-brown, smooth at first becoming shallowly cracked and flaking in maturity. The leaves are quite distinctive, being triangular to ovate, almost straight across the base, with 7–11 lobes on each side and finely serrated margins. Both surfaces are a dull mid-green, covered with fine hair when young. The flowers, produced in late spring, are white, turning rosy-pink and often hidden by the leaves. After flowering, flattened globular, shiny, black, soft fruits appear.

## Oriental Thorn

*Crataegus laciniata*

This beautiful, small, oriental tree has been widely cultivated in parks, gardens and arboreta since its introduction into western Europe in 1810. It is almost always without thorns and has a rounded, sometimes flattened and spreading, canopy of branches, which may become slightly pendulous at the tips in maturity. It is a sun-loving tree, which in the wild is most commonly found growing on woodland edges and in copses. In flower it is one of the loveliest of all hawthorns.

**Identification**: The leaves are deeply lobed, in some instances almost to the midrib, normally with three or four lobes on each side. They are deep green on the upper surface and grey-green with grey hair beneath. The flowers are up to 2.5cm/1in across, white with conspicuous orange-pink anthers, and are borne in much profusion all over the tree in late spring and early summer. They are followed by rounded to oblong fruits, to 2.5cm/1in across, which are bright coral-red or yellow flushed with red.

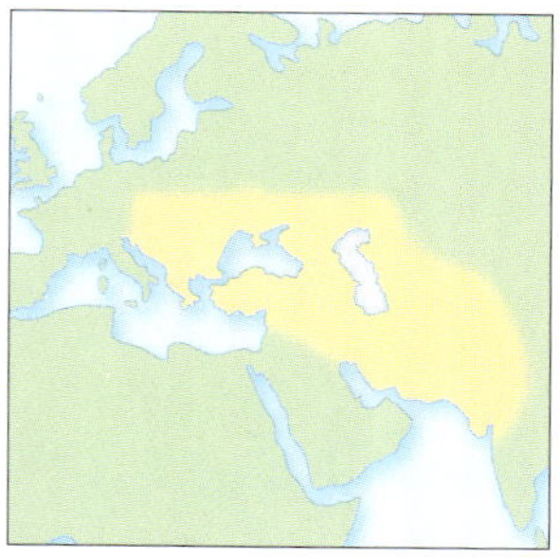

**Distribution**: South-east Europe and south-west Asia.
**Height**: 6m/20ft
**Shape**: Broadly spreading
**Deciduous**
**Pollinated**: Insect
**Leaf shape**: Triangular to diamond-shaped

*Right: The leaves are deeply cut and downy.*

*Left: In autumn, the haws are borne in profusion.*

OTHER SPECIES OF NOTE

**Azarole** *Crataegus azarolus*
Otherwise known as the Mediterranean medlar because of its small, pale yellow to orange edible fruits, which have the flavour of apples, the azarole is native to southern Europe, North Africa and western Asia. It has been cultivated in western Europe, including Britain, since the 17th century. It is a small tree, growing to 9m/30ft, which produces white flowers with purple anthers in dense clusters in mid-summer.

**Hybrid Cockspur Thorn** *Crataegus* x *lavallei*
This attractive, small thorn, 6m/20ft tall and broad, is widely planted in parks and gardens across Europe. It is a hybrid between two American species, *C. crus-galli* and *C. mexicana*, believed to have been raised in France in 1880. It has glossy dark green, elliptic leaves, to 10cm/4in long, and white flowers with pink anthers, which appear in mid-summer. They are followed by rounded fruits that ripen to orange-red, persisting on the tree well into winter.

**Cockspur Thorn** *Crataegus crus-galli*
This small tree, to 8m/26ft, is distinguished by its long, ferociously sharp spines. It is native to eastern and southern North America, from where it was introduced into Europe as early as 1691. In autumn the leaves turn a bright marmalade-orange. It has bright red fruits that persist on the tree long after leaf fall.

## Hawthorn

May *Crataegus monogyna*

Native throughout Europe, the hawthorn is a slow-growing, hardy tree, which can withstand exposure, strong winds and cold better than most northern temperate trees. For centuries, it has been used both to shelter animal stock and to enclose it, particularly in upland areas. A regularly clipped hawthorn hedge is a very effective windbreak and is virtually impenetrable. The other name for hawthorn, may, refers to the month of flowering in Britain, when the tree is at its most conspicuous.

**Distribution**: Europe.
**Height**: 10m/33ft
**Shape**: Broadly spreading
**Deciduous**
**Pollinated**: Insect
**Leaf shape:** Obovate

**Identification**: The bark is dull brown with vertical orange cracks. The leaves are deeply cut, almost to the midrib in some cases, so the outline is not obvious. They are dark green above and paler with some hairs in the vein axils beneath. The creamy white, slightly pungent flowers are produced in profusion in mid-spring.

*Above: The fruit is deep red when ripe.*

*Right: Twigs have vicious thorns.*

## Siberian Hawthorn

*Crataegus dahurica*

This small, extremely hardy hawthorn is native to woodlands in the Transbaikalia region of south-east Siberia, from where it was introduced into western Europe in the early 20th century. It is rare in cultivation and restricted to some botanic gardens and arboreta. It is one of the first hawthorns to come into leaf and flower, regularly producing blossom from mid- to late spring. In autumn it produces attractive orange-red fruits, which persist on the tree well into winter.

**Distribution**: South-east Siberia.
**Height**: 6m/20ft
**Shape**: Broadly spreading
**Deciduous**
**Pollinated**: Insect
**Leaf shape**: Ovate

**Identification**: The bark is brown and smooth at first, becoming finely fissured with some flaking into rectangular plates in maturity. The leaves are broadly ovate, to 5cm/2in long and across, with four or five shallow lobes on each side. They are bright mid-green with some hairs on the underside. The flowers are up to 2.5cm/1in across, white with dark pink anthers. They are slightly fragrant and are borne in clusters of up to ten.

*Below right: The leaves are a main identifying feature.*

*Right: The globular orange-red fruits are 2cm/¾in across.*

## *Crataegus dsungarica*

This small, extremely hardy tree is native to much of central Russia, south-east Siberia and on into northern China from where it is believed to have been introduced into western Europe by the English plant collector Ernest Wilson, in the early years of the 20th century. There is some discussion among taxonomists that it may in fact be a hybrid between two other Asian thorns. It is a small tree with spiny branches and although ornamentally attractive it is seldom found growing outside botanic gardens and arboreta.

**Distribution**: Central Russia.
**Height**: 7m/23ft
**Shape**: Broadly spreading
**Deciduous**
**Pollinated**: Insect
**Leaf shape**: Ovate

**Identification**: It is a handsome thorn with numerous spines up to 2.5cm/1in long borne on all branches. The leaves are triangular, flat at the base with three to seven lobes on each side. The flowers are white, produced in clusters in late spring. They are followed by shiny purple-black fruits, each containing between three and five seeds.

*Right: The fruit is a large, shiny purple-black fruit which contains between three and five seeds.*

## Paul's Scarlet Thorn

*Craetagus laevigata* 'Paul's Scarlet'

This lovely tree is a cultivar of the Midland hawthorn, *C. laevigata*. 'Paul's Scarlet' was produced as long ago as 1858 in England, and since then has become a favourite ornamental species for parks and gardens because of its spectacular, profusely borne, double red flowers.

**Distribution**: Of garden origin.
**Height**: 10m/33ft
**Shape**: Broadly spreading
**Deciduous**
**Pollinated**: Insect
**Leaf shape**: Ovate to obovate

**Identification**: The bark of 'Paul's Scarlet' is grey and smooth, becoming shallowly fissured in maturity. It has ovate to obovate glossy dark green leaves, which are 5cm/2in long and across with shallow lobes around the upper half of each leaf. The double, almost rose-like, flowers are the tree's main feature, being a bright deep pink colour and produced in dense clusters all over the tree in late spring and early summer. They are followed in late summer by bright red, oval-shaped fruits, up to 2.5cm/1in long, which persist long into autumn.

*Right: The flowers are produced in profusion.*

*Right: The leaves have shallow lobes.*

### OTHER SPECIES OF NOTE

**Tansy-leaved Thorn** *Crataegus tanacetifolia*
This small, slow-growing tree is native to Syria and adjacent Middle Eastern countries. It was first cultivated in Europe in 1789. As the name suggests, it has variable, deeply cut leaves, which are reminiscent of some members of the *Tanacetum* genus, a group that includes the common tansy and other feathery-leaved perennials. The tree is normally thornless and produces fragrant white flowers in early summer, followed by yellow fruits, flushed red, which look (and taste) like small apples.

***Crataegus wattiana***
This attractive, small, often thornless, tree originates from central Asia. It has bright red-brown lustrous twigs and large oval leaves, up to 10cm/4in long, which are sharply toothed. White flowers with pale yellow anthers appear in clusters up to 7.5cm/3in across, in late spring and early summer. These are followed by globular, translucent, yellow-orange fruits, which appear in summer but fall to the ground before autumn.

***Crataegus laevigata* 'Rosea Flore Pleno'**
This is one of the oldest of all hawthorn cultivars and still one of the most popular. It is widely cultivated in towns, parks and gardens, and as a street tree. It has beautiful, double, rose-like pink flowers. It is believed to be one of the parents of Paul's scarlet thorn.

## Glastonbury Thorn

*Crataegus monogyna* 'Biflora'

This interesting small tree is planted as a curiosity across western Europe. It has the unique, if rather erratic, habit of producing both new leaves and flowers around Christmas time. A legend attached to the tree suggests that after the crucifixion of Christ, Joseph of Arimathea travelled to England to found Christianity. On arriving at Glastonbury in Somerset, he sat down to rest and stuck his walking staff in the ground. It immediately took root and came into both leaf and flower, even though it was Christmas Day. The tree continued to grow there, flowering at the same time each year, until the early 19th century, when it eventually died. It was of course widely cultivated and its progeny became widespread.

**Distribution**: Of garden origin.
**Height**: 6m/20ft
**Shape**: Broadly spreading
**Deciduous**
**Pollinated**: Insect
**Leaf shape**: Ovate to obovate

*Right: The fruit is a small red haw.*

*Right: The leaves are dark green and deeply cut.*

**Identification**: In most respects this tree reflects its origins, being a variety of the common hawthorn, *C. monogyna*. It has the dull brown bark of that species, which fissures vertically when mature, deeply cut dark green leaves, creamy-white fragrant flowers and small, deep red fruits, which appear from late summer into autumn. Whether the legend itself is true remains open for discussion, but the ability of this tree to produce both leaves and flowers in winter as well as in spring is not in doubt.

# FLOWERING CRABS AND PEARS

*The flowering crab genus,* Malus, *contains over 25 species, mainly native to northern temperate regions. They are hardy, small to medium-sized deciduous trees, widely grown as garden ornamentals for their profusion of spring flowers and late summer fruit. The flowers are similar to those of cherry, except that crab apple flowers have five styles presenting the female stigma for pollination instead of just one.*

## Crab Apple

*Malus sylvestris*

**Distribution**: Europe.
**Height**: 10m/33ft
**Shape**: Broadly spreading
**Deciduous**
**Pollinated**: Insect
**Leaf shape**: Elliptic

*Far right: Crab apples produce a mass of white flowers in spring. The crab apples themselves develop through the summer.*

This tree is known to many as the 'sour little apple', because of its profusion of small, green, inedible apples in late summer. The wild crab apple is not regarded as an important ornamental species; however, it is one of the parents of the domestic orchard apple, *M. domestica*, and of some very attractive ornamental flowering crabs.

**Identification**: The overall appearance is of an uneven, low-domed tree with a head of dense twisting branches, normally weighted to one side. The bark is brown and fissured, even when the tree is relatively young. The leaves are elliptic to ovate, 4cm/1½in long, slightly rounded at the base, deep green above and grey-green beneath with some hair on the leaf veins. The leaf margin is finely, but bluntly, toothed. The flowers are white, flushed with pink, 2.5cm/1in across and carried on short spurs. The fruit is apple-like, up to 4cm/1½in across, green to yellow and sometimes flushed with red.

## Common Pear

*Pyrus communis*

Over 1,000 cultivars of common pear are known, and it is itself believed to be a hybrid that originated in western Asia over 2,000 years ago. This has naturalized throughout Europe, making it very difficult to ascertain what the original species would have looked like. The pear has been cultivated in Europe for centuries, and many named cultivars were raised at Versailles, France, in the 17th century.

**Identification**: A tall columnar to pyramidal crown and distinctive spur shoots (rarely spines) give the tree a slightly angular appearance. The bark is dark grey, cracking into small irregular flakes in maturity. The leaves are mainly rounded at the base, up to 10cm/4in long and 5cm/2in across. The flowers are single white, with five petals and purple anthers, borne in clusters in mid-spring. The fruits are variable, rounded to pear-shaped, containing sweet, edible flesh surrounding small, hard, brown seeds.

**Distribution**: Europe.
**Height**: 15m/50ft
**Shape**: Broadly columnar
**Deciduous**
**Pollinated**: Insect
**Leaf shape**: Ovate to elliptic

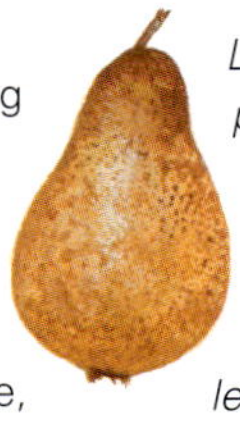

*Left: The fruits of the common pear are russet to yellow, sometimes flushed red.*

*Right: The leaves are a glossy dark but bright green.*

## Hupeh Crab

*Malus hupehensis*

This is one of the most beautiful of all small deciduous trees. The Hupeh crab is a hardy tree growing in the mountainous region of central China, where local people use the leaves to make a drink called 'red tea'. It was introduced to the West in 1900 by the plant collector Ernest Wilson.

**Identification**: The bark is lilac-brown with irregularly shaped plates, flaking from the trunk to reveal orange-brown fresh bark beneath. The leaves are 10cm/4in long, elliptic to ovate and finely toothed around the leaf margin. They are grass green above and pale beneath, with some hairs along the midrib and main veins. The flowers are pink in bud, opening white with a rose flush. They are 5cm/2in across and slightly fragrant. The fruit is greenish-yellow, ripening to red, and normally stays on the tree long after the leaves have fallen.

*Left: The leaf tapers to a fine point.*

**Distribution**: Central and western China.
**Height**: 12m/40ft
**Shape**: Broadly spreading
**Deciduous**
**Pollinated**: Insect
**Leaf shape**: Ovate to elliptic

*Left: The fruit is a rounded 'apple' 1cm/½in across.*

*Right: Flowers open in early spring.*

OTHER SPECIES OF NOTE

***Malus* 'Golden Hornet'**
This hybrid has become one of the most frequently planted of all garden crab apples. In autumn, it produces deep yellow, rounded to egg-shaped fruits, up to 2.5cm/1in long, which stay on the tree long after the leaves have fallen. The fruits shine brightly like little lanterns in winter sun. The parentage is unknown but it was developed in England in the 1940s.

***Malus* 'John Downie'**
Numerous crab apple hybrids are raised and grown for their flowers and fruit. This is one of the most popular and it is widely grown in towns and cities right across Europe. An upright tree, becoming conical as it matures, it has delicate white flowers, which open from soft pink buds, and beautiful egg-shaped fruits, which are bright yellow to orange with a flush of red. They are edible, and when fully ripe, quite sweet-tasting.

**Siberian Crab Apple** *Malus baccata*
As the name suggests, this hardy, broadly spreading 15m/50ft tree is native to north-east Asia and northern China. It was first cultivated in western Europe in 1784 and is well represented in gardens and arboreta. The leaves are reminiscent of pear leaves and the flowers are single, white with yellow anthers, fragrant and borne in mid-spring. The fruits are globular, yellow turning red and borne on a long red stalk.

## Weeping Silver-leaved Pear

Willow-leaved pear *Pyrus salicifolia*

This tree is the most ornamental of all the pears. It was discovered in 1780 by the German botanist and explorer P. S. Pallas, who introduced it to Western cultivation. It is a firm favourite for planting where a small tree with silver foliage is required. Unfortunately, it is not a long-lived species.

*Above left: Blossom appears in clusters.*

*Left: Mature leaves are sage green and smooth.*

**Distribution**: Russia, the Caucasus, from the steppes south into Turkey and northern Iraq.
**Height**: 10m/33ft
**Shape**: Broadly weeping
**Deciduous**
**Pollinated**: Insect
**Leaf shape**: Lanceolate

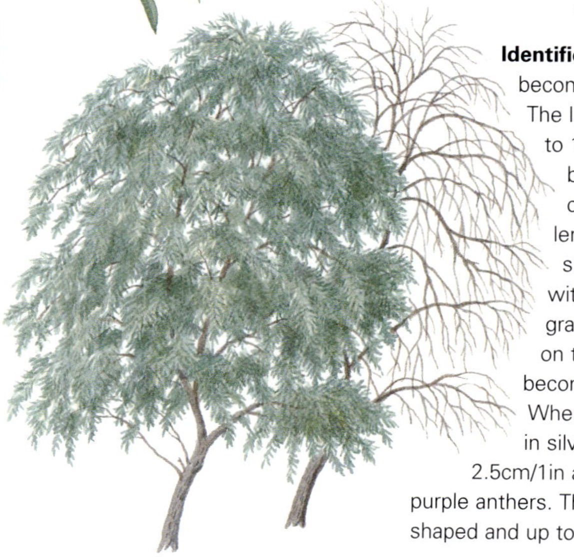

**Identification**: The bark is pale grey, becoming vertically fissured in maturity. The leaves are narrowly lanceolate, up to 10cm/4in long and tapering at both ends. They have a characteristic twist along their length. The young leaves appear silver because they are covered with a silvery white down, which gradually wears off. They are borne on thin, horizontal branches, which become pendulous towards the tip. When young, these are also covered in silvery hairs. The flowers are 2.5cm/1in across, creamy white with purple anthers. The fruit is green, hard, pear-shaped and up to 3cm/1¼in long.

# ROWANS AND WHITEBEAMS

*The rose family, Rosaceae, is one of the largest of all plant families. It encompasses an incredibly diverse range of plants, including cherries, apples, quinces, loquats, cotoneasters, rowans and of course roses. It is also one of the families most commonly represented in cultivation, because of the flowering and fruiting beauty of its members. They include this diverse and beautiful group of trees.*

## Rowan

Mountain ash *Sorbus aucuparia*

This elegant, hardy, small to medium-sized tree grows at elevations in excess of 1,000m/3,300ft on some northern European mountains. Rowan berries are loved by birds and the tree has been associated with providing protection against evil spirits.

**Identification**: The bark varies from silver-grey to purple-grey. It is smooth even in maturity. The pinnate leaves, to 20cm/8in long, comprise up to 15 leaflets, each 5cm/2in long, sharply toothed around the margin, bright green above and blue-green beneath. In autumn, they may turn red or yellow before falling. The flowers, which appear in late spring, are borne in soft clusters, to 15cm/6in across. They are followed in late summer by bright orange-red berries.

*Above: The flowers are creamy-white. and fragrant.*

*Above and right: The berries are 1cm/½in across, borne in large, pendulous clusters.*

**Distribution**: Europe and Asia.
**Height**: 20m/65ft
**Shape**: Broadly conical
**Deciduous**
**Pollinated**: Insect
**Leaf shape**: Pinnate

## Swedish Whitebeam

*Sorbus intermedia*

This is an extremely hardy tree, which thrives in exposed conditions and tolerates localized pollution in city streets. It is often used to transform areas previously affected by industrial despoliation, as it is able to grow on thin, impoverished soils. Swedish whitebeam is very attractive, and makes a handsome garden tree, with its broad, even crown and arching branches, which become pendulous towards the tips.

**Identification**: The bark is grey, smooth at first, cracking into flakes in maturity. The leaves are ovate, sometimes broadly elliptic, glossy dark green above, much paler with grey hair beneath. They are lobed and toothed around the margin, up to 10cm/4in long and 5cm/2in broad. The flowers are individually small but borne in large clusters, up to 12.5cm/5in across.

*Left: The small flowers are dull white with five petals.*

*Right and above left: Deep red, oval berries are borne in pendulous clusters at the end of summer.*

**Distribution**: North-west Europe.
**Height**: 15m/50ft
**Shape**: Broadly columnar
**Deciduous**
**Pollinated**: Insect
**Leaf shape**: Ovate

## Wild Service Tree

Chequer tree *Sorbus torminalis*

This attractive, medium-sized tree has russet-brown, edible fruits, which were at one time used to flavour beer. It has a wide natural range, which extends from northern Europe south to North Africa. Its presence within a wood is considered a good indicator that the woodland is ancient and has suffered little disturbance. At first glance, the leaves are reminiscent of maple or plane; they are stiff to the touch and turn yellow-brown in autumn.

**Identification**: The bark is dark brown and smooth, becoming cracked and flaky in maturity. The winter buds are olive green. The leaves are deeply cut into sharply toothed lobes. They are glossy dark green above, paler green beneath, 10cm/4in broad and long, and held in a distinctive rigid way on the branches. The small, white flowers are borne in flattened clusters, which may be 10cm/4in across, in early summer.

**Distribution**: Europe and North Africa.
**Height**: 20m/65ft
**Shape**: Broadly columnar
**Deciduous**
**Pollinated**: Insect
**Leaf shape**: Broadly ovate

*Left: The fruits, which are small, warty, russet-brown berries, follow the small white flowers.*

OTHER SPECIES OF NOTE

**Chinese Rowan**
*Sorbus hupehensis*
This is a beautiful small tree, native to much of temperate China and cultivated throughout the temperate world as a garden ornamental. It has ash-like, deep green leaves and delightful white berries, flushed with pink.

**Service Tree of Fontainebleau** *Sorbus latifolia*
This interesting medium-sized tree is believed to have originated in a small region in France which has the forest of Fontainebleau at its centre. The tree's existence there has been known since the early 18th century. It probably originally developed as a hybrid of the wild service tree, *S. torminalis*, and the whitebeam, *S. aria*, but is now considered a species in its own right and does come true from seed. Its leaves are like those of the wild service tree, *S. torminalis*.

***Sorbus vilmorinii***
This small, beautiful Chinese tree is one of the most attractive species of the whole *Sorbus* genus and is worthy of space in every garden. It was introduced into Europe in 1889 by Père Jean Marie Delavay, the French Jesuit missionary, and has been widely cultivated as an ornamental tree ever since. It has glaucous, fern-like, pinnate leaves and white flowers in late spring. They are followed by berries, which ripen from deep red through pink to white and persist on the branches like pendulous bunches of white grapes well into winter.

## Whitebeam

*Sorbus aria*

Whitebeam is a tree of calcareous uplands, thriving on thin limestone and chalk soils. Its edible red fruit is collected and made into jam, jelly and wine in some parts of Europe. The timber is dense and hard and at one time was used to make wheels and cogs. In the past it was sometimes referred to as the "weather tree", for when the white underside of the leaf became visible, rain was believed to be on the way.

**Identification**: The bark is smooth silver-grey-brown, even in maturity. By far the most distinguishing feature of this tree is its two-coloured leaves. They are up to 12.5cm/5in long and 6cm/2½in wide, pale green when emerging from the bud, turning a shiny deep green above and white with hairs beneath. When the wind catches the leaves, the effect of flickering green and white over the whole tree is quite remarkable. The flowers are borne in flattened clusters in mid-spring. The fruit is slightly speckled and rough, due to surface lenticels. The leaves do not produce good autumn colour but persist beneath the tree as a grey, crisp covering right through winter.

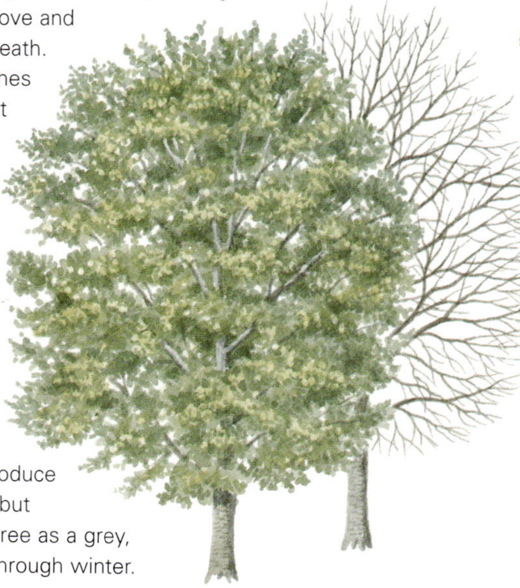

**Distribution**: North, west and central Europe.
**Height**: 15m/50ft
**Shape**: Broadly columnar
**Deciduous**
**Pollinated**: Insect
**Leaf shape:** Ovate

*Above: The flowers are 1cm/½in across.*

*Below: The fruit is a red, round berry.*

## Medlar

*Mespilus germanica*

This small, spreading tree has rather angular branching. It is grown mainly for its fruit, which is an acquired taste. When ripe, the dumpy, pear-shaped fruits have an extremely disagreeable taste and are not edible until they have been exposed to frost and "bletted" (allowed to reach the first stages of decay). Even then the taste is rather acidic. Medlars were popular in the past and were widely cultivated in orchards.

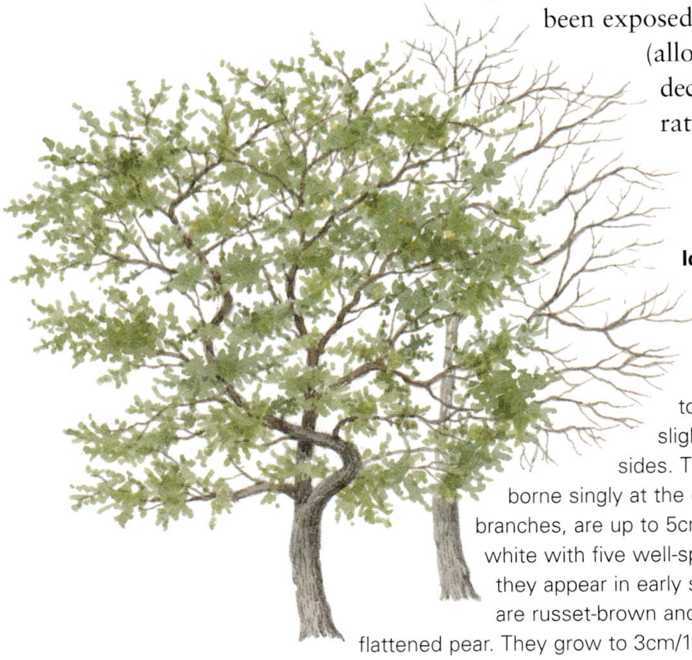

**Identification**: The bark is dull brown, smooth at first, developing fissures in maturity. The bright green, almost stalkless, elliptic leaves are up to 15cm/6in long, minutely toothed around the margin and slightly hairy on both sides. The flowers, borne singly at the end of leaf branches, are up to 5cm/2in across, white with five well-spaced petals; they appear in early summer. The fruits are russet-brown and shaped like a flattened pear. They grow to 3cm/1¼in across and have a slightly open brown top surrounded by a persistent calyx, which gives them a tasselled look.

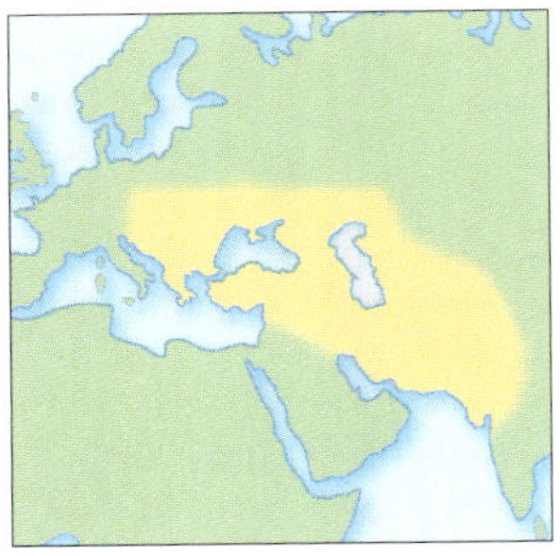

**Distribution**: South-west Asia and south-east Europe.
**Height**: 6m/20ft
**Shape**: Broadly spreading
**Deciduous**
**Pollinated**: Insect
**Leaf shape**: Elliptic to lanceolate

*Left: The flowers open in summer, later than those of many fruit trees.*

*Right: The fruit is small and brown.*

## Quince

*Cydonia oblonga*

Like many trees long cultivated for their fruit, the exact origins of the common quince are unknown. It has certainly been grown around the Mediterranean for at least 1,000 years. In the wild it commonly inhabits shallow limestone soils on mountain slopes. The golden yellow fruit is pear-shaped.

**Identification**: The bark is brownish purple, smooth at first but maturing into irregular plates that flake to reveal orange-brown fresh bark beneath. The leaves are ovate to elliptic, up to 10cm/4in long, dark green and smooth above with cinnamon-grey hairs beneath. The leaves persist on the branches into early winter. The flowers are white flushed with pink and up to 5cm/2in across. They are borne singly at the end of hairy leaf shoots in mid-spring. The fragrant, golden yellow, pear-shaped fruit grows up to 10cm/4in long and is quite waxy to the touch.

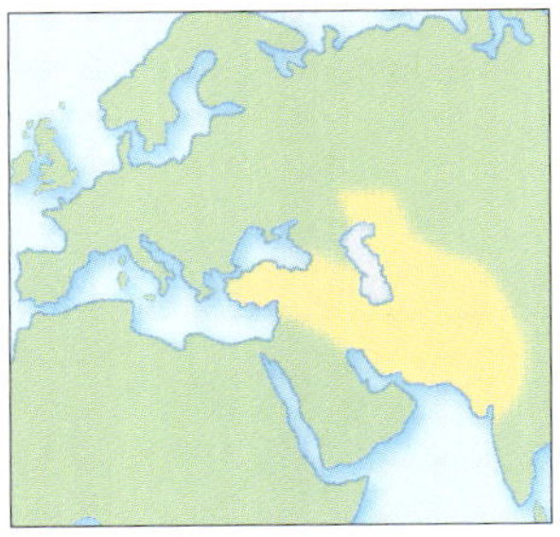

**Distribution**: South-west Asia.
**Height**: 5m/16ft
**Shape**: Broadly spreading
**Deciduous**
**Pollinated**: Insect
**Leaf shape**: Ovate to elliptic

*Right: Quince fruit is fragrant and quite bitter. The leaves have smooth, unserrated edges and stay on the tree until the beginning of winter.*

## Sargent's Rowan

*Sorbus sargentiana*

This magnificent, large *Sorbus*, native to the mountains of Sichuan in south-west China, was collected by Ernest Wilson in 1908, for the Arnold Arboretum in Boston, USA. From there specimens were sent to Europe in 1910. It is one of the best of the genus for autumn colour: the leaves turn brilliant orange-red, it produces large, conspicuous clusters of brilliant red berries and it has large, sticky, mahogany-red winter buds.

**Distribution**: South-west China.
**Height**: 10m/33ft
**Shape**: Broadly columnar
**Deciduous**
**Pollinated**: Insect
**Leaf shape**: Pinnate

**Identification**: The bark is plum-brown and smooth. The pinnate leaves are up to 35cm/14in long and are made up of 11 finely toothed and long tapered leaflets, each of which is up to 10cm/4in long and 5cm/2in broad. They are deep green above, and grey-green beneath with some hair. The flowers, which appear in early summer, are small, creamy-white and borne in large, dense clusters, up to 20cm/8in across. They are followed by large clusters of bright red berries.

*Left: Small white flowers are followed by bright red berries.*

*Right: The sticky winter buds are crimson.*

OTHER SPECIES OF NOTE

***Sorbus arranensis***

A number of rare *Sorbus* species have isolated, or limited, natural ranges, and small populations. Several are variations of whitebeam, *S. aria*, including *S. arranensis*. This is a small tree with a stiff, upright form and deeply lobed ovate leaves that are mid-green above and grey-green with grey hairs beneath. It has rosehip-like red fruits and is restricted to just two glens on the Isle of Arran, off the coast of Scotland.

***Sorbus commixta* 'Embley'**

This is a popular small cultivar of the Japanese rowan, *S. commixta*, raised in the 1930s. It is one of the best for autumn colour, with leaves that turn every shade from scarlet to burgundy, and large bunches of orange-red fruits.

***Sorbus megalocarpa***

The specific name of this small tree, to 7m/25ft tall, roughly translates as "large fruit". The fruits, which appear in late summer, are hard, brown, speckled fawn and the size of a partridge egg. *S. megalocarpa* has oval to obovate leaves that are coarsely toothed around the margin and turn red in autumn. Winter buds are red and sticky.

***Sorbus pohuashanensis***

This small hardy tree, native to the mountains of northern China, was introduced into Europe in 1882. It has a dense canopy and pinnate leaves that are deep green above and hairy grey-green beneath. It is one of the most reliable and showy *Sorbus* for fruit, producing masses of bright red berries in large clusters, the weight of which may cause the branches to bow.

## Kashmir Rowan

*Sorbus cashmiriana*

Although described in 1901 from specimens in Kashmir, this beautiful small tree did not arrive in Europe until the 1930s. Since then it has become a valued species in botanic gardens and arboreta. Despite its warm origins it is perfectly hardy and there are few trees to surpass its splendid display of fruit in autumn and early winter, which for some reason is completely ignored by birds.

**Distribution**: Kashmir, western Himalayas.
**Height**: 8m/26ft
**Shape**: Broadly conical
**Deciduous**
**Pollinated**: Wind
**Leaf shape**: Pinnate

**Identification**: The bark is grey, sometimes red-grey, and smooth even in maturity. The leaves are pinnate, deep green above, paler with some hair beneath, up to 15cm/6in long with 15–17 leaflets, each 5cm/2in long, taper-pointed and sharply toothed. The flowers are produced in clusters, to 10cm/4in across, in late spring. They are followed by pink berries, which ripen to pure white and are borne in loose pendulous clusters, like white marbles, well into winter.

*Left: The flowers are soft pink.*

*Right: The round berries are white, flushed pink.*

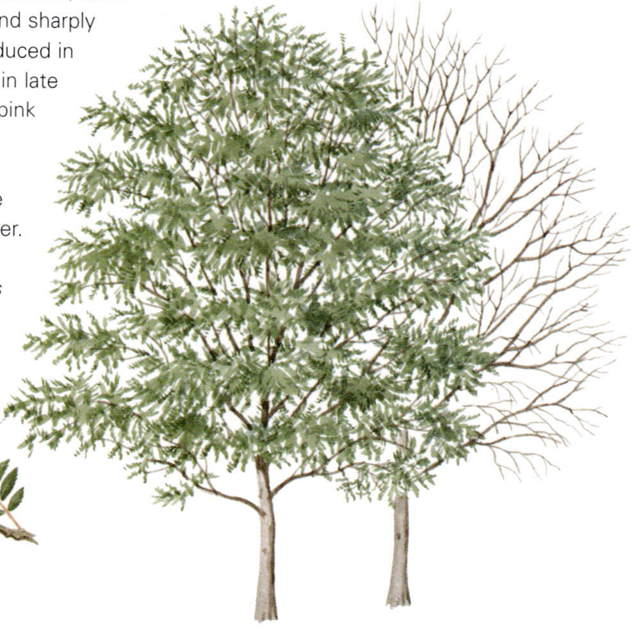

# FALSE ACACIAS, PAGODAS AND TREE OF HEAVEN

*The main trees within this group are members of the pea family. Most have broad crowns, compound leaves, pea-like flowers that grow in clusters and seed pods. The species included here are all grown in Europe as ornamental specimens.*

## Pagoda Tree

*Sophora japonica*

Despite its botanical name, *japonica*, the pagoda tree is thought not to be a native of Japan. However, it has been widely cultivated there for centuries, particularly in temple gardens and places of learning. In China the flower buds were used to make a yellow dye and all parts of the tree, if taken internally, have a strong purgative effect. Flowers on trees grown from seed take anything up to 30 years to appear.

**Identification**: The overall shape is rounded, with branching starting low on the stem. The bark is greenish brown, becoming vertically fissured and ridged in maturity. The leaves are pinnate and up to 25cm/10in long, with up to 15 opposite, untoothed, ovate, pointed leaflets, which are dark green above and glaucous with some hair beneath. The flowers are white, pea-like and fragrant, and are borne in terminal panicles in summer. The fruit is a seed pod up to 7.5cm/3in long, containing up to six seeds. It ripens from green to brown.

**Distribution**: Northern China but could be more widespread.
**Height**: 20m/65ft
**Shape**: Broadly spreading
**Deciduous**
**Pollinated**: Insect
**Leaf shape**: Pinnate

*Left: The white flowers of the pagoda tree are produced in summer in open sprays.*

## Tree of Heaven

*Ailanthus altissima*

This fast-growing Chinese tree, with large ash-like leaves, was introduced into Europe in 1751. It is extremely adaptable and copes admirably with the rigours of city life. However, it does have the habit of producing suckers, which will come up through cracks in city streets, sometimes up to 20m/65ft away from the original tree. The female tree is more commonly grown because the male tree produces flowers that emit a disagreeable odour, which can cause nausea and headaches.

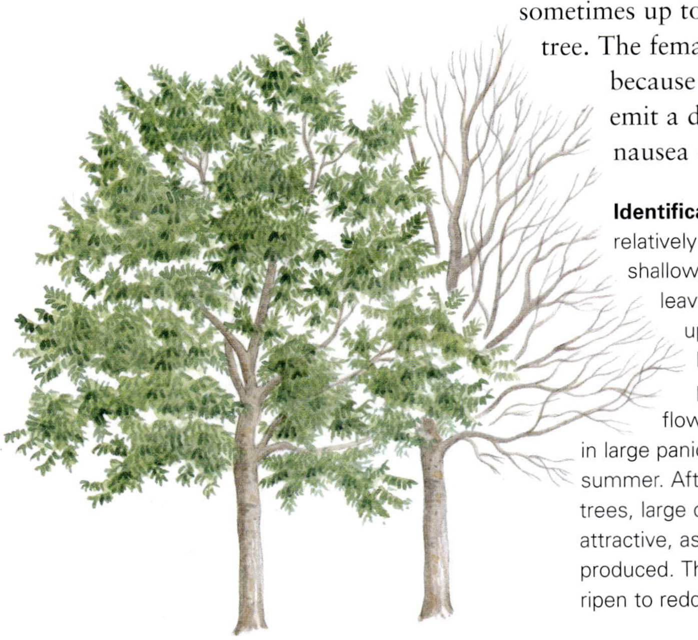

**Identification**: The bark is ash-like – slate grey, relatively smooth at first but developing vertical shallow fissures as the tree matures. The pinnate leaves may be up to 75cm/30in long and made up of 15–20 leaflets, each up to 12.5cm/5in long and ending in a long, tapered point. Both the male and female flowers are yellow-green and are borne in large panicles on separate trees in summer. After flowering on female trees, large conspicuous bunches of attractive, ash-like winged seeds are produced. These start green and ripen to reddish-brown.

**Distribution**: Northern China.
**Height**: 30m/100ft
**Shape**: Broadly columnar
**Deciduous**
**Pollinated**: Insect
**Leaf shape**: Pinnate

*Right: Each pinnate leaf has a tough texture.*

*Left: The conspicuous seed pods are borne in late summer.*

OTHER SPECIES OF NOTE

**Honey Locust** *Gleditsia triacanthos*
This large, spreading tree with frond-like leaves and sharp spines is native to eastern and central North America, from Ontario to Florida. It is also naturalized in parts of southern Europe. It is sometimes known as the sweet locust, because of the sweet, edible flesh that surrounds the seeds in its glossy brown pods, which may be up to 45cm/18in long. The honey locust is extremely tolerant of atmospheric pollution and is widely planted in urban areas.

**Kentucky Coffee Tree** *Gymnocladus dioica*
This elegant tree is native to eastern and central USA, where the seeds were used as a substitute for coffee by early colonial settlers. It is a medium-sized, slow-growing tree, with extremely large compound, bipinnate leaves that can be up to 1m/3ft long. They emerge bronze-pink from the bud in spring, gradually turn dark green in summer and finally become butter yellow before falling in autumn. Small, star-like white flowers appear in early summer.

**Takeda** *Maackia chinensis*
This attractive, small to medium-sized, broad-headed Chinese tree was first cultivated in Europe in 1908. It has pinnate leaves, up to 20cm/8in long, composed of up to 13 ovate leaflets, which when young are silvery blue-green and covered with fine down. Its pea-like flowers, which are dull creamy-white and borne in erect panicles, appear in mid- to late summer.

## *Robinia pseudoacacia* 'Frisia'

This cultivated form of the North American black locust is now one of the most popular trees for ornamental planting in Europe. It was raised at the Jansen Nursery, Zwollerkerspel, Holland, in 1935, and since then has been widely planted both in town and country. It is less vigorous and smaller than the species, but does suffer the same problem of brittle wood, and therefore branches may break in exposed conditions.

**Distribution**: Of Dutch garden origin.
**Height**: 15m/50ft
**Shape**: Broadly columnar
**Deciduous**
**Pollinated**: Insect
**Leaf shape**: Pinnate

**Identification**: The bark is grey-brown, smooth at first becoming deeply and vertically furrowed in maturity. The leaves are pinnate, to 25cm/10in long, with up to 21 elliptic to ovate untoothed leaflets, which are an intense bright, glowing, golden yellow from spring to autumn. 'Frisia' is less floriferous than the species, and any flower that does appear is masked by the brightness of the foliage. Smaller branches carry the occasional sharp spine at the base of some leaves.

*Right: The fruits are pea-like pods.*

*Below: The leaves become greener in summer.*

## Black Locust

*Robinia pseudoacacia*

This is the most widespread of all the locust trees. It originated in the Appalachian Mountains, from Pennsylvania to Georgia, but it is now naturalized over most of the USA and was one of the earliest American trees to reach Europe, in the early 17th century. It was once valued for ships' masts because of its durable, straight timber, and Native Americans used the wood to make bows. Black locust has the habit of suckering from the root system, and on a mature tree suckers can occur up to 10m/33ft from the base of the trunk.

**Distribution**: Eastern USA.
**Height**: 25m/80ft
**Shape**: Broadly columnar
**Deciduous**
**Pollinated**: Insect
**Leaf shape**: Pinnate

**Identification**: The bark is grey-brown, smooth at first, becoming deeply fissured with scaly ridges in maturity. The pinnate leaves, up to 30cm/12in long, have 11–21 ovate to elliptic, untoothed, thin leaflets, each 5cm/2in long and ending in a sharp point. Quite often there are two sharp spines at the base of each leaf. The fragrant flowers are pea-like, white, with a pea-green blotch in the throat.

*Right: The leaves are grass-green above and blue-green below. The fruit is a dark brown bean pod up to 10cm/4in long.*

# EUCALYPTUS

*There are more than 400 species of eucalyptus, or gum trees, all native to the Southern Hemisphere. They are abundant in Australia, Tasmania, New Guinea, the Philippines and Java. Most eucalyptus trees are evergreen and fast-growing, with attractive bark, luxuriant foliage and white flowers. They are widely cultivated for their ornamental qualities and timber in other warm temperate regions of the world.*

## Cider Gum

*Eucalyptus gunnii*

The cider gum is native to the island of Tasmania, where it grows in moist mountain forests up to 1,300m/4,250ft above sea level. It is one of the hardiest of all eucalyptus species and one of the most widely planted around the world.

The glaucous, rounded juvenile foliage is prized by flower arrangers and florists. In order to maintain these attractive leaves, trees have to be regularly coppiced to stimulate the growth of new shoots. Wild trees grow to 30m/100ft tall.

*Right: Mature cider gum leaves are long and slender, and hang down from the branches.*

**Identification**: This potentially large, fast-growing tree has smooth, grey-green to orange bark, peeling to reveal creamy fawn patches. The juvenile leaves are round, 4cm/1½in across, glaucous to silver-blue in colour and borne opposite in pairs. The mature leaves are lanceolate, up to 10cm/4in long, sage-green to silver and borne alternately on the twigs. The flowers are white with numerous yellow stamens, and are borne in clusters of three in the leaf axils during summer. The fruit is a green, woody capsule, open at one end and containing several seeds.

*Right: After pollination in summer, the flowers develop into woody fruit.*

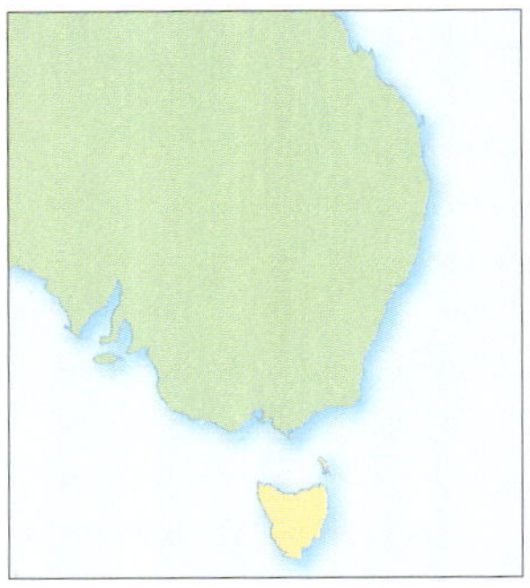

**Distribution**: Tasmania.
**Height**: 30m/100ft
**Shape**: Broadly columnar
**Evergreen**
**Pollinated**: Insect
**Leaf shape**: Juvenile leaves rounded, mature leaves lanceolate

## Tasmanian Snow Gum

Mount Wellington Peppermint *Eucalyptus coccifera*

This is one of the hardiest of all eucalyptus trees. In its native land it grows on mountains up to 1,400m/4,600ft above sea level. The species was introduced into Britain around 1840, and has since been widely planted in temperate regions of Europe. It has striking glaucous foliage and beautiful bark, making it one of the most attractive eucalyptus for ornamental planting. It is a fast-growing tree and requires plenty of space. When the foliage is crushed it emits a peppermint fragrance.

**Identification**: The bark is smooth and grey-white, and peels in vertical strips to reveal bright creamy-white bark beneath. The juvenile leaves are round and mature leaves are lanceolate, to 5cm/2in long and 2cm/¾in broad, running to a fine, tapered tip. Both young and mature leaves are glaucous blue-green. The fruit is a woody, funnel-shaped capsule up to 1cm/½in long.

*Right: The flowers are white with numerous yellow stamens, and are borne in clusters in the leaf axils in early summer.*

**Distribution**: Tasmania.
**Height**: 25m/80ft
**Shape**: Broadly spreading
**Evergreen**
**Pollinated**: Insect
**Leaf shape**: Juvenile leaves round, mature leaves lanceolate

## Mountain Gum

Broad-leaved kindling bark *Eucalyptus dalrympleana*

This attractive, very fast-growing eucalyptus is perfectly hardy in central and western Europe. Near the Wicklow Mountains of Ireland there are specimens that were planted in 1945 and have already reached heights greater than 33m/110ft. In its native habitat in Australasia, the tree is known to reach 45m/145ft tall. However, high in the mountains of New South Wales at 1,300m/4,250ft above sea level, this species grows to only shrub-like dimensions.

**Identification**: The mountain gum has beautiful patchwork bark, which is grey-brown to red-brown, peeling to reveal patches of creamy-white, or blush-white, fresh bark beneath. The juvenile leaves are rounded and borne opposite on the twigs. The mature leaves are lanceolate, to 18cm/7in long. They are bronze when young, turning glaucous as they mature. The flowers are borne in clusters of three in the leaf axils in summer; they are white with yellow stamens. The seeds are contained in a woody brown hemispherical capsule.

**Distribution**: Tasmania and south-east Australia.
**Height**: 33m/110ft
**Shape**: Broadly columnar
**Evergreen**
**Pollinated**: Insect
**Leaf shape**: Juvenile leaves rounded, mature leaves lanceolate

*Left: The mature leaves are lanceolate and up to 18cm/7in long. Bronze when young turning blue-green as they mature.*

### OTHER SPECIES OF NOTE

**Tasmanian Blue Gum** *Eucalyptus globulus*
Native to Tasmania and Victoria in Australia, this fast-growing species can reach 55m/180ft tall. Elsewhere, particularly in Europe, it rarely reaches large proportions. However, one specimen on Jersey in the Channel Islands is supposed to have reached 35m/115ft in just 30 years.

**Silver Top** *Eucalyptus nitens*
Also known as shining gum, this fast-growing blue gum is native to south-east Australia, from where it was introduced into Europe in the early 20th century. It is slightly tender, and although there are some specimens of up to 29m/95ft tall growing in southern Ireland, in colder regions of Europe it is unlikely to reach anywhere near this height. It has long ribbon-like, purple-grey to blue-grey leaves and smooth, silver-white bark, which peels in long vertical strips.

**Silver Gum** *Eucalyptus cordata*
This beautiful tree from Tasmania is one of the most tender eucalyptus to survive in Europe, and is quickly cut to the ground in regions that experience frost. It is also one of the finest for leaf colour, producing vibrant silver-blue-grey, rounded, sometimes heart-shaped, leaves. Quite often in European gardens this tree is grown as bedding and planted just for one summer; in a single season it may grow to 1m/3ft.

## Urn Gum

*Eucalyptus urnigera*

This hardy species is native to the rocky slopes of Mount Wellington in the mountains of south-eastern Tasmania, where it is to be found at elevations of up to 1,000m/3,300ft. Outside Tasmania, including Europe, it has been widely planted as an ornamental species and also for wind protection. It is, in many respects, very similar to the cider gum, *E. gunnii*, but can be distinguished by its smaller fruit and smaller overall size.

**Identification**: This is a small tree or large shrub with horizontal branches that droop at the ends. The bark is pale grey to orange-yellow, shedding vertically in long strips. The juvenile leaves are rounded, up to 5cm/2in in length and width, and are silver-blue with a white bloom. The adult leaves are ovate to lanceolate, waxy to the touch, glossy green and up to 15cm/6in long. The flowers are white with several golden yellow stamens. They grow in clusters of three in the leaf axils in spring. The fruit is urn-shaped (hence the tree's name), about 1.5cm/⅔in long, woody and sharply tapered below the rim. It looks similar to a poppy seed capsule.

**Distribution**: South-east Tasmania.
**Height**: 12m/40ft
**Shape**: Broadly columnar
**Evergreen**
**Pollinated**: Insect
**Leaf shape**: Juvenile leaves rounded, adult leaves lanceolate

*Above: The flowers and urn-shaped seed capsules grow from the leaf axils.*

*Left: The brush-like flowers are held together in clusters of three.*

## Red Flowering Gum

*Eucalyptus ficifolia*

Native to south-west Australia, this small, beautiful but tender eucalyptus was introduced into Europe around 1890. It is widely cultivated under glass for its large corymbs of bright red flowers, which are much valued by florists and flower arrangers. In southern European countries it will grow outside to a height of 12m/40ft. This is one tree which will become more popular for planting in parks and gardens if the climate of western Europe continues to warm up.

**Distribution**: South-west Australia.
**Height**: 12m/40ft
**Shape**: Rounded
**Evergreen**
**Pollinated**: Insect
**Leaf shape**: Ovate to lanceolate

**Identification**: The bark is red with grey-red patches becoming stringy in maturity. The alternate mature leaves are a light sage green colour, evergreen, leathery, lanceolate, up to 10cm (4in) long and very fragrant when crushed. They are borne on reddish-brown shoots. The flowers are highly visible and distinctive being a bright red – sometimes pink – colour and carried in large clusters throughout the summer. These are followed by large oval to rounded woody seed capsules, up to 7.5cm (3in) long, which are green at first ripening to brown.

*Right: Each lanceolate leaf may be up to 10cm/4in long and emits a strong fragrance when crushed.*

*Left: This eucalyptus has showy red-pink flowers which are borne in conspicuous clusters in summer.*

## Small-leaved Gum

*Eucalyptus parviflora*

This extremely hardy, rare, small to medium-sized tree grows wild in just one location in New South Wales, at elevations in excess of 1,500m/4,900ft. Unlike many eucalyptus species, the small-leaved gum will grow well on alkaline soil. It was introduced into Europe in the 1930s and has established itself as far north as Britain, where one tree, at Windsor, west of London, is already over 21m/70ft tall.

**Identification:** The tree is handsome and densely leaved, with attractive smooth, grey, peeling bark. The ovate, juvenile grey-green leaves are borne opposite, in pairs, on short leaf stalks up to 2cm/¾in long. Mature leaves are carried alternately on longer leaf stalks. They are blue-green to glaucous, up to 5cm/2in long and 5mm/¼in wide. The flowers are white and are borne in clusters of four to seven on a short, common stalk in summer. The seed is contained in a woody, grey-green cylinder, which is closed at the base. The hanging seed pods are long and woody.

**Distribution**: New South Wales, Australia.
**Height**: 10m/33ft
**Shape**: Broadly columnar
**Evergreen**
**Pollinated**: Insect
**Leaf shape**: Juvenile ovate, adult lanceolate

*Below: The mature leaves are lanceolate in shape.*

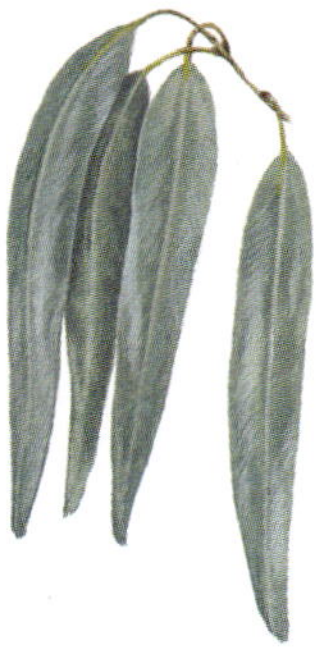

*Left and right: The creamy-white, brush-like flowers are often obscured by leaves.*

## Mountain Swamp Gum

Broad-leaved Sally *Eucalyptus camphora*

This small eucalyptus is able to withstand prolonged water-logging in its native region of south-east Australia. It takes its botanical name from the fact that when the leaves are crushed they emit a strong camphor-like fragrance and are rich in essential oils. The foliage of *E. camphora* is a staple part of the diet of koalas. It has been cultivated in Europe since the early 20th century, but is still an uncommon specimen in parks and gardens.

**Identification**: The bark is rough and dark, normally grey-brown to almost black. It is shed in long vertical ribbons. The juvenile leaves are small and round; the adult leaves are ovate to lanceolate, glaucous blue tinged with red, and up to 10cm/4in long. The flowers, which appear in summer, are small and white, or sometimes pale lemon, with masses of stamens, which give a rather fluffy appearance. They are borne in clusters of around three in the leaf axils.

*Left: The grey-green leaves are broad and long.*

*Left: After flowering, brown woody urn-shaped seed capsules are produced.*

**Distribution**: New South Wales, Australia.
**Height**: 10m/33ft
**Shape**: Broadly spreading
**Evergreen**
**Pollinated**: Insect
**Leaf shape**: Ovate to lanceolate

OTHER SPECIES OF NOTE

**Varnished Gum** *Eucalyptus vernicosa*
This very hardy small tree, or large shrub, is native to mountains in Tasmania, where it is found at heights of up to 1,500m/4,900ft above sea level. It takes its name from its shiny leaves and shoots, which look as if they have been varnished. The juvenile leaves are rounded and thick and have a red margin. The adult leaves are lanceolate, with a red margin, and run to a sharp red point. They are borne on red shoots that point skywards in an erect manner. The tree is often found in European tree collections and botanic gardens.

**Spinning Gum** *Eucalyptus perriniana*
Native to south-east Australia and Tasmania, this small tree, which grows to 10m/33ft, is cultivated in arboreta and botanic gardens across southern and western Europe. It has white bark and both its juvenile and adult leaves are silver. The juvenile leaves, which are rounded, have the curious habit of being attached by the actual leaf blade all around the shoot, and they appear to "spin" in the wind.

**Black Peppermint** *Eucalyptus amygdalina*
As the common name suggests, this small tree has foliage that emits a strong peppermint fragrance when crushed. It is native to eastern Tasmania and, although cultivated in Europe, it is relatively tender and is seldom found in northern or central regions. It has rough, fibrous, dark grey-brown bark. The juvenile leaves are elliptic and grey-green; the adult leaves are stalked, long and narrow with a hooked point at the tip.

## Lemon-scented Gum

*Eucalyptus citriodora*

The lemon-scented gum is native to the central and northern coastal regions of Queensland, Australia, and as such is tender and unlikely to survive in northern or central European winters. However, it is cultivated in southern Europe, particularly in Portugal and Spain, and elsewhere it is sometimes grown as a conservatory plant for its deliciously fragrant foliage. The leaves of this tree yield a lemon-scented oil that is rich in citronella. It is used in the perfume industry and as an antiseptic.

**Identification**: The bark is smooth, normally grey-white, and peels in thin irregular scales to reveal fresh blue-white bark beneath. In maturity the whole crown is a mass of elegant drooping foliage. The aromatic leaves are narrowly lanceolate, up to 20cm/8in long and 2.5cm/1in broad, and are a bright, fresh green. The flowers are borne on short stalks in clusters of three to five. The seeds are contained in a woody brown urn-shaped capsule.

**Distribution**: Queensland, Australia.
**Height**: 40m/130ft
**Shape**: Broadly columnar
**Evergreen**
**Pollinated**: Insect
**Leaf shape**: Lanceolate

*Above: The flowers are white with numerous threadlike stamens.*

*Above right: The leaves yield a lemon-scented oil used in the perfume industry.*

# DOGWOODS AND HANDKERCHIEF TREE

*The trees within this section are some of the most attractive and ornamental of all the trees that grow in Britain and Europe. Consequently they are widely cultivated in parks, gardens and arboreta throughout the region. For some, such as the Japanese strawberry tree and the handkerchief tree, their most beautiful and distinguishing features are the showy white bracts that surround the flowers.*

## Table Dogwood

*Cornus controversa*

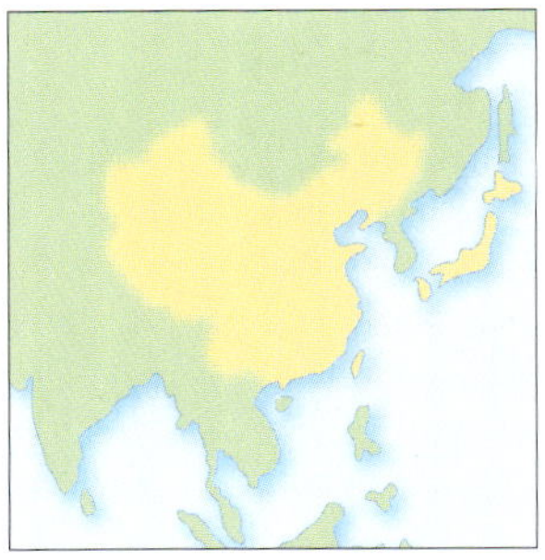

**Distribution**: Japan, China and Taiwan.
**Height**: 20m/65ft
**Shape**: Broadly spreading
**Deciduous**
**Pollinated**: Insect
**Leaf shape**: Ovate to elliptic

This superb ornamental Asian tree has symmetrical, horizontal branches, which become progressively shorter towards the top of the tree in tiers, rather like a wedding cake. It was introduced into Europe before 1880, and is commonly grown in parks, gardens and arboreta throughout western Europe. It is one of only two dogwoods to have alternate leaves (the other being *C. alternifolia*); the rest are all opposite.

**Identification**: The bark is smooth, grey to light brown, becoming slightly fissured with age. The oval leaves are up to 15cm/6in long, dark green and glossy above and blue-green beneath with fine hairs. They are clustered together on thin stalks at the tips of the shoots. In autumn the leaves turn a rich plum-purple before falling. The flowers are small and creamy-white, borne in broad, flattened heads, 15cm/6in wide, along the horizontal branches in early summer. They are followed by small, spherical blue-black berries, also borne in clusters.

*Right: After flowering small blue-black berries are produced.*

## Japanese Strawberry Tree

*Cornus kousa*

This small Japanese tree has been widely planted as an ornamental right across Europe since its introduction in 1875. It has distinctive and decorative creamy-white bracts, sometimes tinged with pink, which surround very small greenish-white flower clusters in early summer. A Chinese variety of this tree, *C. kousa* var. *chinensis*, which was collected by Ernest Wilson in 1907, is also cultivated in Europe and is bigger in overall height, leaves and bracts.

**Distribution**: Japan.
**Height**: 8m/26ft
**Shape**: Broadly columnar
**Deciduous**
**Pollinated**: Wind
**Leaf shape**: Ovate

**Identification**: The bark of this beautiful tree is rich brown-red, and flakes to reveal fresh cream or fawn bark beneath. The leaves are ovate, to 7.5cm/3in long, dark green and slightly shiny above, paler with tiny tufts of rust-coloured hair in the leaf vein axils beneath. They are untoothed and have a conspicuously undulating margin. The tiny flowers are surrounded by four creamy-white, or blush-white, taper-pointed bracts. They are followed by clusters of small, red, strawberry-like, edible, sweet-tasting fruit.

*Left: Small red edible fruits are produced following a hot summer.*

*Left: Four large white or pink-tinged bracts surround the insignificant flowers in summer.*

OTHER SPECIES OF NOTE

**Golden Rain Tree** *Koelreuteria paniculata*
The golden rain tree is native to northern China and southern Mongolia, and was introduced into Europe in 1763. The name *Koelreuteria* commemorates the 18th-century German professor of botany, Joseph Koelreuter, who experimented with plant hybridization. It is a beautiful tree with pinnate leaves and yellow flowers, with bright orange-red stamens, in superb panicles up to 35cm/14in long, which drip from the tree in summer.

**Flowering Dogwood** *Cornus florida*
Native to the eastern USA and introduced into Europe as long ago as 1730, this beautiful small tree grows to 6m/20ft. Despite its name, it is cultivated not for its flowers, which are insignificant, but for the protective bracts that surround them. The bracts are about 5cm/2in long and white with pink blotching around the base. Numerous cultivars have been raised from this species, including 'Cherokee Chief', which has deep rosy-red bracts.

**Pacific Dogwood** *Cornus nuttallii*
This west coast North American dogwood, introduced into Europe in 1835, has some of the largest bracts of any flowering dogwood. Also known as the mountain dogwood, it is a medium-sized tree, to 25m/80ft tall, and has elliptic, deciduous leaves, 15cm/6in long, which turn bright red and yellow in autumn. The insignificant flowers are surrounded by four to seven beautiful creamy-white to blush-white bracts, which may be up to 7.5cm/3in long, in late spring.

## Handkerchief Tree

Dove tree, Ghost tree *Davidia involucrata*

This beautiful tree was introduced to the West from China in 1904 by Ernest Wilson, who had been commissioned by Veitch's nursery to collect propagating material for "this most wondrous of species". All of the tree's common names refer to the spectacular white hanging leaf bracts that appear in late spring.

**Identification**: The bark is orange-brown with vertical fissures, creating flaking, irregular plates. The leaves, up to 15cm/6in long, are sharply toothed with a drawn-out, pointed tip. They are glossy bright green above and paler with some hairs beneath. In times of drought they tend to roll up to reduce water loss. The small flowers, clustered into a ball with conspicuous lilac anthers, appear in late spring, surrounded by the showy white bracts. The fruit is a green-purple husk containing a single hard nut, inside which are up to five seeds.

*Above and below: Surrounding the flowers are two large white bracts of unequal size, up to 20cm/8in long, which flutter in the breeze.*

**Distribution**: Western China.
**Height**: 25m/80ft
**Shape**: Broadly conical
**Deciduous**
**Pollinated**: Insect
**Leaf shape**: Heart-shaped

## Bentham's Cornel

*Cornus capitata*

This beautiful small tree takes its name from the Victorian botanist George Bentham. It has the reputation of being too tender for northern and western European gardens, but in a sheltered location it will survive and indeed flourish. Some of the best specimens are to be found in Cornwall, England: in the latter part of the 19th century there was an avenue of *C. capitata* on the entrance drive to the estate now known as the Lost Gardens of Heligan.

**Identification**: The bark is red-brown and relatively smooth. The evergreen leaves are variably elliptic, tapering at both ends, and are grey-green, leathery and covered with fine hair. They are 7.5–12.5cm/3–5in long and up to 5cm/2in broad. The flowers are very small and clustered together in the centre of four or six beautiful, sulphur-yellow bracts, up to 5cm/2in long.

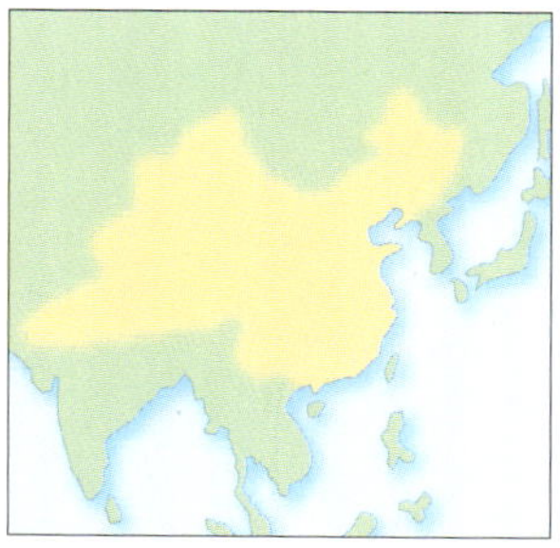

**Distribution**: China and Himalayas.
**Height**: 12m/40ft
**Shape**: Broadly spreading
**Evergreen**
**Pollinated**: Insect
**Leaf shape**: Elliptic

*Left: After flowering, crimson, edible, strawberry-like fruits are produced on long pendulous stalks.*

# HOLLIES

*There are more than 400 species of evergreen and deciduous trees and shrubs in the Aquifoliaceae family; the vast majority belong to the holly, or* Ilex, *genus. Hollies are dioecious and berries are only produced on female trees. The leaves occur alternately on the shoot and the seed is always contained within a berry. Hollies are widely grown as ornamental trees throughout Britain and Europe.*

## Common Holly

*Ilex aquifolium*

Holly is one of the most useful and ornamental trees of the temperate world. It is extremely hardy and its dense foliage provides better shelter in exposed coastal and mountainous localities than just about any other tree. It has long been considered an integral part of Christmas celebrations and its bright berries cheer up the dullest of winter days. Holly timber is dense and hard and has been used for making just about everything from piano keys to billiard cues. The common holly has given rise to numerous attractive garden cultivars.

**Identification**: The bark is silver-grey and smooth even in maturity. The leaves are elliptic to ovate, up to 10cm/4in long, glossy dark green and waxy above, and pale green beneath. They are extremely variable: some leaves have strong spines around the margin; others are spineless. Both the male and female flowers are small and white with a slight fragrance; they appear on separate trees, clustered into the leaf axils in late spring and early summer. The fruits are round, shiny, red berries up to 1cm/½in across, borne in clusters along the shoots in winter.

*Right: The dense foliage of holly makes it a useful hedging plant.*

**Distribution**: Whole of Europe, western Asia and North Africa.
**Height**: 20m/65ft
**Shape**: Broadly columnar
**Evergreen**
**Pollinated**: Insect
**Leaf shape**: Elliptic to ovate

*Right: Holly flowers are scented and appear from spring into summer.*

## Horned Holly

*Ilex cornuta*

This slow-growing, rather shrubby, small tree was discovered by the Scottish plant collector Robert Fortune in China in 1846, and sent to England shortly afterwards. In Europe it is still relatively uncommon in cultivation and is usually found in botanic gardens and arboreta. This is a shame because it is an ideal holly for small gardens, and has distinctive, handsome foliage.

*Right: The rectangular leaves often have a spine in each corner and one at the tip.*

**Identification**: The bark is grey and smooth even in old age. The leaves are almost rectangular and very stiff. They exhibit between three and five spines, but there are normally fewer spines on leaves positioned higher up the tree. The leaves are leathery, glossy dark green, up to 10cm/4in long and 5cm/2in wide. The flowers are small and dull white, and are borne in clusters in spring. The fruits are round and bright red, slightly larger than those of the common holly, *I. aquifolium*.

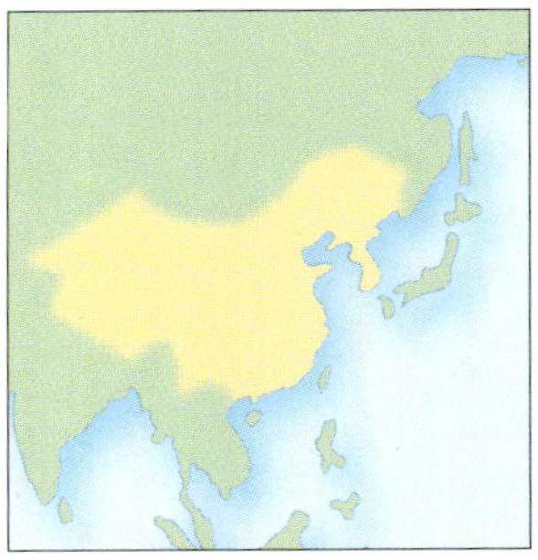

**Distribution**: China and Korea.
**Height**: 4m/13ft
**Shape**: Broadly spreading
**Evergreen**
**Pollinated**: Insect
**Leaf shape**: Rectangular to variable

## Madeira Holly

*Ilex perado*

This elegant, small to medium-sized tree is similar to the common holly, *I. aquifolium*, except for its winged leaf stalks and variable leaf shapes, which may be oval to round, and with or without spines. It has been cultivated in Europe since 1760 but is tender and does not thrive in the colder northern and central regions. It is one of the parents of the ornamental garden cultivars known as the Highclere hollies, *I.* x *altaclarensis*.

**Identification**: The Madeira holly has leathery, flattened leaves, deep green above and paler beneath, that are up to 10cm/4in long and 5cm/2in wide. The leaves may be oval, obovate or rounded. Sometimes they have no spines at all, sometimes just spine-tipped teeth near the tip, and occasionally they are spiny all round the leaf margin. They are borne on short, winged leaf stalks. The bark is silver-grey and smooth, and the wood is creamy-white and extremely dense and hard. The flowers are small, pink in bud and dull white when open, borne in small clusters. The bright, glossy red fruits are pea-sized, egg-shaped to rounded, and are borne in linear clusters along the twigs.

**Distribution**: Madeira.
**Height**: 10m/33ft
**Shape**: Broadly columnar
**Evergreen**
**Pollinated**: Insect
**Leaf shape**: Variable

*Left: The leaves may be oval or rounded and may be spined or have no spines at all.*

OTHER SPECIES OF NOTE

**Yellow-fruited Holly** *Ilex aquifolium* 'Bacciflava'
This handsome and distinctive cultivar of the common holly, *I. aquifolium*, produces large clusters of bright golden fruits in early winter. In all other respects it is virtually indistinguishable from the species. It is sometimes named 'Fructu Luteo' in garden centres and nurseries.

**Hedgehog Holly** *Ilex aquifolium* 'Ferox'
Cultivated since the early 17th century, hedgehog holly was probably one of the first cultivars of the common holly. It is unmistakable and easily identified by the way its small evergreen leaves have short, sharp spines, both around the leaf margin and emanating from the flat, upper surface of the leaf. It is a male cultivar and therefore does not produce fruit.

***Ilex aquifolium* 'J. C. van Tol'**

This is an interesting cultivar of common holly in that it produces both male and female flowers and is therefore self-pollinating, whereas most hollies are of one sex only. It is a handsome tree with large, thick, deep green matt leaves, which are conspicuously veined and normally without spines. Large, bright red berries are produced in profusion.

## Highclere Holly

*Ilex* x *altaclarensis*

This is a hybrid between the common holly, *I. aquifolium*, and the Madeira holly, *I. perado*. It is believed to have been raised in England before 1838, and has given rise to an important group of ornamental garden cultivars, including some of the finest evergreen trees for planting in small gardens. Among them are the varieties 'Golden King', 'Camelliifolia' and 'Hodginsii'. The original hybrid is now rarely planted, having been superseded by its popular offspring.

**Identification**: The bark is silver-grey and smooth, even in maturity. The leaves are variable both in size and shape. They may be oval, round or even oblong and up to 12.5cm/5in long and 7.5cm/3in wide. The leaves normally have a spiny tip, but the spines around the margin may vary. Both male and female flowers are fragrant, small and white, sometimes with a dark pink blush, and are borne in clusters in the leaf axils, on separate trees, in spring. Female trees bear bright red, globular berries in autumn, which are retained into winter.

**Distribution**: Of garden origin.
**Height**: 20m/65ft
**Shape**: Broadly columnar
**Evergreen**
**Pollinated**: Insect
**Leaf shape**: Variable

*Left: Attractive leaf shapes and colours, and a profusion of berries makes these cultivars some of the most ornamental hollies.*

# HOLLIES AND BOX

*Along with hollies, members of the* Buxus *genus, particularly common box, have been widely grown for centuries in gardens throughout Europe. They are hardy small trees, thriving in most soil types and in sun or shade. They respond particularly well to shearing and clipping and consequently have been widely used for topiary.*

## *Ilex corallina*

**Distribution**: Western and south-west China.
**Height**: 10m/33ft
**Shape**: Broadly conical
**Evergreen**
**Pollinated**: Insect
**Leaf shape**: Ovate

This attractive Asian holly was introduced into Europe in 1900 from the province of Hubei in Western China. Since then it has been planted in botanic gardens and arboreta, but is not common in parks and gardens. In cultivation it is a slow-growing, small holly of graceful, slender habit. There are some fine young specimens planted as street trees in the Boskoop region of Holland.

**Identification**: The bark is smooth and light grey, sometimes silver-grey. The wood is hard and dense and a clean creamy-white. The evergreen leaves are glossy dark green above and paler beneath without a sheen. They are ovate-lanceolate, up to 15cm/6in long and 5cm/2in wide with a slender point, and are bluntly serrated around the leaf margin. The juvenile leaves may be prickly. The fruits are small, rounded, bright red berries, borne in clusters from autumn to late winter.

*Right: Evergreen leaves may be up to 15cm/6in long with a slender point.*

## Perny's Holly

*Ilex pernyi*

This small tree, or occasionally large shrub, is native to central and western China, and was discovered by Abbé Paul Perny, a French Jesuit missionary, in 1858. It was another 40 years before it was introduced into Europe. Today, it is common in botanic gardens and arboreta, but still not as widely planted in gardens as might be expected for such a handsome tree.

*Right: The leaves are virtually triangular.*

*Left: The bright red berries persist on the tree well into winter.*

**Identification**: The tree has stiff branches and a slender, conical form. The bark is grey and smooth. The leaves, which are borne on thick leathery stalks, are distinctive in being almost triangular: they are very angular, with between three and seven ferocious spines, small (to 3cm/1¼in long) and dark glossy green. The flowers are cream to pale yellow, borne in tight clusters in the leaf axils. After flowering, clusters of pea-sized, bright red berries are produced.

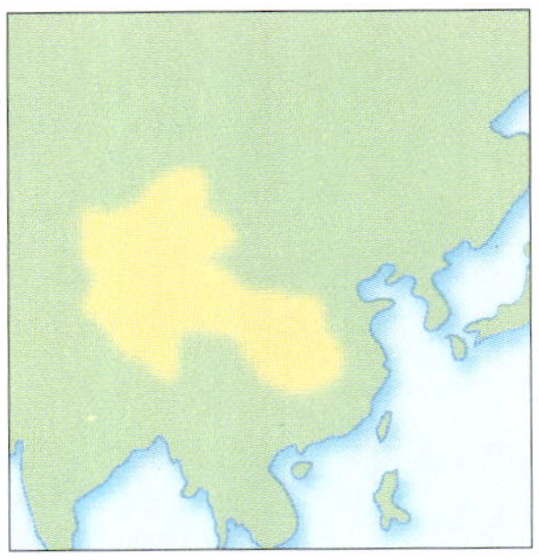

**Distribution**: Central and western China.
**Height**: 10m/33ft
**Shape**: Broadly conical
**Evergreen**
**Pollinated**: Insect
**Leaf shape**: Triangular

### OTHER SPECIES OF NOTE

**Japanese Holly** *Ilex crenata*
Native to both Japan and Korea, this attractive evergreen plant is more a tall shrub than a tree, seldom attaining heights in excess of 4m/13ft. It has stiff, deep green, glossy leaves, which are 1cm/½in long and more akin to those of common box than holly. They are densely borne on reddish-brown shoots, which also carry globular, glossy black berries in winter.

**Himalayan Holly** *Ilex dipyrena*
This large conical tree, to 15m/50ft tall, is native to the eastern Himalayas and western China. It was introduced into Europe in 1840. It has elliptic, dark green leaves, to 10cm/4in long, which have regular, fine, forward-pointing spines around the margin. The leaves are attached to the twigs by short purple leaf stalks. The fruit is a deep red glossy berry.

**Balearic Islands Box** *Buxus balearica*
This slightly tender, Mediterranean evergreen tree, which grows to 10m/33ft, was introduced into western and northern Europe before 1780. It has an upright narrow habit, is densely branched and has large, leathery, matt green, oval leaves, to 5cm/2in long and 2.5cm/1in across. In addition to the Balearic Islands, it is also found growing wild in North Africa and southern Spain. In southern Europe it is used for hedging in much the same way as *B. sempervirens* in northern Europe.

## *Ilex* x *koehneana*

This interesting small evergreen tree is a hybrid between the common holly, *I. aquifolium*, and the Asian tarajo holly, *I. latifolia*, which is said to have arisen in Florence, Italy, around 1890. It is named after the German botanist Bernhard Koehne, who reported its existence. It has large leaves like the tarajo holly, but takes its hardiness from common holly. It is not widely cultivated, but is found in European arboreta and botanic gardens.

**Distribution**: Of garden origin.
**Height**: 6m/20ft
**Shape**: Broadly conical
**Evergreen**
**Pollinated**: Insect
**Leaf shape**: Elliptic to oblong

**Identification**: *Ilex* x *koehneana* has grey smooth bark, olive green twigs and purple young shoots. When the leaves first emerge they are distinctly bronze-coloured. They are elliptic to oblong, 12.5cm/5in long, glossy mid-green above and slightly paler beneath, and they have numerous regular and relatively large spines around the margin. The male and female flowers are small, greenish-white, and borne in clusters in the leaf axils on separate trees in spring. Small clusters of pea-sized, glossy red berries are borne on female trees in winter.

*Left: The leaves have numerous regular and relatively large spines around the margin.*

## Common Box

*Buxus sempervirens*

This small tree or spreading shrub has dense foliage and has been grown for centuries in gardens for hedging, screening and topiary. It is a favourite for use in defining knot gardens and parterres, and clips well. It withstands dense shade and will happily grow beneath the branches of other trees. It has hard, cream-coloured wood, which has been extensively used for wood engraving and turnery.

**Identification**: The bark is fawn or buff-coloured, smooth at first, then fissuring into tiny plates. The leaves are ovate to oblong, 2.5cm/1in long, rounded at the tip with a distinctive notch, glossy dark green above, pale green below and borne on angular shoots. Both male and female flowers are produced in mid-spring; they are small, pale green with yellow anthers and carried in the same clusters on the same trees.

*Left: The fruit is a small woody capsule holding up to six seeds.*

*Right: Flowers are produced in the leaf axils, with several males surrounding one female flower.*

**Distribution**: Europe, North Africa and western Asia.
**Height**: 6m/20ft
**Shape**: Broadly conical to spreading
**Evergreen**
**Pollinated**: Insect
**Leaf shape**: Ovate

# HORSE CHESTNUTS

*The horse chestnut genus,* Aesculus, *contains some of the most popular and easily recognizable ornamental trees in the world. There are just 15 species, all native to northern temperate regions, where they are widely grown in parks, gardens and arboreta for their stately habit and attractive flowers and fruit. All horse chestnuts have compound, palmate leaves and large flowers borne in upright panicles.*

## Indian Horse Chestnut

*Aesculus indica*

**Distribution**: North-western Himalayas and northern India.
**Height**: 30m/100ft
**Shape**: Broadly columnar
**Deciduous**
**Pollinated**: Insect
**Leaf shape**: Compound palmate

This magnificent tree is not as widely known as the common horse chestnut, but equals it in stature and beauty. In the Himalayan forests where it grows wild, Indian horse chestnut regularly exceeds 30m/100ft. Tall flower spikes appear in mid-summer. For many years the white, light timber was used to make tea boxes.

**Identification**: The bark is grey-brown and smooth, even in maturity. The leaves are compound and palmate, with either five or seven leaflets all joining the leaf stalk at a common point. Each leaflet is obovate to broadly lanceolate and up to 25cm/10in long. They emerge bronze-coloured, gradually turning glossy grass-green in summer and then golden yellow in autumn. The white to pale pink flowers are borne in mid-summer on erect, cylindrical panicles, up to 30cm/12in long. Each flower has a yellow or red blotch at the base. The fruit is a rough (but not spiny), green, slightly pear-shaped husk containing up to three dark brown nuts, commonly known as conkers.

*Right: The leaflets spread like fingers from the leaf stalk. Indian horse chestnut seeds are contained in a smooth husk.*

*Right: The flowers appear on long spikes in the middle of summer, poking through the foliage like candles.*

## Red Horse Chestnut

*Aesculus* x *carnea*

**Distribution**: Of garden origin.
**Height**: 20m/65ft
**Shape**: Broadly columnar
**Deciduous**
**Pollinated**: Insect
**Leaf shape**: Compound palmate

This popular tree, planted in parks and gardens throughout Europe, is a hybrid between the common horse chestnut, *A. hippocastanum*, and the American red buckeye, *A. pavia*. Little is known of its origins, but it is possible that it occurred naturally in Germany in the early 1800s. There are several different clones of this hybrid in cultivation and they vary in terms of flower quality, how soon they begin to produce flowers, and their susceptibility to branch-break.

**Identification**: The bark is dull brown, smooth at first, becoming shallowly fissured and flaking in maturity. The leaves are palmately compound, mid-green with pronounced parallel veining. There are normally between five and seven obovate, short-stalked leaflets, each to 25cm/10in long, joined at the base to a long leaf stalk. The flowers, which appear in late spring, range from pink to deep red and are borne in erect panicles up to 20cm/8in long. The fruit is a brown shiny single seed, which may be known as a conker, enclosed in a slightly spiny husk.

*Below: The fruit.*

*Above: Erect panicles of red flowers cover the tree in spring.*

## Common Horse Chestnut

*Aesculus hippocastanum*

*Left: Horse chestnut leaf buds are covered by brown scales. The seed husks have sharp spines.*

This tree is native only to Albania and Greece, but its distribution is often wrongly thought to be larger because of its popularity and widespread planting as an ornamental tree. It came into cultivation outside its natural range as early as 1650, when it was introduced into Vienna, Austria.

**Identification**: The bark is orange-brown to grey, smooth at first, turning shallowly fissured and scaly in maturity. The large winter buds are rich red-brown and covered in sticky resin. The leaves are large: each leaflet can be up to 30cm/12in long, and there are normally five to seven strongly veined, obovate leaflets on each leaf. The nuts are grouped in twos or threes in a husk 5cm/2in across.

**Distribution**: Albania and Greece.
**Height**: 30m/100ft
**Shape**: Broadly columnar
**Deciduous**
**Pollinated**: Insect
**Leaf shape**: Compound palmate

*Left: The flowers are creamy white, blotched with yellow and pink, and borne in large upright, conical panicles, up to 25cm/10in long, in mid-spring.*

OTHER SPECIES OF NOTE

**Sweet Buckeye** *Aesculus flava*
Sometimes called yellow buckeye, this handsome, round-headed tree, to 30m/100ft, was introduced from North America to Europe by 1764. *Flava* means yellow and refers to the yellow flowers that are borne in upright panicles in late spring. The leaves turn brilliant orange-red in mid-autumn.

**Red Buckeye** *Aesculus pavia*
Native to eastern USA, red buckeye grows in moist woods and thickets. It is a small tree with a slightly weeping, pendulous habit in the outer branches. It is one of the parents of the hybrid red horse chestnut, *A.* x *carnea*, to which it gives the red flower. It was introduced into Europe in 1711.

**Ohio Buckeye** *Aesculus glabra*
This small North American horse chestnut, to 10m/33ft, was introduced into Europe around 1812. It has rough bark and palmate leaves with five leaflets. It produces yellow-green flowers in late spring and has bright autumn leaf colour.

**Sunrise Horse Chestnut** *Aesculus* x *neglecta* 'Erythroblastos'
This is a stunningly beautiful cultivar of the hybrid between *A. flava* and *A. sylvatica*. It is grown for its spectacular young leaves, which erupt in mid-spring a brilliant shrimp-pink. The leaves gradually change to pale yellow-green, but the tree is worth growing for those few splendid weeks in spring.

## Japanese Horse Chestnut

*Aesculus turbinata*

In many ways this large tree is similar in appearance to the common horse chestnut, *A. hippocastanum*. It is widely planted in Japan as an ornamental species but is slower-growing than the common horse chestnut and its flowers are not so large or carried in such profusion. Its leaves are much larger, however, and turn bright orange in autumn.

**Identification**: The bark is brown and flaky in maturity. The winter buds are glossy, red-brown and sticky. The leaves are compound and palmate, with five to seven obovate, stalkless, toothed leaflets all attached at the same point to a common leaf stalk. Each rich green leaflet can be up to 40cm/16in long and is heavily veined. The flowers are creamy-white with a red blotch and are borne on upright panicles up to 25cm/10in tall in late spring. The fruit is an egg-shaped, virtually spineless yellow-green husk, ripening brown to reveal two to three shiny brown seeds, or conkers, inside.

**Distribution**: Japan.
**Height**: 30m/100ft
**Shape**: Broadly columnar
**Deciduous**
**Pollinated**: Insect
**Leaf shape**: Compound palmate

*Above: The husks have tiny spines.*

*Below: An elegant flower spike.*

# MAPLES

*There are more than 100 species of maple,* Acer, *in the world and countless cultivars, particularly of the Japanese maples. Mainly deciduous, they are predominantly found throughout northern temperate regions, with a few extending into subtropical Asia. They range in size from mighty American giants to slow-growing Japanese bonsai. Many are cultivated for their attractive foliage and graceful habit.*

## Field Maple

*Acer campestre*

This small, hardy maple has a massively wide natural distribution, which runs from the Atlantic to Siberia and south to the Mediterranean. It is a small to medium-sized tree of woodland edges and hedgerows. It is widely cultivated as a hedging species throughout Europe. Field maple is a long-lived tree, and there are European specimens that are known to be more than 500 years old. It produces good golden-yellow leaf colour in autumn.

**Identification**: The bark is pale grey-brown, smooth at first but becoming very 'corky', particularly on small twigs, which may take on a 'winged' appearance. The leaves are palmately lobed with five lobes, dark green above, paler beneath, to 7.5cm/3in long and much the same across. They are borne on a short stalk that exudes a milky liquid when crushed or broken, particularly in spring. The flowers are small and greenish-yellow, and are borne in erect clusters in spring. These develop into pendulous clusters of winged seeds.

**Distribution**: Europe, south-west Asia and parts of North Africa.
**Height**: 20m/65ft
**Shape**: Broadly spreading
**Deciduous**
**Pollinated**: Insect
**Leaf shape**: Palmately lobed

*Left and right: The leaves turn red and golden brown in the autumn.*

## Sycamore

*Acer pseudoplatanus*

Sycamore is one of the most common northern temperate trees. It has an extensive natural range and has been widely planted, and subsequently naturalized, in Britain and North America. It is hardy and resistant to strong winds and exposure to salt-laden air in coastal areas.

**Identification**: The bark is grey and smooth when young, becoming a delightful greyish-pink in maturity with irregular-sized flaking plates. The leaf buds are lime green. They open in spring to release bronze-yellow leaves, which turn deep green within two weeks of emerging. The leaves are up to 20cm/8in across, palmate, with five coarsely toothed lobes. The small flowers are borne in dense, yellow-green, pendulous clusters as the leaves emerge in spring. The fruit is the familiar two-winged seed. Each wing is 2.5cm/1in long. The seeds are grouped in pendulous clusters from early summer; they are red-green in colour, ripening to brown in mid-autumn.

*Right: Sycamore seeds are easily recognized by their paired wings. The leaves are typically maple-shaped.*

**Distribution**: Europe, from the Pyrenees in Spain to the Carpathians in the Ukraine.
**Height:** 30m/100ft
**Shape**: Broadly columnar
**Deciduous**
**Pollinated**: Insect
**Leaf shape**: Palmate

## Norway Maple

*Acer platanoides*

This fast-growing, handsome, hardy maple has been cultivated as an ornamental species for centuries. It has a large, spreading crown with upswept branches and is as much at home in parkland settings as in woodland. Recently, smaller cultivars have been developed, which are being planted in great numbers alongside roads.

**Identification**: The bark is grey and smooth when young, becoming vertically ridged and fissured in maturity. The leaves are rather like the leaf on the Canadian flag – palmate with five lobes, each ending in several sharp teeth and a slender point. Each leaf is bright green and up to 15cm/6in in both length and width; it is borne on a long, slender, pink-yellow leaf stalk. The flowers are bright yellow, sometimes red, and are borne in conspicuous drooping clusters in spring as the leaves emerge.

**Distribution**: South-west Asia and Europe, north to southern Norway.
**Height**: 30m/100ft
**Shape**: Broadly columnar
**Deciduous**
**Pollinated**: Insect
**Leaf shape**: Palmate

*Right: Flowers may be either yellow or red.*

*Left: The fruit is a pair of green-yellow winged seeds, borne in clusters. Each is up to 5cm/2in long.*

*Right: The foliage is a fresh bright green.*

OTHER SPECIES OF NOTE

**Paperbark Maple**
*Acer griseum*
This beautiful small tree was discovered in China in 1901 and almost immediately became a European garden favourite. It has cinnamon-coloured, wafer thin bark, which flakes to reveal fresh orange bark beneath. Its distinctive trifoliate leaves turn burgundy-red and orange in autumn.

**Père David's Maple** *Acer davidii*
This beautiful small tree, to 15m/50ft, is native to China, from where it was introduced into Europe in 1879, when it was named after the great French Jesuit missionary and plant collector. *A. davidii* is a snake-bark maple and has smooth, olive-green bark, beautifully marked with narrow, vertical white stripes. The leaves are ovate, to 15cm/6in long and 10cm/4in broad, and turn a beautiful orange-red in autumn.

**Hers's Maple** *Acer grosseri* var. *hersii*
This superb Chinese maple, growing to 10m/33ft and introduced into Europe around 1923, is widely planted as an ornamental. It has green bark that has beautiful long, narrow, linear white markings. The leaves are ovate, normally with three to five lobes, and turn a rich marmalade-red in autumn. The green flowers are borne in conspicuous pendulous racemes in late spring.

## Cappadocian Maple

*Acer cappadocicum*

This stately, handsome tree was introduced into western Europe in 1838, and has since been a popular addition to parks, gardens and arboreta. It makes a relatively large, broad but round-headed tree, normally with a clean trunk of 3–4m/10–13ft, though some root suckering may occur around the base. In autumn the leaves turn a rich butter-yellow before falling. If the leaf or leaf stalk is torn, a milky, sticky sap is extruded.

**Identification**: The bark is similar to that of ash, light grey and quite smooth. The palmate leaves, to 10cm/4in long and 15cm/6in across, have five to seven pointed lobes. They emerge a warm red before fading to grass-green above and paler beneath, with some hair in the leaf vein axils. The small, greenish-yellow flowers appear in spring. The fruits are winged seeds carried in pairs, in clusters, turning from green to light brown as they ripen.

**Distribution**: Turkey, Iran, the Caucasus and into western China.
**Height**: 20m/65ft
**Shape**: Broadly spreading
**Deciduous**
**Pollinated**: Insect
**Leaf shape**: Palmately lobed

*Above: Flowers are in erect clusters.*

*Right: The seeds are winged pairs.*

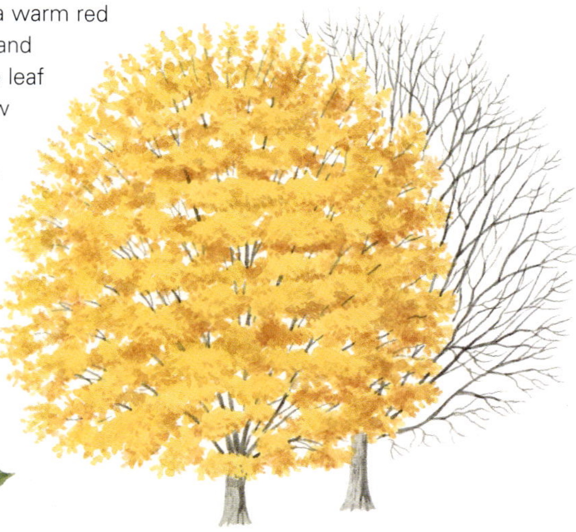

## Smooth Japanese Maple

*Acer palmatum*

The smooth Japanese maple was discovered in 1783 and introduced into the West in 1820. Surprisingly, however, it was almost another 80 years before it became popular and began to be widely planted. The famous Acer Glade at Westonbirt Arboretum, in Gloucestershire, England, was planted in 1875. Today there are literally hundreds of cultivars of smooth Japanese maple. In the wild the species grows within, or on the edge of, mixed broad-leaved woodland, providing dappled shade and shelter.

**Distribution**: China, Taiwan, Japan and Korea.
**Height**: 15m/50ft
**Shape**: Broadly spreading
**Deciduous**
**Pollinated**: Insect
**Leaf shape**: Palmate

**Identification**: The overall shape of the tree is like a large natural bonsai, with horizontal, spreading, meandering branches forking from the main stem quite close to the ground. The bark is grey-brown and smooth, even in maturity. The leaves are palmate with between five and seven deep, pointed lobes that have forward-facing serrations around the margin. They are up to 10cm/4in across. The flowers are burgundy-red with yellow stamens. They are borne in upright or drooping clusters as the leaves emerge in spring. The fruits are green to red winged seeds carried in pairs; each wing is up to 1cm/½in long and up to 20 seeds are clustered together on the branch.

*Below left: In autumn, the leaves turn red and gold before falling.*

*Right: The new leaves emerge bright green.*

## Full Moon Maple

Japanese maple *Acer japonicum*

The full moon maple is native to the islands of Hokkaido and Honshu in Japan, from where it was introduced into Europe in 1864. Despite this relatively early introduction it is hardly ever seen as a species in gardens, having been superseded by a number of beautiful cultivars produced from it, including 'Vitifolium' and 'Aconitifolium'. The species has lime green foliage, which turns red, orange and yellow in autumn.

**Distribution**: Japan.
**Height**: 10m/33ft
**Shape**: Broadly spreading
**Deciduous**
**Pollinated**: Insect
**Leaf shape**: Palmately lobed

**Identification**: The bark is silver grey and smooth, even in maturity. The leaves are palmately rounded, to 12.5cm/5in across, with 7–11 toothed, taper-pointed lobes. The leaves have some fine hair on both their upper and lower surfaces. The flowers are small and purple-red, and are borne in conspicuous pendulous clusters, which appear before the tree comes into leaf. These are followed by winged seeds, to 2.5cm/1in long, which are borne in pairs. They are green at first, ripening to light brown in late summer.

*Right: Clusters of small, reddish-purple flowers open in spring.*

*Below: The lobed leaves are rounded and mid-green in summer, turning red before they fall in autumn.*

OTHER SPECIES OF NOTE

**Silver Maple** *Acer saccharinum*
This is one of the fastest-growing North American maples and is widely planted as an ornamental specimen in parks and gardens across Europe. Although at first glance it is similar in outline to sycamore, it is altogether a much more elegant tree, with a light, open crown and bi-coloured leaves that are bright green above and silver-green beneath.

**Oregon Maple** *Acer macrophyllum*
The specific name of this large North American tree, *macrophyllum*, translates as "large-leaved", which is an appropriate description. The dark green leaves, which are palmate with large, coarsely toothed lobes, may be up to 25cm/10in long and 30cm/12in broad, and are carried on long straw-coloured leaf stalks. They turn yellow and orange in autumn.

**Ash-leaved Maple** *Acer negundo*
Also known as box elder, this medium-sized, 20m/65ft, North American maple is not immediately recognizable as a maple, having pinnate leaves, rather like ash, with up to seven leaflets. *A. negundo* has a light airy look, with long, widely spaced branching, silver-grey bark and bright green leaves. Yellow-green flowers hang, tassel-like, from the outer twigs in spring.

## Snake-bark Maple

*Acer capillipes*

This beautiful small tree, native to Japan, was introduced into the USA by Charles Sargent in 1892. From there saplings were sent to Kew Gardens, London, in 1894. It is now widely planted in parks and gardens right across western Europe. It is a graceful tree with arching branches and foliage that turns bright orange and red in autumn.

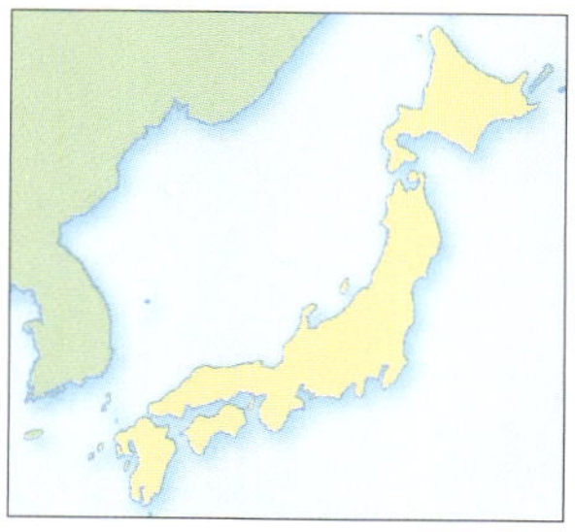

**Distribution**: Japan.
**Height**: 10m/33ft
**Shape**: Broadly spreading
**Deciduous**
**Pollinated**: Insect
**Leaf shape**: Lobed

**Identification**: The bark is grey-green and marked with vertical, narrow white stripes, flushed with red when young, and carried through from the main stem to the secondary branching and twigs. The leaves are up to 15cm/6in long and 10cm/4in across, serrated around the margin and with three to five lobes. The central lobe is the largest and runs to an extended tip. The leaf, borne on a red leaf stalk, has a grass-green upper surface and is paler beneath. The flowers are small, green and carried in pendulous clusters in spring. The fruits are winged seeds, carried in pairs.

*Left: The flower clusters are to 10cm/4in.*

*Right: The bark is marbled with vertical white stripes.*

## Forrest's Maple

*Acer pectinatum* subsp. *forrestii*

This beautiful snake-bark maple takes its name from the plant collector George Forrest, who discovered it in south-west China in 1906. Considering its beauty it is surprising that it is not more widely cultivated in Europe. However there are two splendid specimens at Caerhays Castle, Cornwall, England, which are believed to be plants originally introduced by Forrest.

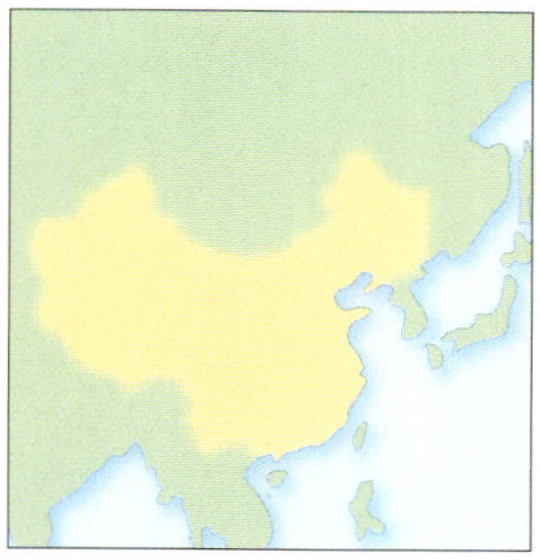

**Distribution**: China.
**Height**: 15m/50ft
**Shape**: Broadly spreading
**Deciduous**
**Pollinated**: Insect
**Leaf shape**: Cordate lobed

**Identification**: This is a beautiful maple with stem bark and young branches that are purple-red, sometimes green, striated with white. It is a sparsely and gracefully branched tree. The leaves, to 12.5cm/5in long and 5cm/2in wide, are mostly heart-shaped with three to five lobes; the central lobe is drawn out into a long slender point and the side lobes are triangular. The leaves are carried on coral-red leaf stalks. The flowers are small and purple-red, and are borne in pendulous clusters as the leaves emerge in spring. They are followed by winged seeds, carried in pairs, in clusters, in late summer to early autumn. The leaves turn bright yellow or orange-red before falling in autumn.

*Left: Small purple-red flowers are borne in pendulous clusters in spring.*

*Left: Winged seeds are borne in late summer into autumn as the leaves begin to turn colour.*

# MAPLE CULTIVARS

*Over the centuries hundreds of maple cultivars have been developed for their beauty and then grown as ornamental trees in gardens right across the world. Many of the original Japanese maple cultivars were developed in the Japanese temple gardens during the 17th and 18th centuries. The vast majority produce stunning leaf colour changes in autumn.*

## Vine-leaved Japanese maple

*Acer japonicum* 'Vitifolium'

The origins of this beautiful maple cultivar have been lost in time, although it is known to have been in cultivation by around 1882. The name refers to the fact that its leaves, which are more deeply divided than those of the species, resemble the leaves of a grape vine (*Vitis*). It is one of the mainstays of autumn leaf colour displays throughout Europe, producing leaves of several colours on the same tree at the same time: the autumn foliage ranges from green to burgundy, scarlet, orange, gold and yellow.

**Identification**: It is a wide-spreading tree with many low horizontal branches, which sweep upwards towards the tip. The bark is grey-brown and smooth, even in maturity. The leaves are heart-shaped at the base, rounded with a serrated margin. They are up to 12.5cm/5in long and 15cm/6in wide, with 9–11 lobes, separated almost halfway to the leaf centre and ending in sharp points. The flowers are small and green to red-purple, borne in conspicuous clusters in the spring.

**Distribution**: Of garden origin.
**Height**: 10m/33ft
**Shape**: Broadly spreading
**Deciduous**
**Pollinated**: Insect
**Leaf shape**: Roundly lobed

*Above: The leaves resemble those of the grape vine (Vitis).*

## *Acer palmatum* 'Katsura'

In Japan and in the USA this maple cultivar is considered a dwarf form; however, in Europe it may reach 7m/23ft in height. It is widely cultivated for its leaf shape and spring colour. New leaves, when they emerge from bud, are a beautiful apricot-yellow with a darker margin, which is usually orange. As the season progresses the leaves turn a bright golden-green.

**Identification**: The tree has rather shrubby, dense growth, with strongly ascending branching. The bark is grey and smooth. The leaves are small, seldom more than 5cm/2in long and broad. They have five lobes, which are lanceolate and taper to a long point, with the centre lobe being the longest. The margins are shallowly toothed and the lobes divide the leaf almost to the centre. Each leaf is attached to the twig by a 1cm/½in leaf stalk. In autumn the leaves revert from green to bright apricot-yellow and orange before falling.

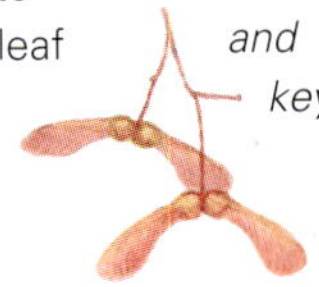

*Above and below: The autumn leaf and key.*

**Distribution**: Of garden origin.
**Height**: 7m/23ft
**Shape**: Broadly columnar
**Deciduous**
**Pollinated**: Insect
**Leaf shape**: Palmately lobed

## *Acer palmatum* 'Beni komachi'

Everything about this small tree is attractive. It is one of the very best dwarf Japanese maple cultivars and probably the best of all red-leaved cultivars. Even the name is delightful, roughly translating to "the beautiful, red-haired little girl". It was raised in the USA in the 1960s and is still relatively uncommon in European gardens, although its beauty is sure to make it more popular in the future.

**Identification**: The leaves, which are up to 5cm/2in long and broad, are deeply divided, almost to the midrib and leaf stalk, by five long, narrow lobes. The margins of each lobe are finely toothed and taper to a blunt point. Each lobe tends to curve down and to the side. When the leaves emerge from bud in spring they are bright scarlet-red. They then fade through late spring and summer to a blue-green red and then in autumn they return to brilliant scarlet.

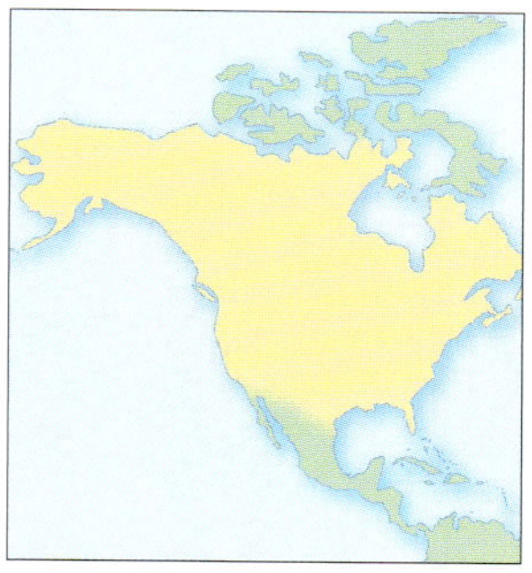

**Distribution**: Of North American garden origin.
**Height**: 2m/6ft
**Shape**: Broadly columnar
**Deciduous**
**Pollinated**: Insect
**Leaf shape**: Palmately lobed

*Left: The leaves turn a bright scarlet colour.*

OTHER SPECIES OF NOTE

***Acer palmatum* 'Linearilobum'**
This is an elegant cultivar that is said to have been raised in the Netherlands in 1867. However, there is some suggestion that it is a natural form of *A. palmatum*, which was being grown in Japan before then. It is a very popular maple, widely cultivated across Europe. It has bright grass-green leaves, with seven very narrow lobes, often little more than broadened midribs. It is vigorous and may eventually reach 5m/16ft tall.

***Acer palmatum* 'Crimson Queen'**
This outstanding, dissected, purple-leaved Japanese maple was raised in the USA in the 1960s, and is now very popular in gardens in Europe. It has fern-like dissected leaves, which emerge dark purple-red and stay that way throughout summer, unlike some cultivars, which turn a muddy green or bronze as they mature. In autumn the leaves turn bright red. It reaches an eventual height of around 3m/10ft.

***Acer palmatum* 'Burgundy Lace'**
'Burgundy Lace' is an American cultivar, which was raised in Washington in the 1950s but was not introduced into Europe until 1972. It is grown in European gardens but not so widely as 'Crimson Queen'. It is a small tree, eventually attaining a height of 4–5m/13–16ft, and has deeply dissected palmate leaves, to 7.5cm/3in long and wide. Throughout spring and early summer the leaves are the colour of Burgundy wine; in late summer they turn bronze-green.

## *Acer palmatum* 'Filigree'

This is one of the most beautiful of all dissected maples. Its delicate foliage and pendulous habit make it a real showstopper. It is a small tree, ideal for planting in a confined space, and is widely cultivated in European gardens, parks and arboreta. It is a fairly recent cultivar, having being raised by Joel Spingarn in New York, around 1955. It is very slow-growing and eventually forms a neat mushroom shape. The name refers to the lace-like quality of the dissected foliage.

**Identification**: The overall form resembles an overgrown bonsai specimen, with a multitude of twisted branches making the tree a striking feature in winter when the leaves have fallen. The bark is grey and smooth, becoming silvery green, with white elongated flecks on the twigs and shoots. Each leaf may be up to 10cm/4in across but is deeply dissected all the way to the centre by seven lobes; these are also deeply dissected, and up to 7.5cm/3in long.

**Distribution**: Of garden origin.
**Height**: 2m/6ft
**Shape**: Broadly spreading
**Deciduous**
**Pollinated**: Insect
**Leaf shape**: Palmately lobed

*Left: The leaves are light yellow-green in spring, darkening through summer, then turning fiery gold in autumn.*

## Coral-barked Maple

*Acer palmatum* 'Sangokaku'

Sometimes referred to as 'Senkaki', this is one of the original Japanese maple cultivars and was widely cultivated in Japan in the 19th century, and probably earlier. It was introduced into Europe and was being cultivated in Britain around 1920. It is a stunning tree, grown for its brilliant red stems, which are at their most vibrant in winter. The Japanese name 'Sangokaku' translates as "coral tower".

**Distribution**: Of garden origin.
**Height**: 11m/36ft
**Shape**: Broadly columnar
**Deciduous**
**Pollinated**: Insect
**Leaf shape**: Palmately lobed

**Identification**: This relatively large and vigorous Japanese maple is easily identified by its striking red bark. The deepest colour is on the new growth, but even old branches and the main stem maintain a red-green colouring. The palmate leaves are 5cm/2in across and long, normally with five lobes, each serrated around the margin and ending in a sharp point. They are a fresh green, becoming a subtle but beautiful apricot colour in autumn.

*Right: The bright green leaves turn yellow in autumn, creating a contrast with the reddish bark.*

## Fern Leaf Maple

*Acer japonicum* 'Aconitifolium'

This outstanding Japanese maple has been cultivated since the 1880s and is grown for its beautiful aconite-like leaves and glorious autumn colours. It can make a large tree for a Japanese maple and is often as wide as it is tall. It is a strongly structured plant, not at all thin and wispy like some *A. palmatum* cultivars. It has leaves that tend to lie in the horizontal plane, giving the tree a Far-Eastern appearance.

**Distribution**: Of garden origin.
**Height**: 8m/26ft
**Shape**: Broadly spreading
**Deciduous**
**Pollinated**: Insect
**Leaf shape**: Palmately lobed

**Identification**: The bark is smooth and grey-brown. The leaves are deeply cut, almost to the base, with 7–11 lobes that are themselves divided on each side with irregular cuts extending to the lobe midrib. Each lobe runs to an extended but blunt point. The leaf size is variable, from 7.5–15cm/3–6in long and broad. The foliage is deep green, becoming clear scarlet with gold, orange or purple tints in autumn.

*Below: The leaves resemble those of the aconite plant and turn from rich green to scarlet and orange in autumn. The seeds appear inside keys.*

OTHER SPECIES OF NOTE

**Red-leaved Norway Maple** *Acer platanoides* 'Crimson King'
This large, 25m/80ft, handsome tree is the best red-leaved Norway maple cultivar and is particularly attractive in spring when the leaves and flowers combine. It was raised in 1937 in a nursery in Belgium. Since then it has become a popular tree for planting in towns and cities right across Europe and North America. The flowers are bright yellow flecked with red.

***Acer platanoides* 'Drummondii'**
This striking variegated Norway maple cultivar was raised in a Scottish nursery in 1903. It is a medium-sized tree, seldom exceeding 12m/40ft tall, with a broad rounded crown, light brown bark and strongly lobed leaves, to 15cm/6in long and 17.5cm/7in broad. The leaves are deep green in the centre with a broad creamy-yellow to creamy-white margin.

***Acer pseudoplatanus* 'Brilliantissimum'**
This sycamore cultivar is like a bright star that burns itself out too soon. In spring when the leaves emerge from bud they are a glorious terracotta-orange-pink, but within days they fade through yellow to light green and the show is over. It is a compact small tree of slow growth.

***Acer palmatum* 'Seiryu'**
This is the only upright cultivar of the cut-leaved or dissected Japanese maple. That alone makes it worth growing, but it also has beautiful dissected foliage that emerges in spring with reddish flecks to the tips of each fresh green leaf. The autumn colour ranges from rich marmalade to translucent apricot.

## Golden Moon Maple

*Acer shirasawanum* 'Aureum'

When this graceful tree is given the right soil and moisture conditions it makes one of the finest ornamental trees for garden use, providing a glorious canopy of subtle golden foliage. It was for a long time thought of as a cultivar of *A. japonicum*, which it is not, but old labels on specimens in botanic gardens may refer to it as such. It has been in cultivation in Japan for at least 200 years, and was introduced into Europe in 1865.

**Distribution**: Of garden origin.
**Height**: 8m/26ft
**Shape**: Broadly spreading
**Deciduous**
**Pollinated**: Insect
**Leaf shape**: Orbicular

**Identification**: The bark is grey-brown and smooth, even in maturity. The leaves are round, to 10cm/4in across, with 9–11 lobes divided a third of the way to the centre. They are a clear golden yellow, sometimes with a reddish margin. In autumn they turn old gold before falling. The flowers are small and red, and are borne in conspicuous erect spikes in spring. They are followed by winged seeds, borne in pairs, which are green ripening to brown in late summer.

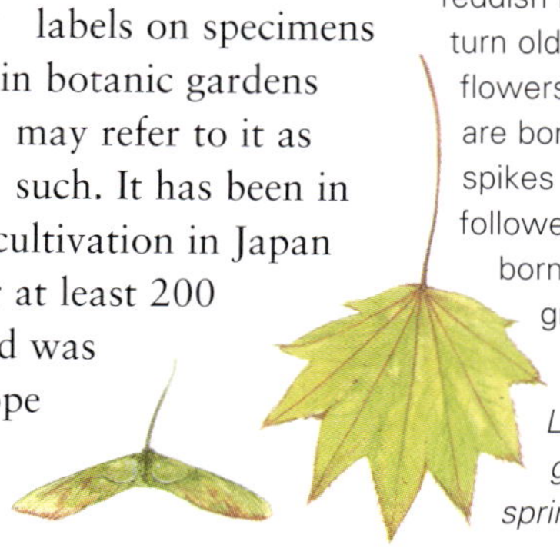

*Left: The leaves are a golden lime colour from spring to late summer.*

## *Acer palmatum* 'Osakazuki'

Undoubtedly the best-known Japanese maple cultivar of all and widely planted in parks, gardens and arboreta right across Europe, this maple has been listed in nursery catalogues since the 1850s. 'Osakazuki' takes on intense crimson-red autumn leaf colouring reliably every year, producing the best colour if it is planted in a sunny location. Although the tree grows rapidly in the first few years, it slows down considerably after ten years or so, and becomes a densely branched, round-topped tree reaching a maximum of 10m/33ft tall.

**Distribution**: Of garden origin.
**Height**: 10m/33ft
**Shape**: Broadly spreading
**Deciduous**
**Pollinated**: Insect
**Leaf shape**: Palmately lobed

**Identification**: The bark is grey-brown and smooth. The leaves are palmate, normally up to 10cm/4in long and broad, with seven ovate lobes, which are finely serrated around the margin and run to a long narrow tip. The two lobes on each side of the leaf stalk are normally much smaller than the other five. When the leaves first emerge in spring they are olive-orange; they turn slowly to bright grass-green in late spring and summer before finally changing to intense crimson in autumn.

*Below: In autumn, when the leaves turn, the seeds turn scarlet, becoming part of a very showy display from this tree.*

*Below: The leaves will reliably turn a good crimson-red colour every autumn.*

# ASHES

*There are about 65 species within the ash genus,* Fraxinus. *All have pinnate leaves and are found in temperate regions of the world, primarily North America, Europe and Asia. They are hardy, fast-growing deciduous trees that tolerate exposure, poor soils and atmospheric pollution. Included within this genus are members of the beautiful flowering ashes, which includes the manna ash from southern Europe.*

## Common Ash

*Fraxinus excelsior*

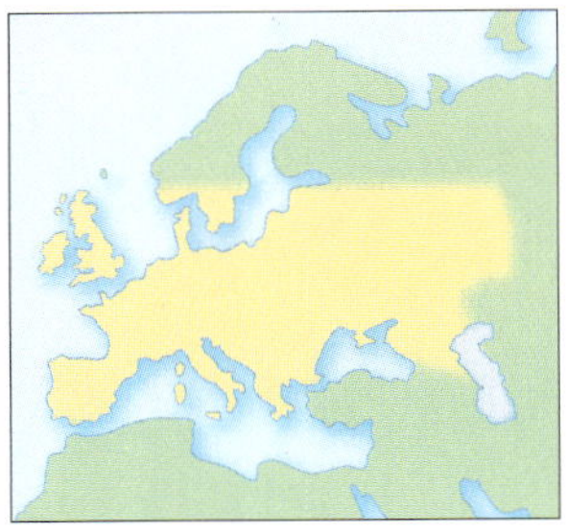

**Distribution**: Europe.
**Height**: 40m/130ft
**Shape**: Broadly columnar
**Deciduous**
**Pollinated**: Insect
**Leaf shape**: Pinnate

One of the largest of all European deciduous trees, the common ash is found growing wild from the Pyrenees to the Caucasus. Ash grows particularly well on calcareous limestone soils. It produces strong, white timber that has long been used where strength and durability are required, along with impact resistance, such as in coach building and for items such as ladders and tool handles.

**Identification**: The ash has a light airy crown, with a trunk that tends to be straight and long with little branching. The bark is pale fawn when young, becoming grey and fissured with age. One distinguishing characteristic of ash is its velvet-black winter buds. The fruits, flattened, winged seeds known as 'keys', 4cm/1½in long, are borne in clusters throughout the winter.

*Right: The leaves are pinnate and up to 30cm/12in long. Each may have up to 12 pairs of shallow toothed, rich green leaflets.*

*Left: Both male and female flowers are produced in profusion in early spring.*

## Narrow-leaved Ash

*Fraxinus angustifolia*

This elegant tree has, as its name suggests, the narrowest leaves of any ash. These give the tree an open, feathery look. It is a fast-growing tree, which was introduced into western Europe in 1800. There are several cultivars of *F. angustifolia*, including 'Raywood', which has leaves that turn plum-purple in autumn.

**Identification**: The bark is grey-brown with vertical fissures. Older trees may have been grafted on to the rather incompatible, slower-growing *F. excelsior*, which results in a prominent horizontal banding effect at the graft union. Winter buds are dark brown. The small, inconspicuous, green or purple flowers are borne on bare twigs in early spring. The fruits are flattened, winged seeds, up to 4cm/1½in long, borne in hanging fawn clusters that persist well into winter.

*Above: Narrow-leaved ash is a graceful tree with well-spaced branches and light, airy foliage.*

**Distribution**: Southern Europe, North Africa and western Asia.
**Height**: 25m/80ft
**Shape**: Broadly columnar
**Deciduous**
**Pollinated**: Insect
**Leaf shape**: Pinnate

*Right: The leaves are pinnate, with up to 13 lanceolate, sharply toothed, glossy dark-green leaflets, up to 10cm/4in long.*

OTHER SPECIES OF NOTE

**American Ash** *Fraxinus americana*
Otherwise known as white ash, this North American tree with rust-brown winter leaf buds and olive-green shoots was introduced into Europe in 1724. It is a fast-growing tree, which can eventually attain heights up to 25m/80ft. It has pinnate leaves, small purple flowers and single winged seeds, which are borne in dense pendulous clusters.

***Fraxinus angustifolia* 'Raywood'**
This is a handsome tree and one that deserves to be even more widely planted in European towns and cities than it already is. Raised in 1928, in the Raywood Gardens, Adelaide, Australia (hence its name), it is a fast-growing tree, with a compact, dense crown of upward arching branches and pinnate leaves, which turn a distinctive and handsome, vibrant plum-purple in autumn.

***Fraxinus holotricha***
This medium-sized ash, uncommon in cultivation, is native to the Balkan peninsula of south-eastern Europe. It was cultivated in Germany in the early 20th century, but is seldom seen growing outside botanic gardens and arboreta. It has pinnate grey-green leaves, normally with 11 or 13 narrow leaflets, and downy shoots.

## Manna Ash

*Fraxinus ornus*

This beautiful flowering ash grows wild in south-western Asia and southern Europe and has been widely cultivated throughout central and western Europe since around 1700. It produces, rather unusually for ash, large panicles of creamy-white, fragrant flowers, which hang in fluffy clusters from the branches in late spring. Manna sugar, a form of sweetener tolerated by diabetics, is derived from the sap of this tree.

**Distribution**: Southern Europe and south-west Asia.
**Height**: 20m/65ft
**Shape**: Broadly spreading
**Deciduous**
**Pollinated**: Insect
**Leaf shape**: Pinnate

**Identification**: The bark is grey and smooth and the winter leaf buds are dark grey. The pinnate leaves, to 20cm/8in long, have five to nine sharply toothed and tapered leaflets, each up to 10cm/4in long and 5cm/2in broad. They are matt mid-green above and slightly paler beneath. The fruit is a single, flat, winged seed, up to 4cm/1½in long, green at first ripening to pale brown.

*Below: Hanging clusters of creamy-white flowers are produced in late spring.*

## Chinese Flowering Ash

*Fraxinus mariesii*

This beautiful flowering ash, which is occasionally planted in gardens and arboreta throughout Europe, is named after the English plant collector Charles Maries, who introduced propagation material to James Veitch's nursery in 1878. It is a small, slow-growing tree, which forms a rounded, bushy head of branches in maturity. It is ideal for medium-sized gardens and should be more widely planted for its beauty.

**Identification**: The bark is light grey-brown and smooth, and the light grey winter leaf buds are covered with fine down, which makes them look as if they are covered with frost. The pinnate leaves, up to 17.5cm/7in long, have up to seven oval to ovate leaflets, each up to 10cm/4in long and 5cm/2in wide. They are dark green above and silver-green beneath, with a purple tinge to the leaf stalks. The flowers are creamy-white, borne in pendulous clusters, followed in late summer by attractive deep purple fruits.

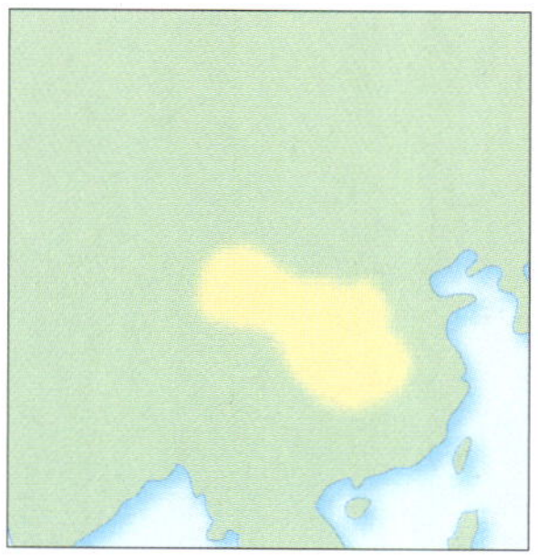

**Distribution**: Central China.
**Height**: 7m/23ft
**Shape**: Broadly spreading
**Deciduous**
**Pollinated**: Insect
**Leaf shape**: Pinnate

*Left: Each leaf has between five and seven leaflets.*

*Right: Purple-brown single-winged seeds appear in late summer.*

## Afghan Ash

*Fraxinus xanthoxyloides*

This small shrubby ash is native to dry valleys in the north-western Himalayas and Afghanistan. It was first cultivated in central and western Europe in the 1870s. It is not common in cultivation, but can be found in some botanic gardens and arboreta, where it is quite often grafted on to the common ash, *F. excelsior*. It is not immediately recognizable as an ash, having leaflets of variable shape and size.

**Identification**: The bark is dull grey-brown and smooth, even in maturity. The leaves, leaf stalks and young shoots may be covered in a fine white down on some trees and be entirely hairless on others. Each leaf has between five and nine leaflets – sometimes as many as 13. The leaflets are usually lanceolate or narrowly elliptic, up to 5cm/2in long and 2cm/¾in broad. The flowers are borne in short, dense clusters in the leaf axils in early spring.

**Distribution**: North-west Himalayas and Afghanistan.
**Height**: 8m/26ft
**Shape**: Broadly spreading
**Deciduous**
**Pollinated**: Insect
**Leaf shape**: Pinnate

*Right: Each leaf is made up of smaller leaflets.*

## Chinese Ash

*Fraxinus sogdiana*

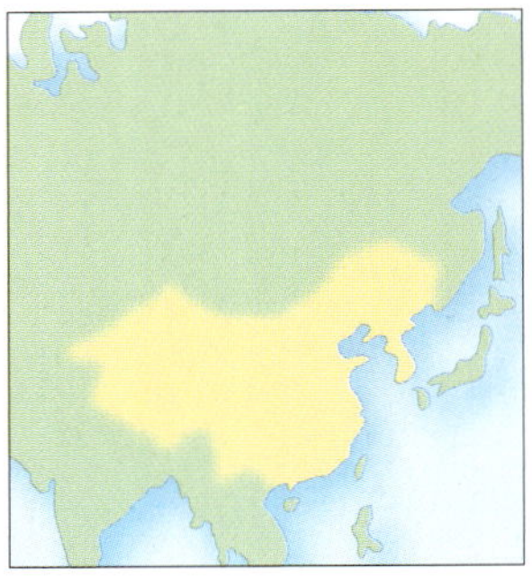

**Distribution**: Central Asia
**Height**: 10m/33ft
**Shape**: Broadly spreading
**Deciduous**
**Pollinated**: Insect
**Leaf shape**: Pinnate

This small tree, a native of Turkistan, was introduced into the St Petersburg Botanic Garden in 1891, and from there into much of central, and eventually western, Europe. It is an elegant small tree, growing to 10m/33ft, with green shoots and pinnate leaves with between seven and eleven lanceolate, toothed leaflets, which are attached to the midrib by a short leaf stalk.

**Identification**: The bark is silver-grey and smooth even in maturity. The shoots are bright green. The leaves are produced in whorls of three towards the tips of the branches. Each leaf is made up of between seven and eleven leaflets. Each leaflet is dull olive-green and bears distinctive teeth around the leaf margin. Each leaf is attached to the shoot by a short leaf stalk. The flowers are white, borne in clusters up to 5cm/2in long. In the autumn they turn rich butter yellow before falling.

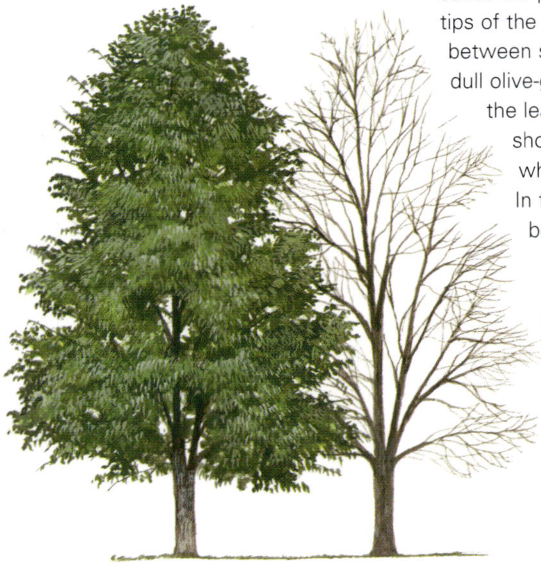

*Left: The leaflets are typical of the species in their size and shape.*

*Left: The seeds appear in the autumn.*

### OTHER SPECIES OF NOTE

**Syrian Ash** *Fraxinus angustifolia* subsp. *syriaca*
This rare small ash, first cultivated in Europe in 1880, is sometimes still referred to as *F. syriaca*. It has bright, apple-green pinnate leaves, densely borne in whorls of three. Each leaf is made up of between three and seven lanceolate, sharply toothed leaflets. This ash is recognizable from a distance by its crowded, dense foliage.

**Golden Ash** *Fraxinus excelsior* 'Jaspidea'
This is a fast-growing and popular cultivar of the common ash, *F. excelsior*, which was raised in the 1870s. It produces bright orange-yellow shoots, which are very conspicuous when the tree is bare of leaves in winter, and golden yellow autumn leaf colour. It is sometimes wrongly named as *F. excelsior* 'Aurea', which is, in fact, a golden-leaved dwarf ash.

**Weeping Ash** *Fraxinus excelsior* 'Pendula'
Originally discovered growing wild in Cambridgeshire, England, in the 18th century, this beautiful structural tree with weeping branches, which form a spreading umbrella-like canopy, is widely planted in parks and gardens throughout Europe. Most trees are grafted on to the common ash, *F. excelsior*, at 3–5m/10–16ft above the ground. It does not grow much above the point of grafting.

## Griffith's Ash

*Fraxinus griffithii*

This handsome ash, native to northern India, Burma and south-west China, and slightly tender, was introduced into Europe in 1900 by the plant collector Ernest Wilson. It has a tidy form and bright green shiny leaves, which are evergreen in warm winter climates and semi-evergreen to deciduous in colder regions of northern and western Europe. It has a leaf that is reminiscent of Chinese privet, *Ligustrum lucidum*.

**Identification**: The bark is pale grey and smooth and the young shoots are angular and bright green. The leaves are pinnate, with five to eleven ovate, glossy, bright green leathery leaflets, each up to 7.5cm/3in long, and tapering to a blunt point at the tip. Griffith's ash has large, fluffy, open panicles of creamy-white, slightly fragrant flowers, which are borne in late spring. They are followed by loose clusters of single winged seeds, which ripen to brown in the autumn and persist on the tree into winter.

**Distribution**: Northern India and south-east Asia.
**Height**: 11m/36ft
**Shape**: Broadly spreading
**Semi-evergreen**
**Pollinated**: Insect
**Leaf shape**: Pinnate

*Left: The shiny green leaves are smaller than typical ash.*

## *Fraxinus paxiana*

**Distribution**: Northern India, Himalayas and western China.
**Height**: 20m/65ft
**Shape**: Broadly spreading
**Deciduous**
**Pollinated**: Insect
**Leaf shape**: Pinnate

This distinctive, medium-sized Asian ash is a member of the Ornus section of the genus – the group of about 15 species that bear terminal, rather than lateral, inflorescences – also known as flowering ashes. As such, it produces large clusters of white flowers in late spring. It was introduced into Europe from western China in 1901. It has large winter buds and stiff young shoots. It is uncommon in European parks and gardens but normally represented in botanic gardens and arboreta.

**Identification**: The bark of *F. paxiana* is light grey and smooth. The winter leaf buds are conspicuously large and are covered with velvety brown down, resembling moleskin. The leaves are pinnate, up to 30cm/12in long, with between seven and nine toothed, lanceolate leaflets, each up to 15cm/6in long and 5cm/2in wide, and running to a slender point. Panicles of creamy-white flowers, up to 25cm/10in across, are carried on the tree in late spring and early summer.

*Left: After flowering, clusters of brown single-winged seeds are produced.*

*Left: With each individual leaflet up to 15cm/6in long, the leaves can become huge.*

# SPINDLE TREES

*This genus of small trees and large shrubs is extremely diverse. They thrive in almost any soil including chalk. They are widely grown as ornamental trees in European gardens for their autumn leaf colour and fruits. Spindles are found growing wild right across the Northern Hemisphere from Japan to Great Britain and are common in gardens, parks and in hedgerows.*

## Winged Spindle

*Euonymus alatus*

This distinctive, slow-growing small tree, or large shrub, has been planted throughout Europe since its introduction from China in 1860. It has conspicuous corky bark and angular branches that develop thin, corky wings. It is one of the best spindles for autumn colour, turning deep scarlet-pink.

**Identification**: Instantly recognizable by its corky winged branches, the tree has a rather stiff habit and is often wider than it is tall. It has narrow dark green leaves, finely toothed around the margin, up to 7.5cm/3in long and 2.5cm/1in broad. The flowers are small, greenish-yellow and insignificant. The fruits, which appear in late summer, are purple-red; four small pods, joined at the base, open when ripe to reveal bright orange seeds.

*Left: The vividly coloured fruits ripen in autumn.*

*Right: The dark green leaves turn brilliant red.*

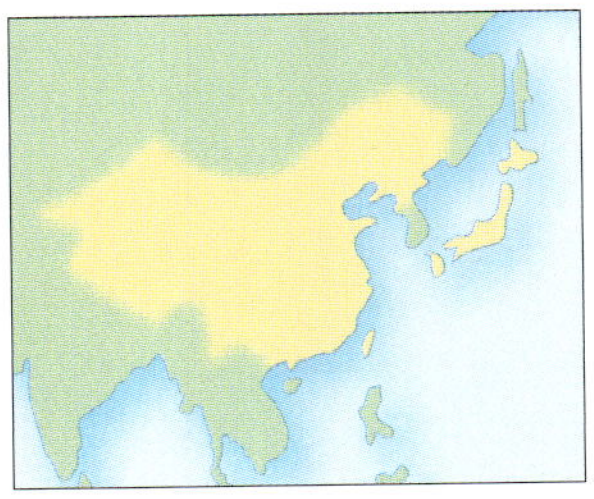

**Distribution**: China and Japan.
**Height**: 3m/10ft
**Shape**: Broadly spreading
**Deciduous**
**Pollinated**: Insect
**Leaf shape**: Oval to obovate

## Common Spindle Tree

*Euonymus europaeus*

This common small tree inhabits woodland edges and hedgerows throughout most of Europe. The name "spindle" comes from the fact that the tree has very hard, dense wood, which was at one time used to make spindles, skewers, charcoal and clothes pegs (pins). In the Victorian era it was commonly called skewerwood. The spindle tree is one of the best small European trees for autumn colour.

**Identification**: The stem bark is smooth and grey. However, young twigs are angular and may have a covering of thin, corky bark. The leaves are pale green, elliptic to ovate, to 7.5cm/3in long and 2.5cm/1in wide, finely toothed and running to a short point at the tip. In autumn they turn purple-red and yellow before falling. The flowers are small and greenish-white, borne in clusters in the leaf axils in early summer. The fruits, which appear in autumn, are bright pink, and open like parasols to reveal bright orange-coated seeds.

**Distribution**: Europe and western Asia.
**Height**: 6m/20ft
**Shape**: Broadly spreading
**Deciduous**
**Pollinated**: Insect
**Leaf shape**: Elliptic to ovate

*Right: The yellow-green leaves turn purple-yellow in autumn.*

*Left: The seed is contained within a bright pink or pink/red capsule.*

## Euonymus latifolius

This small tree, or large spreading shrub, which has a rather lax crown, has been in cultivation in western Europe since 1730. It is planted in gardens and parks for its vibrant autumn displays of fruit, seeds and foliage. In many ways this is a more ornamental tree than its European cousin, *E. europaeus*, having larger fruit and a more graceful habit.

**Identification**: The bark is smooth and grey and the young shoots are angular. The leaves are oval, sometimes oblong, a dull grass-green, to 12.5cm/5in long and 5cm/2in broad, finely and evenly toothed and running to a short point. Their autumn colours range from wine-purple to light pink and orange. The fruits are pendulous and bright scarlet, opening to reveal bright orange-coated seeds.

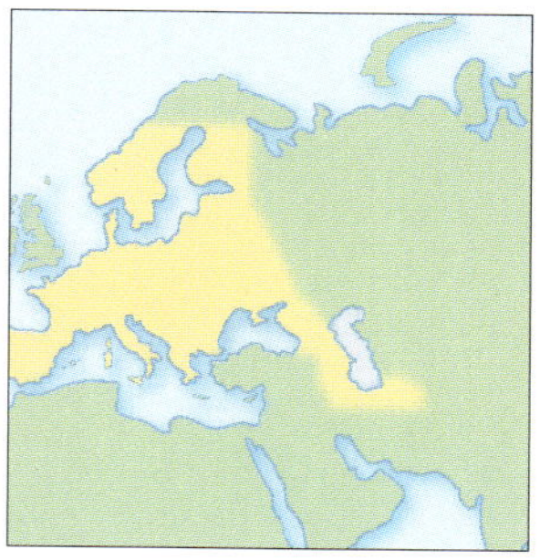

**Distribution**: Europe, the Caucasus and northern Iran.
**Height**: 6m/20ft
**Shape**: Broadly spreading
**Deciduous**
**Pollinated**: Insect
**Leaf shape**: Oval to oblong

*Right: The scarlet fruits have four or five winged lobes.*

OTHER SPECIES OF NOTE

***Euonymus verrucosus***
A native of Eastern Europe and western Asia, cultivated in Austria since 1763, this small, densely branched tree, to 3m/10ft tall, has conspicuous and distinctive "warts" along the younger branches. It has ovate leaves, which turn red and lemon in autumn. The fruit capsule may be yellow or red and contains orange-coated seeds.

***Euonymus oxyphyllus***
This handsome, slow-growing small tree, or large shrub, which is native to Japan, Korea and China, has been widely planted in parks and gardens since its introduction into Europe in 1895. It produces wine-red leaves in autumn at the same time as rich maroon-pink pendulous fruits containing bright orange-coated seeds.

***Euonymus illicifolius***
This unusual euonymus has evergreen, holly-like leaves, which are thick, spined and glossy green. It is native to central China and although introduced into Europe in 1930 it has never been widely cultivated. It is rather tender and will not grow outdoors in northern Europe. It produces round, white seed capsules that contain orange-coated seeds.

***Euonymus europaeus* 'Red Cascade'**
This popular cultivar of the common spindle was raised in England just after World War II. It is similar in size and leaf shape to the species but has graceful, long, arching branches, which in autumn weep under the weight of a profusion of bright red fruits.

## Chinese Spindle Tree

*Euonymus hamiltonianus*

The sight of a good specimen of *E. hamiltonianus*, its branches dripping with bright fruits set against beautiful coloured foliage, is one of the joys of autumn. It is a Himalayan species, which grows in China and Japan and was introduced into Europe in the early 20th century by Ernest Wilson. There are several varieties and sub-species, some of which are now more common in cultivation than the species.

**Identification**: The bark is dark grey-brown and smooth. The leaves are variable in shape and anywhere between 5cm/2in and 15cm/6in long. In autumn they turn copper-red on the top side, but beneath may be pale fawn. The fruits are pale pink capsules, borne on long stalks and ripening to reveal deep pink-orange coated seeds.

**Distribution**: Asia.
**Height**: 9m/30ft
**Shape**: Broadly spreading
**Deciduous**
**Pollinated**: Insect
**Leaf shape**: Variable, ovate to obovate

*Above: The pink fruits.*

*Right: The leaves are arranged opposite.*

# BEANS, FOXGLOVE AND VARNISH TREES

*The catalpas make up a genus of eleven species of beautiful flowering trees. Along with the foxglove trees they produce some of the largest leaves of any temperate deciduous tree. Catalpas have a strong resistance to atmospheric pollution and are commonly planted in towns and cities across warmer regions of Europe. The genus* Rhus *includes the very popular Stag's horn sumach.*

## Varnish tree

*Rhus verniciflua*

The sap of this handsome Asian tree is the source of the varnish used to give a high-gloss finish to Chinese and Japanese lacquerware. Oil extracted from the fruit is also used in China to make candles. However, almost all parts of the tree are poisonous and if they contact the skin they may cause irritation and blistering. *R. verniciflua* was cultivated in Europe before 1862.

**Identification**: This is an open-branched tree with light grey-brown bark, smooth at first, becoming vertically and shallowly fissured in maturity. The pinnate leaves, up to 60cm/24in long, are divided into 7–19 broadly ovate leaflets, each to 15cm/6in long. These are bright green above, paler beneath with some hairs. Large drooping panicles of small yellow-white flowers are produced in summer. On female trees they are followed by pea-sized yellow fruits.

*Left: The flowers are pea-sized and appear in panicles in late summer.*

*Right: The leaves were traditionally used to extract tannin.*

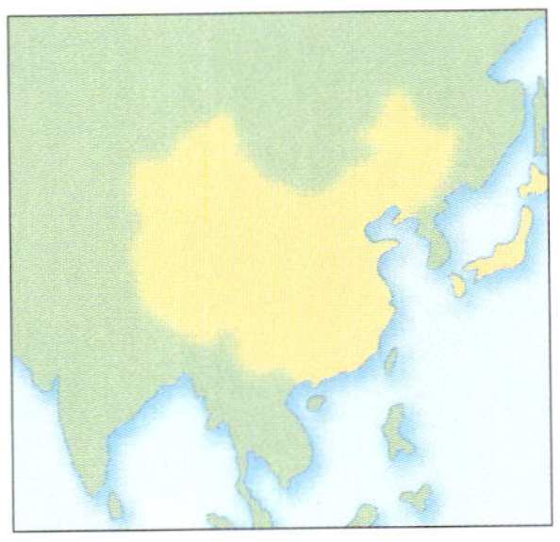

**Distribution**: China, Japan and eastern Himalayas.
**Height**: 20m/65ft
**Shape**: Broadly spreading
**Deciduous**
**Pollinated**: Insect
**Leaf shape**: Pinnate

## Farge's Catalpa

*Catalpa fargesii*

This beautiful flowering Chinese tree is named after the French Jesuit missionary Paul Farges, who discovered it in western China in 1896. Since then it has been widely grown in both Europe and North America for its delicate pink flowers, which appear in early to mid-summer, several weeks earlier than its American counterpart, *C. bignonioides*. The form *duclouxii* is the most common in cultivation.

*Above: Each broadly ovate leaf may be up to 15cm/6in long.*

**Identification**: The bark is dark grey, becoming scaly and peeling to reveal grey-pink fresh bark beneath. The leaves are broadly ovate, with the occasional small side lobe (not always present), and up to 15cm/6in long and 12.5cm/5in wide. They are bronze when first emerging from buds, becoming bright green with a slight sheen by late spring. Farge's catalpa has bell-shaped flowers, which are light pink with maroon spots and a yellow blotch at the entrance to the flower throat. These are followed by pendulous seed pods, up to 45cm/18in long.

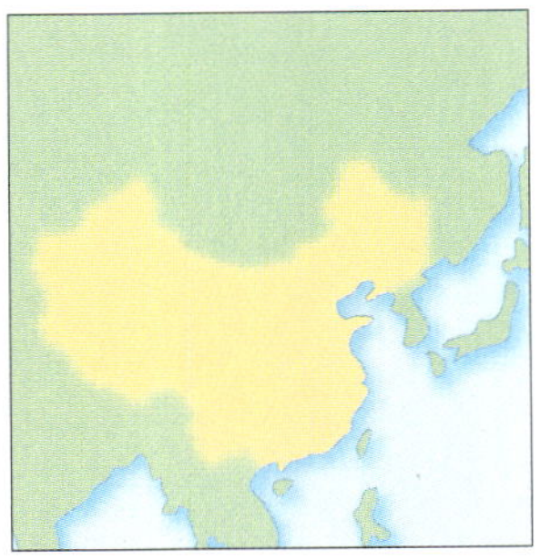

**Distribution**: China.
**Height**: 20m/65ft
**Shape**: Broadly columnar
**Deciduous**
**Pollinated**: Insect
**Leaf shape**: Broadly ovate

*Right: Long brown-black pendulous seed pods appear in autumn.*

## Foxglove Tree

Princess tree *Paulownia tomentosa*

This beautiful flowering tree is native to central China, and is perfectly hardy in Europe. It takes its genus name from Anna Paulownia, the daughter of Tsar Paul I of Russia. The spectacular pale purple, foxglove-like flowers appear on spikes in late spring. The timber has a resinous quality and was used in China and Japan to make a stringed instrument similar to a lute. *Paulownia* is quite often coppiced to enhance its foliage, which on juvenile shoots can be up to 45cm/18in across.

**Identification**: The bark is rather like that of beech, being grey and smooth, even in maturity. The leaves are ovate, up to 45cm/18in wide and long, heart-shaped at the base with two large, but normally shallow, lobes on each side. They are dark green, with hair on both surfaces and shoots. The shoots are soft and pithy. Each trumpet-shaped, pale purple flower, blotched inside with dark purple and yellow, is 5cm/2in long. The flowers are on upright panicles up to 45cm/18in tall.

**Distribution**: Central and eastern China.
**Height**: 20m/65ft
**Shape**: Broadly columnar
**Deciduous**
**Pollinated**: Insect
**Leaf shape**: Ovate

*Left: The fruit is a green, pointed, egg-shaped, woody capsule containing several winged seeds. The purple flowers resemble foxgloves.*

OTHER SPECIES OF NOTE

**Stag's-horn sumach** *Rhus typhina*
This North American, spreading and sparsely branched tree, to 10m/33ft, was first cultivated in Europe as early as 1629. It is now widely planted in gardens throughout Europe. It is quite distinctive, with downy thick shoots, large pinnate leaves and upright, candle-like, red hairy fruits.

**Potanin's sumach** *Rhus potaninii*
This sumach, first cultivated in Europe in 1902, grows to 20m/65ft in its native China, with a neat, round-topped form. In cultivation it is usually a smaller, multi-stemmed tree. It is grown for the autumn colours of its large pinnate leaves, which may be burgundy, bright red or orange.

***Rhus trichocarpa***
From mountainous country in China, Japan and Korea, this is a small, broadly spreading tree, to 8m/26ft, with large pinnate leaves made up of 17 taper-pointed leaflets, each 10cm/4in long. It is widely cultivated in Europe for its beautiful autumn leaf colour, which is rich marmalade-orange.

**Empress Tree** *Paulownia fargesii*
This perfectly hardy, fast-growing, handsome Chinese tree, to 20m/65ft tall, produces fragrant foxglove-like white flowers, with dark purple speckles in the throat. It appears to be hardier than the foxglove tree, *P. tomentosa,* and flowers at an early age. It has large ovate leaves, which on pruned stems may be up to 60cm/24in across.

## Indian Bean Tree

*Catalpa bignonioides*

The Indian referred to in the name is in fact the Native American who used to wear the seeds. This is one of the last trees in its region to flower, and is normally at its best in mid-summer. It is hardy throughout Europe. It tolerates atmospheric pollution well and has become a firm favourite for planting in towns and cities, despite its broadly spreading crown.

**Identification**: The tree has grey-brown bark, becoming loose and flaking in patches in maturity. The leaves are broadly ovate, up to 25cm/10in long and 15cm/6in wide, rarely lobed and heart-shaped at the base. On emerging from the bud they are bronze-coloured, gradually turning grass-green with some hair beneath. Each leaf is borne on a long, lax leaf stalk. The branches are quite brittle and prone to breakage in summer.

**Distribution**: South-east USA.
**Height**: 20m/65ft
**Shape**: Broadly spreading
**Deciduous**
**Pollinated**: Insect
**Leaf shape**: Ovate

*Right: The seed pods are 40cm/16in long.*

*Below: Each of the trumpet-shaped flowers is up to 5cm/2in long.*

# TAMARISKS

*The tamarisks are a beautiful group of small, elegant trees that thrive in warm, sunny locations. They are found growing wild throughout southern Europe and are often planted in coastal locations because they withstand exposure and salt spray extremely well. They have graceful slender branches, plume-like foliage and small pink flowers, which are often produced in profusion.*

## French Tamarisk

*Tamarix gallica*

Like most tamarisks, French tamarisk thrives in coastal locations and copes with salt spray, exposed conditions and periods of drought. Originally native to an area running from coastal north-west France southwards to North Africa, this species has become widely naturalized elsewhere in Europe, including the south coast of Britain. If regularly pruned, tamarisk makes a good windproof hedge; unpruned it soon becomes wide-spreading and straggly.

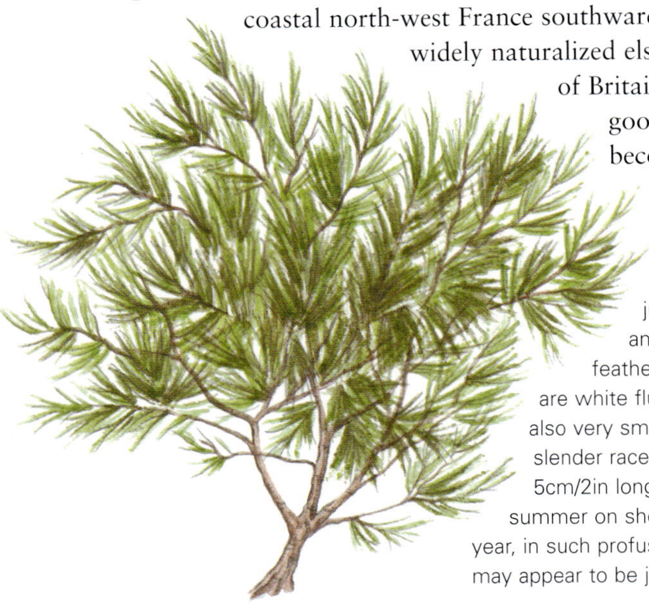

**Identification**: The bark on the main stems is brown; on branches and new growth it is purple-brown. The leaves are scale-like, like those of juniper. They are blue-green, without any hairs, and very small, creating an overall feathery appearance. The flowers, which are white flushed with pink on the outside, are also very small and crowded on lax, slender racemes, which may be up to 5cm/2in long. They appear in late summer on shoots from the previous year, in such profusion that the whole crown may appear to be just flowers.

*Far left: The tiny leaves and flowers give the whole plant a soft, feathery*

*Left: The individual flowers are star-shaped and soft pink.*

**Distribution**: South-west Europe.
**Height**: 8m/26ft
**Shape**: Broadly spreading
**Deciduous**
**Pollinated**: Insect
**Leaf shape**: Scale-like

## *Tamarix ramosissima*

**Distribution**: Western and central Asia.
**Height**: 6m/20ft
**Shape**: Broadly spreading
**Deciduous**
**Pollinated**: Insect
**Leaf shape**: Scale-like

Otherwise known as *T. pentandra*, this rather sprawling, multi-stemmed small tree has a wide natural distribution that runs from southern Russia south through most of temperate Asia. It has been cultivated in Europe since 1885 and is a popular garden tree, particularly in coastal locations. Pruning stems to the ground quickly stimulates new growth. There are two main cultivars that have been developed from the species and are commonly found in gardens, 'Rubra' and 'Rosea'.

*Left: Tiny pink flowers are borne in long slender racemes throughout the summer. They are as fine as the leaves.*

**Identification**: The bark of each mature stem is reddish-brown; it is smooth at first, becoming finely fissured in maturity. The current year's shoots are green-yellow, darkening to reddish-brown as they mature. The leaves are bright green, without any hairs. They are scale-like, like those of juniper, and very small, giving the tree an overall feathery appearance. The very small flowers, which are pink, are borne in slender racemes up to 5cm/2in long. They are carried in profusion, on the current year's shoots, in late summer.

## Tamirix parviflora

This beautiful, spreading, multi-stemmed small tree is native to the coastal regions bordering the Aegean Sea, through the Balkans and possibly even in North Africa, where it thrives in dry, infertile and salty soils. It has been cultivated elsewhere in Europe, particularly in parks, since at least 1853, and has become naturalized in central and southern Europe. It differs from most other tamarisk species in that the flowers appear on the old wood in spring, rather than on the current year's growth in late summer.

**Identification**: The mature bark of *T. parviflora* is brown or purplish-brown. The bark on new growth is purple. The branches are long and have a graceful arching habit. Each shoot is covered in very small, bright-green, scale-like deciduous leaves. The flowers, which are small and appear in late spring, are deep pink, sometimes stained purple, and carried in much profusion, in lax racemes about 5cm/2in long, which are densely arranged along the branches.

*Right: Each shoot is covered in small, bright green scale-like leaves.*

**Distribution**: South-east Europe.
**Height**: 6m/20ft
**Shape**: Broadly spreading
**Deciduous**
**Pollinated**: Insect
**Leaf shape**: Scale-like

*Left: The flowers of* T. parviflora *have only four petals.*

OTHER SPECIES OF NOTE

**Canary Tamarisk** *Tamarix canariensis*
As its name suggests, this shrubby, small tree is native to the Canary Islands and is to be found growing wild on all the islands except El Hierro. It is multi-stemmed and has red-brown bark, small, scale-like grey-green leaves, which are arranged alternately on the branches, and thin, spike-like racemes of pink-white flowers. In the Canaries it is cultivated as a windbreak for crop protection.

**Chinese Tamarisk** *Tamarix chinensis*
This native of eastern and central Asia has been cultivated in Europe since at least the 1830s. It is a small tree, or large shrub, of dense habit, with distinctive, very thin branches. It has small, scale-like, pale green foliage. The flowers, which are dark pink in bud, become pale pink on opening, in late spring, on the previous year's wood.

***Tamarix ramosissima* 'Rosea'**
This beautiful tamarisk was raised in Orleans, France, around 1883, and since then has become a popular small tree for planting in parks and gardens. The flowers are a bright rosy-pink and are densely borne on slender branching racemes, to 10cm/4in long, in late summer. Such is the profusion of flower that the foliage is completely hidden.

## Tamarix tetrandra

This shrubby tamarisk is native to southern Russia, the eastern Balkans and into Iran. It has been cultivated in Europe since 1821, but is not as widespread as some of the other species. Like *T. parviflora*, it flowers on the previous year's wood in late spring or early summer, rather than on new growth in late summer. In some old collections trees labelled *T. tetrandra* var. *purpurea* are in fact *T. parviflora*.

**Identification**: *T. tetrandra* is unique among tamarisk species in having bark that is very dark brown, almost black. This is a distinctive feature and is helpful in identifying this particular species. The overall appearance of the tree is of an open crown with lax, rather sparse, long, dark-coloured branches. The leaves are scale-like, small and bright green. The flowers are very small and light pink. They are clustered together on racemes that may be up to 7.5cm/3in long. These racemes are densely carried on the branches.

**Distribution**: South-east Europe and western Asia.
**Height**: 5m/16ft
**Shape**: Broadly spreading
**Deciduous**
**Pollinated**: Insect
**Leaf shape**: Scale-like

*Below:* T. tetrandra *is a perfect specimen plant for a sunny spot.*

*Right: The light pink flowers are stunningly pretty in spring.*

# PALMS AND TREE FERNS

*There are about 150 genera of palms in the world. Thirty years ago, the species detailed below would have only grown in southern Europe. Recent climate warming across the region means that some now survive outside as far north as the Netherlands, Estonia and Scotland. In particular, chusan palms and tree ferns are becoming increasingly popular for planting in parks and gardens.*

## Canary Island Date Palm

*Phoenix canariensis*

This is the most majestic and stately of all palms that thrive outside the tropics, and is able to withstand several degrees of frost. Although native only to the Canary Islands, it is widely planted as an ornamental species in warm temperate regions throughout the world, commonly on sea fronts and in formal avenues. Specimens over 100 years old thrive in the gardens of Tresco in the Scilly Isles.

*Far right: The fruits ripen to purple-brown in autumn.*

**Identification**: The Canary Island date palm has a long, straight golden-brown fibrous stem with reptilian scales formed by the shedding of previous fronds. It may be up to 1.5m/5ft in diameter at the base, which tends to splay out just above ground level. The leaves are evergreen, comb-like fronds, up to 5m/16ft long, which gracefully arch skywards before drooping towards the tip. Golden yellow flower spikes, up to 2m/6½ft long, are borne in spring, followed in warm climates on female trees by bunches of purple-brown fruits.

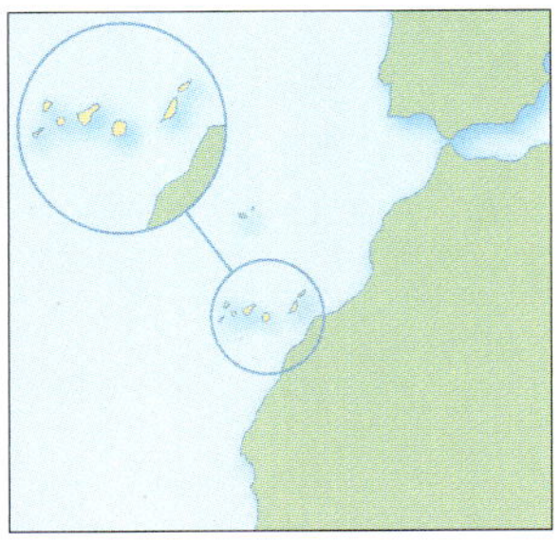

**Distribution**: Canary Islands.
**Height**: 12m/40ft
**Shape**: Palm-like
**Evergreen**
**Pollinated**: Insect
**Leaf shape**: Pectinate

## Chusan palm

Chinese windmill palm *Trachycarpus fortunei*

Despite this tree's image as a tender desert island native, it does, in fact, originate from the mountains of China, and is perfectly hardy in warm regions of Europe. It was introduced into Europe in 1830 by the German botanist Philipp von Siebold, though not from China: Siebold sent seeds home from a tree he had found growing in Japan. Today, the chusan palm is commonly found as an ornamental, planted in coastal parks and gardens throughout Europe, including coastal regions of south-west Britain.

**Identification**: The bark is covered with grey-brown fibrous hairs and is clearly marked with discarded leaf scars. The leaves are fan-shaped, stiff, and blue-green, up to 1.2m/4ft across, and divided almost to the base into approximately 40 linear and pointed strips. They are joined to the tree by a stiff leaf stalk. The flowers are small, fragrant and golden yellow, borne in large drooping panicles in early summer.

*Far left: The flowers are golden yellow and fragrant.*

*Left and right: The blue-green leaves are fan shaped.*

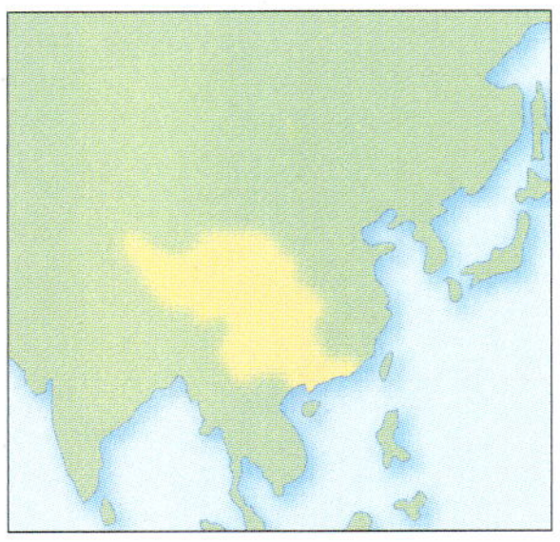

**Distribution**: Central and southern China.
**Height**: 12m/40ft
**Shape**: Palm-like
**Evergreen**
**Pollinated**: Insect
**Leaf shape**: Fan-shaped

## Tree Fern

Australian tree fern *Dicksonia antarctica*

Since their introduction into Europe in 1880, tree ferns have become a favourite for planting in warm, wet regions, where they thrive and bring a touch of the exotic to temperate gardens. Strictly speaking they are not trees at all but true ferns, but their interest lies in the fact that they reach tree-like proportions. The trunk is made up of the fibrous remains of old roots. New roots grow down the trunk each year, to enable the fern to produce a new set of fronds. Growth is slow, normally less than 5cm/2in in height each year. Some fronds may live for two years.

**Identification**: The surface of the trunk is soft, fibrous and chestnut-brown. Under this surface covering is a black, bone-like woody core. In winter the tree resembles little more than an upright log. In spring new fronds unfurl from the top of the trunk. Each frond is like a typical fern leaf but larger, up to 4m/13ft long. The upper surface of the leaf is a dark but rich green, the lower surface paler and duller, and close inspection reveals light brown spores at regular intervals along each frond.

**Distribution**: South-eastern Australia and Tasmania.
**Height**: 7m/23ft
**Shape**: Palm-like
**Semi-evergreen**
**Pollinated**: Wind
**Leaf shape**: Fern-like

*Right: Each leaf is like a fern frond but may be up to 4m/13ft long.*

OTHER SPECIES OF NOTE

**Jelly Palm** *Butia capitata*
Otherwise known as *Cocos capitata*, and also commonly known as the pindo palm, this beautiful, small palm is native to Brazil, but has been widely cultivated in southern coastal areas of the USA, and is planted in south-west Europe. It has grey-green to silver arching fronds and fragrant yellow flowers, which are tinged with purple. These are followed by yellow or orange round fruits, with a flavour reminiscent of apricots, pineapple and bananas, which can be used to make jelly or wine.

**Cabbage Palm** *Cordyline australis*
This slow-growing, palm-like tree, to 10m/33ft tall, is in fact a member of the lily family. It is native to New Zealand and has been grown in Europe since 1823. New Zealand Maoris used to eat the tender tips of the shoots, hence its common name. It is widely cultivated right across the warmer coastal regions of Europe, including south-western Britain. In early summer the tree produces masses of fragrant, creamy-white flowers.

**Cabbage Palmetto** *Sabal palmetto*
This tall palm, to 25m/80ft, is native to south-eastern USA and the West Indies, but has been widely cultivated in parks and gardens across the Mediterranean region of Europe. It has fan-shaped leaves that are palmately divided and may be up to 3m/10ft long, with a prominently arching midrib. Creamy to yellowish-white flowers are borne in drooping clusters in summer.

## Dwarf Fan Palm

European fan palm *Chamaerops humilis*

This is the only member of the palm family that is truly native to mainland Europe. It is found growing wild in Spain, Gibraltar, Italy, Sardinia, Sicily, Algeria and Morocco, where it inhabits mountainsides in coastal regions. It is widely cultivated in mild coastal regions in other parts of Europe, including southern and western Britain, but is not as hardy as the chusan palm, *Trachycarpus fortunei*. It has been in cultivation since 1731. Most specimens do not exceed 2m/6½ft tall.

**Identification**: The overall appearance of this palm tree is of a dense semicircular mass of stiff, grey-green, fan-shaped leaves on a short trunk, which is covered with stiff dark grey-brown fibres towards the top. The large fan-shaped leaves are green to grey-green. They can measure up to 90cm/36in across, and are divided nearly to the base into stiff, pointed segments, which may be up to 45cm/18in long. Each leaf is attached to the trunk by a thick leaf stalk, of variable length, which is armed with sharp, forward-pointing spines. The flowers are small and yellow and are borne in a stiff upright panicle, to 15cm/6in long.

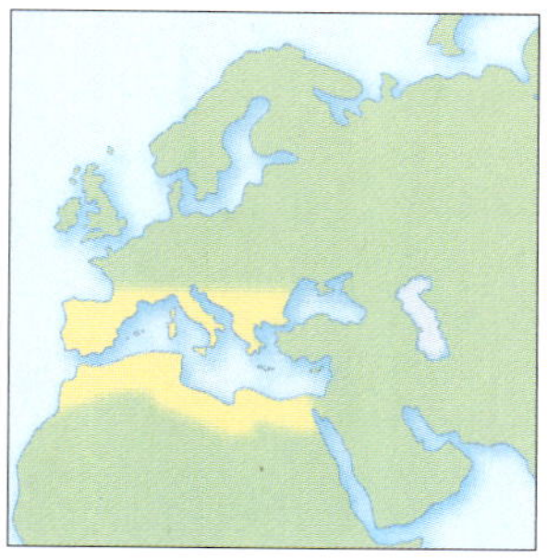

**Distribution**: Southern Europe and North Africa.
**Height**: 2.5m/8ft
**Shape**: Palm-like
**Evergreen**
**Pollinated**: Insect
**Leaf shape**: Fan-shaped

*Above: The yellow flowers are stiff upright panicles.*

*Below: The pointed leaf segments*

# GLOSSARY

**Anther** The terminal part of the stamen in which the pollen matures.
**Aril** A fleshy and colourful appendage to the seed capsule.
**Axil** The upper angle between the stalk and the leaf.

**Bast** The outer fibrous part of the trunk.
**Bipinnate** (of leaves). Having leaflets which are also divided in pinnate manner.
**Bract** A small leaf or scale placed below the calyx.

**Cambium** A layer of cells just within the outer coating of the trunk of the tree from which annual growth of bark and wood occurs.
**Catkin** A cylindrical cluster of male or female flowers. Catkins usually hang down from the tree.
**Chlorophyll.** Green colouring matter of plants.
**Chloroplast** Part of the tree cell containing chlorophyll.
**Class** A collection of orders of trees containing a common characteristic.
**Columnar** Refers to a tall, thin upright tree shape.
**Compound** (of leaves). With leaf divided into leaflets.
**Conical** Cone-like shape of a tree.
**Conifer** Cone-bearing tree. Can be evergreen or deciduous.
**Cordate** (of leaves). Heart-shaped.

*Below: Canadian yew,* Taxus canadensis.

*Above: Chinese tulip tree,* Liriodendron chinese.

**Cotyledon** The first leaf or leaf-pair within seed.
**Cultivar** A variety of tree produced from a natural species and maintained in existence by cultivation.
**Cuticle** The protective surface film on leaves.
**Cutting** A tree propagated from part of original tree. The cut edge of the twig may form roots and grow into a small tree once potted in suitable compost.

**Deciduous** Shedding leaves annually or seasonally.
**Dicotyledon** Plant with double leaf or leaf-pair within seed.
**Dioecious** Having male and female cones or flowers on separate trees.
**Drupe** Fleshy fruit containing stony seed-cover, for example plum.

**Elliptic** (of leaves). Oval in shape, with widest point at midsection.
**Epidermis** Protective layer of cells on leaves and stalks.
**Evergreen** Bearing leaves all year round, although each leaf has a limited life span.

**"False"** (of species). Trees that have characteristics that superficially resemble a specific species.
**Family** A collection of genera of trees sharing a common characteristic.
**Fastigiate** (of trees). Having a conical or tapering outline.
**Fissure** Splits, cracks and grooves, usually in the surface of the bark.
**Form** A group of trees distinguished from others by a single characteristic.

**Genera** Plural of genus.
**Genus** A taxonomic group into which a family is divided and usually containing one or more species.
**Glaucous** (of leaves). Blue-green colouring.
**Graft** A plant produced by joining dissimilar plant materials.
**Gymnosperm** A plant bearing seed that is unprotected by seed vessels, for example conifer.

**Hardwood** Deciduous trees.
**Heartwood** Dense wood within inner core of tree trunk.
**Hybrid** A new species of plant that results from a cross between two genetically dissimilar plants, often growing in close vicinity.

**Indumentum** The outer coating of down or hair on the surface of a leaf.
**Inflorescence** The arrangement of flowers on the flower-bearing stalks.

**Lanceolate** (of leaves). Narrow oval shape, tapering to point.
**Layer** (of propagation). A shoot or branch that grows roots while still attached to the parent plant.
**Lenticel** An aeration pore in bark.

*Below: Cedar of Lebanon,* Cedrus libani.

*Above: Silver maple,* Acer saccharinum.

**Linear** (of leaves). Having a narrow, elongated shape.
**Lobed** (of leaves). Having rounded indentations around the leaf edge.

**Meristem** Growing tissue in trees.
**Monocotyledon** A plant with a single leaf or leaf-pair within a seed.
**Monoecious** Having male and female flowers on the same tree.
**Monotypic** A tree that is the only species in the genus. These are quite unusual.

**Needle** A slender, elongated leaf.

**Oblong** (of leaves). Being longer than broad, with parallel sides.
**Obovate** (of leaves). Egg-shaped, with the broadest end growing furthest from the stem.
**Orbicular** (of leaves). Round.
**Osmosis** Transfer of solutions between porous partitions; process whereby liquid moves from one cell to another.
**Ovate** (of leaves). Egg-shaped, with the broadest end nearest the stem.
**Ovoid** (of flowers). Egg-shaped.
**Ovule** Female reproductive structure which develops into a seed after being fertilized.

**Palmate** (of leaves). With three or more leaflets arising from the same point.
**Panicle** A head of stalked flowers.
**Petiole** Leaf stalk.
**Phloem** Soft tissue within the tree trunk.
**Photosynthesis** Use of sunlight to create nutrients within leaves.
**Phreatophyte** Tree or other plant with long taproots.
**Pinna(e)** Primary division of pinnate leaf.
**Pinnate** (of leaves). Having leaflets in pairs on either side of petiole.
**Primates** An order of mammals typically with flexible hands and feet, a highly developed brain and good eyesight.
**Prostrate** With a trunk that grows along the ground. Could be caused by adverse weather.
**Pubescence**; pubescent. A layer of short, fine hairs; downy.

**Raceme** A flowerstem from which flowerheads grow, with the oldest flowers closest to the base.

**Samara** A winged fruit, for example ash key.
**Sapwood** Soft wood between heartwood and bark.
**Scale** Small, modified leaf.
**Scale-like** Plate-like covering.
**Sessile** (of leaves). Without stalks.
**Simple** (of leaves). Leaves which are not divided into leaflets.
**Softwood** Coniferous trees.
**Species** A group into which the members of a genus are divided. Can contain forms, varieties and subspecies.
**Stomata** Pores in the epidermis of leaves.
**Subclass** A subdivision of a class.
**Subspecies** A subdivision of a species.

**Tepal** The outer part of the flower that is clearly not part of the corolla and the calyx.

*Above:* Prunus *'Tai Haku'*

**Terminal inflorescence** The final flowerhead on the stem.
**Transpiration** Loss of moisture through evaporation.
**Trifoliate** (of leaves). Having three leaflets to make up one leaf.
**Tripinnate** (of leaves). Having three or more pinnae.

**Variety** A group of trees with distinct characteristics, but insufficiently different from the true species to be recognized as a true species of its own.

**Xerophyte** Tree or other plant capable of conserving and storing water.
**Xylem** Woody tissue within tree trunk.

## Acknowledgements

The publisher would like to thank the following picture libraries for permission to use their photographs:
**Ardea** p63b p122b, p176tr.
**Edward Parker** p7 br, p8–9, p15tl and bc, p16 bl, p28 bl, p29 tr, p30 t, p32 tl, p33, p36 bl, p37 t, p41 tr, p42 tr, p47 all, p84, p85 bl and b.
**The Garden Picture Library** p50 tr, p51 tl, p52 br, p53 bl, and bc, p55 bl and br, p56 bl, p57 tr, p59 all, p108b, 118b, p127tl, p139tr, p234tr.
**Garden World Images** p 132b, p168br, p182bl, p186bl, p227tr.
**Oxford Scientific Films** p45 br, p65 all, p69, p75.
**Peter Anderson** p31b, p73t.
**Tony Russell** p34 tr, p51 bl, p93 both, p95b, p145bl, p232tr, p233tl, p235.

b= bottom, t = top, r = right, c = centre, l = left

*Left: Temple juniper,* Juniperus rigida.

# INDEX